MW00452960

# SELECTED SOLU

## EDWARD DERRINGH

*Wentworth Institute*
*Boston, MA*

to accompany

## VOLUMES ONE AND TWO EXTENDED

# PHYSICS

## FOURTH EDITION

### DAVID HALLIDAY

*Professor of Physics, Emeritus*
*University of Pittsburgh*

### ROBERT RESNICK

*Professor of Physics*
*Rensselaer Polytechnic Institute*

### KENNETH S. KRANE

*Professor of Physics*
*Oregon State University*

## JOHN WILEY & SONS, INC.

*New York* • *Chichester* • *Brisbane* • *Toronto* • *Singapore*

Copyright © 1992 by John Wiley & Sons, Inc.

All rights reserved.

Reproduction or translation of any part of
this work beyond that permitted by Sections
107 and 108 of the 1976 United States Copyright
Act without the permission of the copyright
owner is unlawful.  Requests for permission
or further information should be addressed to
the Permissions Department, John Wiley & Sons.

ISBN 0-471-51860-3

Printed in the United States of America

10 9 8 7 6 5

# PREFACE

This solutions supplement is written for use by students. It contains solutions to just over 25% of the problems appearing at the end of the chapters. But keep your textbook handy: to save space the problem statements are not reproduced here, and there are many references to equations, figures, tables and sample problems in the textbook. Note that we observe the rules for significant figures in giving the final answers. We try to solve the problems in the most 'obvious', down-to-earth, straightforward manner, avoiding any use of 'tricks' or sophisticated reasoning. Our aim is to make it easier for you to solve, on your own, similar problems that do not appear in this supplement, whether they be assigned homework problems, or those on exams.

July 3, 1992

Edward Derringh
*Wentworth Institute of Technology*
*Boston, MA 02115*

# CONTENTS

## 1-3

From Table 2, we see that the prefix 'micro' equals a factor of $10^{-6}$ so that 1 microcentury = $(1 \times 10^{-6})$(century). Since 1 century = 100 y, 1 microcentury = $(1 \times 10^{-6})(100$ y$)$ = $1 \times 10^{-4}$ y. Now we need to find the number of minutes in one year so that we can compare this with the standard lecture period of 50 min. One year consists of 365.25 days, so we have

$$1 \text{ y} = (365.25 \text{ d})(24 \text{ h/d})(60 \text{ min/h}) = 5.26 \times 10^5 \text{ min.}$$

Therefore,

$$1 \text{ microcentury} = (1 \times 10^{-4})(5.26 \times 10^5 \text{ min}) = 52.6 \text{ min.}$$

The percent difference from Fermi's 50 min approximation is

$$\% \text{ diff} = \frac{52.6 - 50}{50}(100\%) = 5.2\%.$$


## 1-8

Let $v$, $t$, $L$ stand for the average speed, time, and distance for each runner. Since there are two runners, we use subscripts 1 and 2 to distinguish between them. That is, we have

$$v_1 t_1 = L_1,$$
$$v_2 t_2 = L_2.$$

If the distances are laid out accurately, we should have $L_1 = L_2$. But, we could have $t_1 < t_2$ and $v_1 < v_2$ (i.e., the slower runner with the shorter time!) if $L_1$ is sufficiently less than $L_2$. With this in mind, let us write $L_1 = L_2 - x$, where $x$ is the error in laying out the distance $L_1$. The two equations above now become

$$v_1 t_1 = L_2 - x,$$
$$v_2 t_2 = L_2.$$

Now use your algebra skills to combine these last two equations to get

$$v_1 = v_2 \left(\frac{t_2}{t_1}\right)\left(1 - \frac{x}{L_2}\right).$$

This equation tells us that in order to conclude that $v_1 > v_2$ with the error $x \neq 0$, the error must meet the requirement

$$x < L_2\left(1 - \frac{t_1}{t_2}\right).$$

Let us suppose that neither runner runs a distance greater than 1 mile = 5280 ft. Converting the running times to seconds (why?) and putting $L_2$ = 5280 ft gives for the maximum error $x$ = 3.325 ft = 4 in.

1-12

The last day of the twenty centuries is longer than the first day by the amount

(20 centuries)(0.001 s/century) = 0.020 s.

Thus, the average day during the twenty centuries is $\frac{1}{2}$(0 + 0.020) = 0.010 s longer than the first day. The total cumulative effect $T$ is given by

$$T = \text{[average difference][number of days]},$$
$$T = \text{[0.010 s/average day][(365.25 day/y)(2000 y)]},$$
$$T = 7305\ s = 2\ h\ 1\ \min\ 45\ s.$$

1-17

"Ignore the curvature of the Earth" means to assume that Antarctica is flat. In this case, the volume equals the area times the thickness or depth $D$ of the ice cover. However, before calculating the volume, we must express all dimensions in the same distance unit and, since the volume in $cm^3$ is demanded, it makes sense to use cm throughout. The radius $R$ of the semicircle becomes

$$R = (2000\ km)(1 \times 10^5\ cm/km) = 2 \times 10^8\ cm.$$

The thickness D is

$$D = (3000\ m)(100\ cm/m) = 3 \times 10^5\ cm.$$

The area of a circle is $\pi R^2$, so that the area of a semicircle must be $\frac{1}{2}\pi R^2$. Therefore, the volume $V$ of the ice is $V = \frac{1}{2}\pi R^2 D$; numerically,

$$V = \frac{1}{2}\pi(2 \times 10^8 \text{ cm})^2(3 \times 10^5 \text{ cm}) = 1.88 \times 10^{22} \text{ cm}^3.$$

## 1-24

The distance $D$ sought must satisfy the requirement

$$\log D = \frac{1}{2}[\log(1 \times 10^{-15}) + \log(2 \times 10^{26})] = 5.6505,$$

$$D = 10^{5.6505} = 4.47 \times 10^5 \text{ m} = 447 \text{ km}.$$

Using natural logarithms will give the same result:

$$\ln D = \frac{1}{2}[\ln(1 \times 10^{-15}) + \ln(2 \times 10^{26})] = 13.011,$$

$$D = e^{13.011} = 4.47 \times 10^5 \text{ m}.$$

The reader can surely identify a distance of about 450 km.

## 1-31

We have $N_A = 6.022 \times 10^{23}$ sugar cubes. Each cube has a volume of $(1 \text{ cm})^3 = 1 \text{ cm}^3$, so that the volume of 1 mole of cubes must be $6.022 \times 10^{23} \text{ cm}^3$. The volume of a cube is $a^3$, where $a$ is the edge length. Hence,

$$a = (6.022 \times 10^{23} \text{ cm}^3)^{1/3} = 8.44 \times 10^7 \text{ cm} = 844 \text{ km}.$$

## 1-33

The mass of water in the container is $(5700 \text{ m}^3)(1000 \text{ kg/m}^3) = 5.7 \times 10^6$ kg. The mass flow rate is mass/time. We need the time in seconds to get the rate in kg/s. But $12 \text{ h} = (12 \text{ h})(3600 \text{ s/h}) = 43,200 \text{ s}$. Thus, the mass flow rate is

$$\text{rate} = (5.7 \times 10^6 \text{ kg})/(43,200 \text{ s}) = 132 \text{ kg/s}.$$

## 1-40

Writing $[K]$ for the units of hydraulic conductivity $K$, we can find $[K]$ by inserting SI units for the other quantities into Darcy's law and solving algebraically:

$$\text{m}^3/\text{s} = [K](\text{m}^2)(\text{m/m}),$$

$$[K] = \text{m/s}.$$

3

2-2

We begin by converting the average velocity from km/h to m/s, the SI base unit. From Appendix G,

$$160 \text{ km/h} = (160)(0.2778 \text{ m/s}) = 44.45 \text{ m/s}.$$

Therefore, by Eq. 5,

$$\Delta t = \frac{\Delta x}{\overline{v}} = \frac{18.4 \text{ m}}{44.45 \text{ m/s}} = 0.414 \text{ s},$$

or 414 ms.

2-6

We will work the problem in the metric distance units and use hours for the time unit. It is necessary to calculate the two travel times for the two speeds and then subtract. By Eq. 5,

$$\Delta t_1 = 700/104.6 = 6.692 \text{ h},$$
$$\Delta t_2 = 700/88.5 = 7.910 \text{ h}.$$

The time saved is 7.910 - 6.692 = 1.218 h = 1 h (0.218)(60 min) = 1 h 13 min.

2-10

We are not told the distance the car travels in going up the hill, so let us call this distance $\Delta x$; the distance traveled when coming down the hill is the same. If we use km and h for the units, then we have $\Delta t_{up} = \Delta x/40$ and $\Delta t_{down} = \Delta x/60$. For the round trip we know that the total distance = $2(\Delta x)$ and the total time is the sum of the times to travel up and down; therefore, for the round trip,

$$\overline{v} = \frac{2(\Delta x)}{\dfrac{\Delta x}{40} + \dfrac{\Delta x}{60}} = \frac{(2)(40)(60)}{60 + 40} = 48 \text{ km/h}.$$

This is less than the, perhaps expected, answer of 50 km/h since the car spends more time traveling up the hill than it does in traveling down.

## 2-13

(a) $\Delta x = x(3) - x(2)$. Now $x(3) = 9.75 + 1.50(3)^3 = 50.25$ cm; also $x(2) = 9.75 + 1.50(2)^3 = 21.75$ cm. $\Delta t = 3 - 2 = 1$ s. Therefore,

$$\overline{v} = \frac{\Delta x}{\Delta t} = \frac{50.25 - 21.75}{1} = 28.5 \text{ cm/s}.$$

(b) The instantaneous velocity is

$$v = \frac{dx}{dt} = 4.5t^2.$$

Therefore $v(2) = 4.5(2)^2 = 18.0$ cm/s.

(c) Using the equation for $v$ found in (b) we find $v(3) = 4.5(3)^2 = 40.5$ cm/s.

(d) Again, using the result found in (b), $v(2.5) = 4.5(2.5)^2 = 28.1$ cm/s.

(e) Refer to (a) to find that the desired position of the particle is $x = \frac{1}{2}(50.25 + 21.75) = 36$ cm. Now we need to find when (value of $t$) the particle is at this position; using the given equation for $x(t)$ we have

$$36 = 9.75 + 1.50t^3 \rightarrow t = 2.596 \text{ s}.$$

Finally, using the relation for $v(t)$ found in (b) we have $v(2.596) = 4.5(2.596)^2 = 30.3$ cm/s.

## 2-16

Use Eq. 13 to obtain immediately

$$\overline{a} = \frac{\Delta v}{\Delta t} = \frac{(-30) - (+18)}{2.4} = -20 \text{ m/s}^2.$$

## 2-23

(a) We need $x(3) = 50(3) + 10(3)^2 = 240$ m, and $x(0) = 0$. We also have $\Delta t = 3 - 0 = 3$ s. Therefore

$$\overline{v} = \frac{\Delta x}{\Delta t} = \frac{240 - 0}{3} = 80 \text{ m/s}.$$

(b) The instantaneous velocity is $v = dx/dt = 50 + 20t$. Therefore $v(3) = 50 + 20(3) = 110$ m/s.

(c) Use the expression for $v$ found in (b) above to calculate the instantaneous acceleration $a = dv/dt = 20$ m/s$^2$. Since $a$ does not depend on the time, this value of 20 m/s$^2$ applies at $t = 3$ s, as at all other times.


## 2-29

"From rest" means that the initial velocity $v_0 = 0$ at the origin $x_0 = 0$. Converting the data to SI base units, we have $v = 360$ km/h $=$ $(360)(0.2778$ m/s$) = 100$ m/s and $x = 1.8$ km $= 1800$ m. To find the acceleration $a$, Eq. 20 is most suitable:

$$v^2 = v_0{}^2 + 2a(x - x_0),$$

$$100^2 = 0^2 + 2a(1800 - 0),$$

$$a = 2.8 \text{ m/s}^2.$$


## 2-33

The length units in the data are mixed: let us convert the distance $x$ to meters: $x = 1.2$ cm $= 0.012$ m. Now use Eq. 20:

$$v^2 = v_0{}^2 + 2a(x - x_0),$$

$$(5.8 \times 10^6)^2 = (1.5 \times 10^5)^2 + 2a(0.012 - 0),$$

$$a = 1.4 \times 10^{15} \text{ m/s}^2.$$

Since the data was inserted into the equation in SI base units, the units of the answer will be the appropriate SI base units, in this case, the units of acceleration.


## 2-36

(a) We use the SI base unit data: no unit conversions are needed. We have $a = -4.92$ m/s$^2$ ("deceleration" implies negative acceleration, opposite to the direction of the initial velocity), $v_0 = 24.6$ m/s, $v = 0$ (come "to rest"). The time $t$ is sought: use Eq. 15:

$$v = v_0 + at,$$

$$0 = 24.6 + (-4.92)t,$$

$$t = 5.00 \text{ s.}$$

(b) The brakes are applied at $x_0 = 0$; the distance traveled in coming to rest is represented by $x$. To obtain $x$ using only the original data of the problem, use Eq. 20:

$$v^2 = v_0^2 + 2a(x - x_0),$$

$$0^2 = (24.6)^2 + 2(-4.92)(x - 0),$$

$$x = 61.5 \text{ m.}$$

## 2-39

(a) Put $x_0 = 0$ at the point where the speed of the train is $v_0 = 33$ m/s. Then $v = 54$ m/s at $x = 160$ m. Use Eq. 20:

$$v^2 = v_0^2 + 2a(x - x_0),$$

$$54^2 = 33^2 + 2a(160 - 0),$$

$$a = 5.71 \text{ m/s}^2.$$

(b) To find the time $t$ from the original data, use Eq. 21:

$$x = x_0 + \tfrac{1}{2}(v + v_0)t,$$

$$160 = 0 + \tfrac{1}{2}(54 + 33)t,$$

$$t = 3.68 \text{ s.}$$

(c) It is convenient to reset the equations since now we focus our attention on the interval between the train starting from rest until it reaches a speed of 33 m/s; the numerical value of the acceleration found above still applies since the acceleration is assumed to be constant. The train starts from rest, $v_0 = 0$, at $x_0 = 0$; at a point a distance $x$ down the track the speed is $v = 33$ m/s. To find the time $t$ needed to reach this speed, use Eq. 15:

$$v = v_0 + at,$$

$$33 = 0 + (5.71)t,$$

$$t = 5.78 \text{ s.}$$

(d) To find the distance $x$ we could use any of Eqs. 19, 20, 21, 22. We choose Eq. 19:

$$x = x_0 + v_0 t + \tfrac{1}{2}at^2 = 0 + 0 + \tfrac{1}{2}(5.71)(5.78)^2 = 95.4 \text{ m.}$$

Again, we have not inserted the units with the data, but the data is entirely in SI base units, so the units of the various answers must be the corresponding SI base units. If you are in doubt in any or all of these cases, you should insert the units and see that they work out as asserted above.

The average speed (which equals the average velocity in this case since the motion is entirely in one direction), is given by Eq. 5:

$$\overline{v} = \frac{\Delta x}{\Delta t} = \frac{x - x_0}{t} .$$

Combining this with Eq. 21, we find

$$\overline{v} = \frac{1}{2}(v_0 + v) .$$

In our problem, this last equation gives

$$90 \text{ km/h} = \frac{1}{2}(120 \text{ km/h} + v) \rightarrow v = 60 \text{ km/h}$$

as the needed maximum speed upon crossing the second strip. Now use Eq. 20:

$$v^2 = v_0^2 + 2a(x - x_0),$$

$$(60 \text{ km/h})^2 = (120 \text{ km/h})^2 + 2a(0.110 \text{ km}),$$

$$a = -4.909 \times 10^4 \text{ km/h}^2 = -13.6 \text{ km/h} \bullet \text{s},$$

the last step since 1 h = 3600 s.

(a) Choosing the metric data, we must first convert the initial velocity to m/s: $v_0 = 56(0.2778 \text{ m/s}) = 15.6 \text{ m/s}$. The brakes are applied at a point that we can call the origin $x_0 = 0$; at time $t = 4$ s, the car hits the barrier located at $x = 34$ m with a speed $v$ that we cannot assume is zero. Use Eq. 19 to find $a$:

$$x = x_0 + v_0 t + \frac{1}{2}at^2,$$

$$34 = 0 + (15.6)(4) + \frac{1}{2}a(4)^2 \rightarrow a = -3.6 \text{ m/s}^2.$$

We expect a negative acceleration since the car is slowing down. (b) Now we want to find the impact speed $v$; to do this using only the original data [so that we can get the correct answer even if we made an error in (a) above], use Eq. 21:

$$x = x_0 + \frac{1}{2}(v_0 + v)t,$$

$$34 = 0 + \frac{1}{2}(15.6 + v)(4) \rightarrow v = 1.4 \text{ m/s}.$$

## 2-52

(a) "Dropped" implies that the wrench was released with initial velocity $v_0 = 0$. We let $y_0 = 0$ be the point from which the wrench was dropped. We are told that $v = -24$ m/s at impact (recall that up is positive and the wrench is falling downward). Eq. 25 yields

$$v^2 = v_0^2 - 2g(y - y_0),$$

$$(-24)^2 = 0^2 - 2(9.8)(y - 0) \quad \rightarrow \quad y = -29.4 \text{ m},$$

that is, the ground is 29.4 m below (negative sign) the release point.

(b) To find $t$ using only the original data, invoke Eq. 23:

$$v = v_0 - gt,$$

$$-24 = 0 - (9.8)t \quad \rightarrow \quad t = 2.45 \text{ s}.$$

## 2-54

(a) Put $y_0 = 0$ at the point the rock was dropped; its initial speed is $v_0 = 0$ ("dropped"). Up is positive, so after falling 50 m the position of the rock is $y = -50$ m (i.e., 50 m below the point it was dropped). We are to find the time of fall $t$: Eq. 27 applies:

$$y = y_0 + v_0 t - \tfrac{1}{2}gt^2,$$

$$-50 = 0 + 0 - \tfrac{1}{2}(9.8)t^2 \quad \rightarrow \quad t = 3.19 \text{ s}.$$

(b) After falling an additional 50 m the position of the rock is $y = -100$ m. The time elapsed in reaching this location is found as in part (a); the result obtained is 4.52 s. This is the time to fall 100 m. Since 3.19 s was occupied in falling the first 50 m, the time needed to fall the second 50 m is 4.52 - 3.19 = 1.33 s.

## 2-58

(a) The launching point of the ball is taken as the origin $y_0 = 0$. The data given is $y = +36.8$ m when $t = 2.25$ s. The unknown sought is $v_0$. Apply Eq. 24:

$$y = y_0 + v_0 t - \tfrac{1}{2}gt^2,$$

$$36.8 = 0 + v_0(2.25) - \tfrac{1}{2}(9.8)(2.25)^2 \quad \rightarrow \quad v_0 = 27.4 \text{ m/s}.$$

(b) The desired velocity is $v$; use Eq. 27:

$$y = y_0 + vt + \tfrac{1}{2}gt^2,$$

$$36.8 = 0 + v(2.25) + \tfrac{1}{2}(9.8)(2.25)^2 \quad \rightarrow \quad v = 5.33 \text{ s}.$$

(c) Find the greatest height $y$ the ball will reach by setting $v = 0$; use the value of $v_0$ found in (a) in Eq. 25:

$$v^2 = v_0^2 - 2g(y - y_0),$$

$$0^2 = (27.4)^2 - 2(9.8)(y - 0) \rightarrow y = 38.3 \text{ m}.$$

Hence, the ball will travel an extra distance $38.3 - 36.8 = 1.5$ m above the point that was mentioned in the problem statement.

2-62

(a) Apply Eq. 25 between points $A$ and $B$, taking $y_0 = 0$ at point $A$:

$$v^2 = v_0^2 - 2g(y - y_0),$$

$$(\tfrac{1}{2}v)^2 = v^2 - 2(9.8)(3 - 0) \rightarrow v = 8.85 \text{ m/s}.$$

(b) Now apply the same equation between points $B$ and the highest point reached (at which $v = 0$). The velocity at point $B$ is now $v_0 = \tfrac{1}{2}(8.85) = 4.425$ m/s. Therefore, we get

$$0^2 = 4.425^2 - 2(9.8)(y - y_0) \rightarrow y - y_0 = 0.999 \text{ m},$$

or 99.9 cm above point $B$.

2-65

Call the speed of the ball upon striking the floor $v$ and the speed of the ball as it leaves the floor $u$. Up is positive so that the associated velocities are $-v$ and $+u$. The average acceleration is given, by definition, by Eq. 13:

$$\overline{a} = \frac{+u - (-v)}{t - 0} = \frac{u + v}{t}.$$

As this is positive, the average acceleration is directed upward. Since the speed with which the ball leaves the floor is the same as the speed with which it would strike the floor if dropped from a height of 1.9 m, both $u$ and $v$ can be found from Eq. 25 with $v_0 = 0$ and $y_0 = 0$; that is, $v^2 = -2gy$, using $y = -2.2$ m for $v$ and $y = -1.9$ m for $u$, $g = 9.8$ m/s² for each. Solving gives $v = 6.57$ m/s and $u = 6.10$ m/s, so that $\overline{a} = (6.10 + 6.57)/(0.096) = 132$ m/s², directed upward.

2-67

(a) Let $y$ be the height of fall and $t$ the time to fall this distance. Since the object falls from rest $v_0 = 0$ and Eq. 24 gives

$$y = \tfrac{1}{2}gt^2,$$

$$\tfrac{1}{2}y = \tfrac{1}{2}g(t - 1)^2.$$

Eliminating $y$ between these two equations using our algebra skills gives

$$t = \sqrt{2}(t - 1) \quad \rightarrow \quad t = 3.41 \text{ s}.$$

(b) The height of fall is

$$y = \tfrac{1}{2}gt^2 = \tfrac{1}{2}(9.8)(3.41)^2 = 57.0 \text{ m}.$$

## 2-70

(a) Choose the origin $y_0 = 0$ at the balloon at the instant of release of the package. The initial velocity of the package, presuming that it is not tossed from the balloon, is the same as the velocity of the balloon at that instant; i.e., $v_0 = +12.4$ m/s. Note that the ground is at $y = -81.3$ m (81.3 m below the origin). By Eq. 25

$$v^2 = v_0{}^2 - 2g(y - y_0),$$

$$v^2 = (12.4)^2 - 2(9.8)(-81.3 - 0) \quad \rightarrow \quad v = \pm 41.8 \text{ m/s}.$$

Since the package is heading downward, we pick $v = -41.8$ m/s (up is positive); the speed is 41.8 m/s.
(b)  By Eq. 23, and using the result of (a),

$$v = v_0 - gt,$$

$$-41.8 = 12.4 - (9.8)t \quad \rightarrow \quad t = 5.53 \text{ s}.$$

## 2-75

At the top of its path, the speed of the flowerpot is zero. Times to travel corresponding distances moving up or down are equal. Focus on the downward trip. Let $y_0 = 0$ at the highest point reached, at which point $v_0 = 0$ also. Let $D$ be the distance from the highest point to the top of the window. Applying Eq. 24 to travel from the top of the path to the top of the window, and from the top of the path to the bottom of the window yields in turn (bearing in mind that down is negative),

$$-D = -\tfrac{1}{2}(9.8)t^2,$$

$$-(D + 1.1) = -\tfrac{1}{2}(9.8)(t + 0.37)^2.$$

We use 0.37 s since $0.37 = \tfrac{1}{2}(0.74)$. Eliminating $D$ between these two equations gives

$$4.9t^2 + 1.1 = 4.9t^2 + 2(0.37)(0.49)t + (4.9)(0.37)^2,$$

$$t = 0.118 \text{ s}.$$

Therefore, the distance sought is given by $D = (4.9)(0.118)^2 = 0.068$ m $= 6.8$ cm.

3-3

(a) See the sketch below. Measure the length of the displacement vector **d**; applying the scale this length, or magnitude, is about 375 m; measure the angle with a protractor to find that $\theta \approx 57°$ east of north.

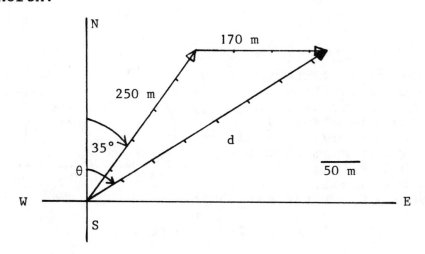

(b) The distance the woman actually walks is 250 m + 170 m = 420 m, whereas the magnitude of her displacement is 375 m, as shown above.


3-10

(a) Use Eqs. 5:

$$a_x = a\cos\phi = (7.34)\cos252° = -2.27,$$

$$a_y = a\sin\phi = (7.34)\sin252° = -6.98.$$

(b) To determine magnitude and direction from the components, use Eqs. 6:

$$a = \sqrt{a_x^2 + a_y^2} = \sqrt{(-25)^2 + (43)^2} = 50,$$

$$\phi = \tan^{-1}(a_y/a_x) = \tan^{-1}(+43/-25) = 120°.$$

We write 120° rather than the -60° our calculator gave, because the vector is in the second quadrant. (Our calculator cannot tell, once the division has been done, whether the minus sign goes with the x-component or the y-component of the vector; indeed, in this problem, the calculator has attached the minus sign to the y-component, rather than the x, with the result that it puts the vector in the fourth quadrant.)

<u>3-11</u>

(*a*) If we picture the *x* axis as horizontal and directed to the right and the *y* axis as vertical and directed upward, then the height sought is the *y* component of a vector of magnitude 13 m and angle 22°. By Eq. 5, then, the vertical distance raised = 13sin22° = 4.9 m.
(*b*) Reasoning as in (*a*), the horizontal distance moved = 13cos22° = 12 m.


<u>3-16</u>

Since the wheel rolls without slipping, the straight line distance PQ equals one-half the circumference of the wheel, for the wheel rolled through one-half a revolution. The circumference of the wheel is $2\pi R$, where $R$ is the radius of the wheel. Hence, the displacement of point P is

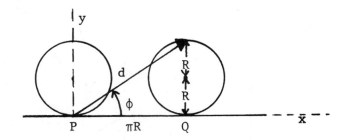

$$\mathbf{d} = \tfrac{1}{2}(2\pi R)\mathbf{i} + 2R\mathbf{j} = 141\mathbf{i} + 90\mathbf{j}.$$

Use Eqs. 6 to find the magnitude $d$ and angle $\phi$:

$$d = \sqrt{[141^2 + 90^2]} = 170 \text{ cm}; \quad \phi = \tan^{-1}(90/141) = 33°.$$


<u>3-18</u>

(*a*) Add corresponding components: $\mathbf{a} + \mathbf{b} = (5 - 3)\mathbf{i} + (3 + 2)\mathbf{j} = 2\mathbf{i} + 5\mathbf{j}$; this vector is in the first quadrant since both components are positive.
(*b*) Use Eqs. 6: $|\mathbf{a} + \mathbf{b}| = \sqrt{[2^2 + 5^2]} = \sqrt{29} = 5.39$; $\phi = \tan^{-1}(5/2) = 68.2°$.


<u>3-23</u>

(*a*) Refer to Fig. 26 in HRK; since $\mathbf{r} = \mathbf{a} + \mathbf{b}$, $r_x = a_x + b_x$ and $r_y = a_y + b_y$. The components of **a** are $a_x = 12.7\cos28.2° = 11.19$ and $a_y = 12.7\sin28.2° = 6.001$. Now, the angle counterclockwise from the +*x* axis to vector **b** is 28.2° + 105° = 133.2°. Hence, the components of **b** are $b_x = 12.7\cos133.2° = -8.694$ and $b_y = 12.7\sin133.2° = 9.258$. Therefore $r_x = 11.19 + (-8.694) = 2.50$ and $r_y = 6.001 + 9.258 = 15.3$, so that $\mathbf{r} = 2.50\mathbf{i} + 15.3\mathbf{j}$.
(*b*) The magnitude of **r** is $r = \sqrt{[2.50^2 + 15.3^2]} = 15.5$.
(*c*) The angle for **r** is $\phi = \tan^{-1}(15.3/2.50) = 80.7°$.

3-25

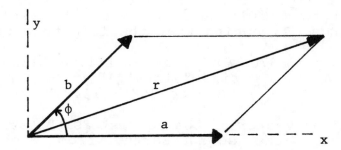

Orient the coordinate axes so that one of the vectors lies along one of the vectors lies along one of the axes; in the sketch above, for example, the vector **a** is positioned on the $x$ axis. Then we have for **a** and **b**,

$$\mathbf{a} = a\mathbf{i}; \quad \mathbf{b} = (b\cos\phi)\mathbf{i} + (b\sin\phi)\mathbf{j},$$

and therefore

$$\mathbf{a} + \mathbf{b} = (a + b\cos\phi)\mathbf{i} + (b\sin\phi)\mathbf{j}.$$

Recalling that $\sin^2\phi + \cos^2\phi = 1$, we obtain for the magnitude of this vector

$$r = |\mathbf{a} + \mathbf{b}| = \sqrt{(a + b\cos\phi)^2 + (b\sin\phi)^2} = \sqrt{a^2 + b^2 + 2ab\cos\phi}.$$

3-26

Use the properties of the dot product: $\mathbf{a} \cdot \mathbf{a} = a^2$; $\mathbf{a} \cdot \mathbf{b} = \mathbf{b} \cdot \mathbf{a}$; $\mathbf{a} \cdot \mathbf{b} = 0$ if **a** and **b** are perpendicular. Now let **a** and **b** be the two vectors referred to in the problem and proceed as follows:

$$(\mathbf{a} + \mathbf{b}) \cdot (\mathbf{a} - \mathbf{b}) = 0,$$
$$\mathbf{a} \cdot \mathbf{a} - \mathbf{a} \cdot \mathbf{b} + \mathbf{b} \cdot \mathbf{a} - \mathbf{b} \cdot \mathbf{b} = 0,$$
$$a^2 - \mathbf{b} \cdot \mathbf{a} + \mathbf{b} \cdot \mathbf{a} - b^2 = 0,$$
$$a^2 - b^2 = 0,$$
$$a = b.$$

3-28

(a) Picture the origin of the coordinate system shown in Fig. 7 as being at the center of the Earth and the vector **a** to point to Washington. The magnitude of **a** is the radius of the Earth, which equals 6370 km. Adopting the hint, we have, by Eqs. 7,

$$a_x = 6370 \sin51°\cos77° = 1113.6 \text{ km},$$
$$a_y = 6370 \sin51°\sin77° = 4823.5 \text{ km},$$
$$a_z = 6370 \cos51° = 4008.8 \text{ km}.$$

Now go back to Fig. 7 and picture **a** pointing to Manila; let us call the vector **A** to avoid confusion with the vector to Washington. Again apply Eqs. 7, noting that the longitude of Manila is east:

$$A_x = 6370 \sin75°\cos(-121°) = -3169.0 \text{ km},$$
$$A_y = 6370 \sin75°\sin(-121°) = -5274.1 \text{ km},$$
$$A_z = 6370 \cos75° = 1648.7 \text{ km}.$$

The displacement vector of the tourist is the straight line vector, passing through the solid Earth, from Washington to Manila (obviously not the actual route taken). This displacement vector **D** is given by **a** + **D** = **A**, so that **D** = **A** - **a**. Subtracting corresponding components as given above yields, in km,

$$\textbf{D} = -4282.6\textbf{i} - 10{,}097.6\textbf{j} - 2310.1\textbf{k}.$$

(b) The magnitude of the displacement vector found above is

$$D = \sqrt{[(-4282.6)^2 + (-10{,}097.6)^2 + (-2310.1)^2]} = 11{,}220 \text{ km}.$$

## 3-33

(a) The angle between **r** and **s** is 320° - 85° = 235°. By Eq. 13,

$$\textbf{r}\cdot\textbf{s} = (4.5)(7.3)\cos235° = -19.$$

(b) Picking the smaller angle between the vectors and applying Eq. 16 gives for the magnitude of the vector product

$$|\textbf{r} \times \textbf{s}| = (4.5)(7.3)\sin(360° - 235°) = 27.$$

The direction of the vector product is given by the right hand rule, Fig. 17(a), and is in the +z direction.

## 3-38

The magnitudes of the vectors are $a = \sqrt{[3^2 + 3^2 + 3^2]} = 5.196$ and $b = \sqrt{[2^2 + 1^2 + 3^2]} = 3.742$. By Eq. 13, $\textbf{a}\cdot\textbf{b} = (5.196)(3.742)\cos\phi$. Now apply Eq. 15: $\textbf{a}\cdot\textbf{b} = (3)(2) + (3)(1) + (3)(3) = 18$. Hence we have

$$\phi = \cos^{-1}[18/(5.196)(3.742)] = 22.2°.$$

## 3-41

(a) See Fig. 28 in HRK. The vectors **a** and **b** are at a right angle (90°) so that $\textbf{a}\cdot\textbf{b} = 0$.
(b) The three vectors form a closed figure, so that **a** + **b** + **c** = 0 and therefore we have

$$\textbf{a}\cdot\textbf{a} + \textbf{a}\cdot\textbf{b} + \textbf{a}\cdot\textbf{c} = 0,$$

16

$$a^2 + 0 + \mathbf{a} \cdot \mathbf{c} = 0$$
$$\mathbf{a} \cdot \mathbf{c} = -a^2 = -16.$$

(b) This time, take $\mathbf{b} \cdot (\mathbf{a} + \mathbf{b} + \mathbf{c}) = 0$ to find

$$\mathbf{b} \cdot \mathbf{a} + \mathbf{b} \cdot \mathbf{b} + \mathbf{b} \cdot \mathbf{c} = 0,$$
$$0 + b^2 + \mathbf{b} \cdot \mathbf{c} = 0,$$
$$\mathbf{b} \cdot \mathbf{c} = -b^2 = -9.$$

## 3-45

See Fig. 29. The area of a triangle is area = $\frac{1}{2}$(base)(height). From the figure, we see that the most convenient base to use is that formed by the vector $\mathbf{a}$, of length $a$. The height is the dashed line perpendicular to $\mathbf{a}$ to the opposite angle. But, from the small right triangle to the left in the parallelogram, this height is $b\sin\phi$. Hence, area = $\frac{1}{2}(a)(b\sin\phi)$ = $\frac{1}{2}(ab\sin\phi)$ = $\frac{1}{2}|\mathbf{a} \times \mathbf{b}|$, by Eq. 16.

## 3-48

(a) By Eq. 15, $\mathbf{a} \cdot \mathbf{b} = (3.2)(0.5) + (1.6)(4.5) + (0)(0) = 8.80$. The magnitudes of the vectors are $a = \sqrt{[3.2^2 + 1.6^2]} = 3.58$ and $b = \sqrt{[0.5^2 + 4.5^2]} = 4.53$. Hence,

$$\mathbf{a} \cdot \mathbf{b} = ab\cos\phi,$$
$$8.80 = (3.58)(4.53)\cos\phi \rightarrow \phi = 57°.$$

(b) Write $\mathbf{c} = c_x\mathbf{i} + c_y\mathbf{j}$. In essence, we are told that $c_x^2 + c_y^2 = 25$ and $\mathbf{c} \cdot \mathbf{a} = 0$. By Eq. 15, this last requirement becomes

$$3.2c_x + 1.6c_y = 0 \rightarrow c_y = -2c_x.$$

Thus,

$$c_x^2 + (-2c_x)^2 = 25 \rightarrow c_x = \pm 2.2.$$

By the second equation above, we have then $c_y = -2(\pm 2.2) = -(\pm 4.4)$. That is, either we have $c_x = 2.2$, $c_y = -4.4$, or $c_x = -2.2$, $c_y = 4.4$.

4-2

(a) Into the given equation for **r**, we substitute $t = 2$ s to find

$$\mathbf{r} = [2(2)^3 - 5(2)]\mathbf{i} + [6 - 7(2)^4]\mathbf{j} = 6\mathbf{i} - 106\mathbf{j}.$$

(b) First find **v**$(t)$ from $\mathbf{v} = d\mathbf{r}/dt = (6t^2 - 5)\mathbf{i} - 28t^3\mathbf{j}$. Then, by substituting $t = 2$ we get $\mathbf{v}(2) = 19\mathbf{i} - 224\mathbf{j}$.
(c) Use the expression for **v**(2) found in (b); the acceleration is $\mathbf{a} = d\mathbf{v}/dt = 12t\mathbf{i} - 84t^2\mathbf{j}$. Now put $t = 2$ to find a(2) = 24**i** - 336**j**.

4-8

(a) At the maximum $x$ coordinate $v_x = 0$. From the given expressions for **v**$_0$ and **a**, we have $v_{x0} = 3.6$ m/s and $a_x = -1.2$ m/s$^2$. Hence

$$v_x = v_{x0} + a_x t,$$

$$0 = 3.6 + (-1.2)t \quad \rightarrow \quad t = 3 \text{ s}.$$

(b) The velocity in m/s is found from Eq. 11:

$$\mathbf{v} = \mathbf{v}_0 + \mathbf{a}t,$$

$$\mathbf{v}(3) = 3.6\mathbf{i} + (-1.2\mathbf{i} - 1.4\mathbf{j})(3) = -4.2\mathbf{j}.$$

(c) Use Eq. 12; we have **r**$_0$ = 0 since the particle is at the origin at time $t = 0$. Therefore,

$$\mathbf{r} = \mathbf{v}_0 t + \tfrac{1}{2}\mathbf{a}t^2,$$

$$\mathbf{r}(3) = (3.6\mathbf{i})(3) + \tfrac{1}{2}(-1.2\mathbf{i} - 1.4\mathbf{j})(3)^2 = 5.4\mathbf{i} - 6.3\mathbf{j}.$$

4-11

(a) Rolling off a horizontal surface means that $\phi_0 = 0$. The floor is at $y = -4.23$ ft (i.e. below the origin at the point where the ball rolls off the table) and $g = 32$ ft/s$^2$. By Eq. 22:

$$y = v_0 \sin\phi_0 t - \tfrac{1}{2}gt^2,$$

$$-4.23 = 0 - \tfrac{1}{2}(32)t^2 \quad \rightarrow \quad t = 0.514 \text{ s}.$$

(b) Now use Eq. 21 with $x = 5.11$ ft:

$$x = (v_0 \cos\phi_0)t,$$

$$5.11 = v_0(1)(0.514) \quad \rightarrow \quad v_0 = 9.94 \text{ ft/s}.$$

## 4-14

(a) Since the rifle is fired horizontally $\phi_0 = 0°$ and $\sin 0 = 0$. Up is positive. By Eq. 22 (recalling that 1 in. = 1/12 ft),

$$y = v_0 \sin\phi_0 t - \tfrac{1}{2}gt^2$$

$$(-0.75)(1/12) = 0 - \tfrac{1}{2}(32)t^2 \quad \rightarrow \quad t = 0.0625 \text{ s.}$$

(b) For the horizontal motion use Eq. 21:

$$x = v_0 \cos\phi_0 t,$$

$$130 = v_0 \cos 0°(0.0625) \quad \rightarrow \quad v_0 = 2080 \text{ ft/s.}$$

## 4-18

(a) We have $v_0 = 15$ m/s, $\phi_0 = 360° - 20° = 340°$, $t = 2.3$ s. By Eq. 21, $x = v_0 \cos\phi_0 t = (15)(\cos 340°)(2.3) = 32.4$ m.

(b) Use Eq. 22:

$$y = v_0 \sin\phi_0 t - \tfrac{1}{2}gt^2 = (15)(\sin 340°)(2.3) - \tfrac{1}{2}(9.8)(2.3)^2,$$

which gives $y = -37.7$ m. The negative sign for $y$ indicates that, as expected, the ball is below the point of projection at this time.

## 4-23

(a) We are given that $v_0 = 120$ ft/s and $\phi_0 = 62°$. The height $h$ sought corresponds to the value of $y$ at time $t = 5.5$ s. By Eq. 22,

$$y = (120)(\sin 62°)(5.5) - \tfrac{1}{2}(32)(5.5)^2 = 99 \text{ ft.}$$

(b) Use Eq. 13 written in scalar form:

$$v^2 = v_0^2 - 2gy,$$

$$v^2 = (120)^2 - 2(32)(99) \quad \rightarrow \quad v = 90 \text{ ft/s.}$$

(c) Here the $t = 5.5$ s no longer applies since that is the time to point A, not the time to reach maximum height which is clearly less. At the highest point, however, $v_y = 0$ so that $v^2 = v_x^2 = v_{x0}^2$ here. Using this in the scalar form of Eq. 13 gives

$$v^2 = v_0^2 - 2gy,$$

$$(120\cos 62°)^2 = 120^2 - 2(32)y \quad \rightarrow \quad y = H = 175 \text{ ft.}$$

Throughout, where necessary, we put $x_0 = y_0 = 0$ since we pick the origin at the point of projection.

## 4-28

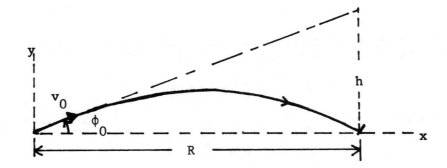

We presume the target to be at the same horizontal level as the bullet as it leaves the rifle. We need the angle $\phi_0$ at which the bullet emerges from the rifle with respect to the horizontal: see the sketch above. If we hit the target, then the horizontal range $R$ must equal the distance to the target. By Eq. 24,

$$R = (v_0^2/g)\sin2\phi_0,$$

$$150 = (1500^2/32)\sin2\phi_0 \rightarrow \phi_0 = 0.0611°.$$

From the sketch, we see that the desired aiming height $h = R\tan\phi_0$ = (150 ft)(tan0.0611°) = 0.160 ft = 1.92 in.


## 4-33

(a) As usual the origin is put at the point of projection, in this case at the summit of Mt. Fuji. By Eq. 21, $v_0t = x/\cos\phi_0$; substitute this into Eq. 22 to find

$$y = v_0\sin\phi_0 - \tfrac{1}{2}gt^2 = x\tan\phi_0 - \tfrac{1}{2}g(x/v_0\cos\phi_0)^2,$$

$$-3300 = 9400\tan35° - \tfrac{1}{2}(9.8)(9400/v_0\cos35°)^2 \rightarrow v_0 = 256 \text{ m/s.}$$

(b) Go back to Eq. 21:

$$9400 = 256\cos35°t \rightarrow t = 44.8 \text{ s.}$$


## 4-37

(a) Refer to Fig. 31. The origin is at the plane as the decoy is released. The angle $\phi_0$ (counterclockwise from the +x axis) is 360° -27° = 333°. Also $v_0$ = 180 mi/h = (180)(1.467 ft/s) = 264 ft/s (the initial speed of the decoy equals the speed of the plane at the moment of release). By Eq. 21,

$$x = (v_0\cos\phi_0)t,$$

$$2300 = 264\cos333°t \rightarrow t = 9.78 \text{ s.}$$

(b) With the time $t$ found, Eq. 22 can be used to determine the height:

$$y = (264\sin333°)(9.78) - \tfrac{1}{2}(32)(9.78)^2 = -2700 \text{ ft;}$$

i.e., the ground is 2700 ft below the point at which the decoy was released, so that the height of the plane above the ground is 2700 ft at this instant.


4-42

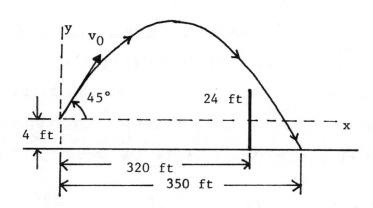

The situation, presuming the ball clears the fence, is shown in the sketch. The origin is at the point the ball is struck, so the ground is at $y = -4$ ft. The initial speed of the ball is unknown and must be found from the range of 350 ft that the ball will reach if there is no fence to (possibly) interfere with its flight. By Eq. 23, then, we have

$$-4 = (\tan45°)(350) - (16)(350/v_0\cos45°)^2 \quad \rightarrow \quad v_0 = 105.23 \text{ ft/s.}$$

Now use Eq. 23 again, but this time set $x = 320$ ft and solve for $y$; we find $y = 24$ ft. This means that at 320 ft from the plate the ball is 24 ft above the point it was struck, or 28 ft above the ground. Since the fence is 24 ft high, the ball clears the fence by about 4 ft.


4-43

Put the origin at the point where the ball is kicked. We use Eq. 23. Note that $gx^2/2v_0^2 = (9.8)(50)^2/2(25)^2 = 19.6$ m. Therefore, Eq. 23 becomes

$$3.44 = 50\tan\phi_0 - (19.6)/\cos^2\phi_0.$$

Now $1/\cos^2\phi_0 = \sec^2\phi_0 = \tan^2\phi_0 + 1$ (see Appendix H). Using this, the equation above can be written in the form

$$(19.6)\tan^2\phi_0 - (50)\tan\phi_0 + 23.04 = 0.$$

This is a quadratic equation with the unknown $\tan\phi_0$; use the quadratic formula to find $\tan\phi_0 = 1.9474$ and $0.6036$. These solutions correspond to angles of $\phi_0 = 63°$ and $31°$ and these are the angular limits sought.

## 4-49

The origin is put at the antitank gun. Find the time of flight $t$ of the shell from Eq. 22:

$$y = v_0\sin\phi_0 t - \tfrac{1}{2}gt^2,$$

$$-60 = 240\sin10°t - \tfrac{1}{2}(9.8)t^2,$$

$$4.9t^2 - 41.676t - 60 = 0 \rightarrow t = -1.255 \text{ s, } 9.760 \text{ s.}$$

Hence $t = 9.760$ s (the negative solution corresponds to motion before $t = 0$; the shell is not in free fall before firing so this solution has no meaning). Now compute the horizontal range of the shell by Eq. 21:

$$x = v_0\cos\phi_0 t = (240\cos10°)(9.760) = 2307 \text{ m.}$$

For a hit, the tank must be at $x = 2307$ m also. Thus, the tank must travel $2307 - 2200 = 107$ m if it wants to be hit. The time $t$ needed for the tank to travel this distance is given from

$$x = \tfrac{1}{2}a_{\text{tank}}t^2,$$

$$107 = \tfrac{1}{2}(0.9)t^2 \rightarrow t = 15.42 \text{ s.}$$

Hence, the gun crew should wait $15.42 - 9.76 = 5.66$ s before firing.

## 4-52

(a) We have $a = 6.8(9.8) = 66.6$ m/s$^2$ and $r = 5.2$ m. Use Eq. 28: $v = \sqrt{[ar]} = \sqrt{[(66.6)(5.2)]} = 18.6$ m/s.
(b) Since the speed is constant, the time required for one revolution is $t = 2\pi r/v$, since the distance that is traveled in one revolution is the circumference of the circular path. Putting in the numbers gives $t = 2\pi(5.2)/(18.6) = 1.76$ s. Therefore, the number of revolutions completed in 1 min = 60 s is $60/1.76 = 34.1$.

## 4-55

(a) This distance is the circumference of the circular path, and this equals $2\pi r = 2\pi(0.15) = 0.94$ m.
(b) The time required for one revolution is $(60 \text{ s})/1200 = 0.05$ s. Hence the speed is $v = (0.94)/(0.05) = 19$ m/s.
(c) Find the acceleration from Eq. 28: $a = v^2/r = 19^2/(0.15) = 2400$ m/s$^2$.

<u>4-57</u>

(*a*) The time needed for one revolution is evidently 1 second. Therefore, the speed is $v = 2\pi r/t = 2\pi(20 \times 10^3 \text{ m})/(1 \text{ s}) = 1.26 \times 10^5 \text{ m/s}$.
(*b*) By Eq. 28, $a = v^2/r = (1.26 \times 10^5 \text{ m/s})^2/(20 \times 10^3 \text{ s}) = 7.94 \times 10^5 \text{ m/s}^2$.


<u>4-60</u>

We can find the time of flight of the stone from the moment the string breaks until it strikes the ground from Eq. 22 with $\phi_0 = 0$; we find $t = \sqrt{[2h/g]} = \sqrt{[2(1.9)/(9.8)]} = 0.623$ s (we insert all data in SI base units). From Eq. 21 we find the speed of the stone when the string broke: $v = x/t = 11/(0.623) = 17.7$ m/s. Finally, from Eq. 28, $a = v^2/r = (17.7)^2/(1.4) = 224 \text{ m/s}^2$.


<u>4-63</u>

See the sketch at the right; The radius $r$ of the circle traveled during the day is given by $r = R\cos L$, where $R = 6370$ km is the radius of the Earth and $L$ is the latitude of the person. The time required for one revolution is one day = 86,400 s. Therefore, the speed $v$ with which the person is moving due to the rotation of the Earth is, since the speed is constant,

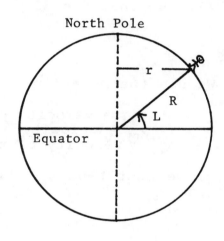

$$v = 2\pi r/t = 2\pi R\cos L/t,$$

$$v = 2\pi(6370 \times 10^3 \text{ m})\cos 40°/(86,400 \text{ s}) = 355 \text{ m/s}.$$

The acceleration is $a = v^2/r = v^2/R\cos L$, so we have

$$a = (355 \text{ m/s})^2/[(6370 \times 10^3 \text{ m})\cos 40°] = 0.026 \text{ m/s}^2 = 2.6 \text{ cm/s}^2.$$


<u>4-67</u>

The person can walk at a speed of $(15 \text{ m})/(90 \text{ s}) = 0.167$ m/s. The escalator moves at a speed of $(15 \text{ m})/(60 \text{ s}) = 0.250$ m/s. Thus, the speed of the person walking on the moving escalator, relative to the building, is $0.167 + 0.250 = 0.417$ m/s. Therefore, the time needed to cover the length of the escalator is $(15 \text{ m})/(0.417 \text{ m/s}) = 36.0$ s. To see that the length of the escalator does not enter, redo the calculation assuming that the escalator is, say, 35 m long, or do the problem algebraically.

Use the following notation, in the spirit of Eq. 43:

$$v_{sc} = \text{velocity of snow relative to car,}$$
$$v_{sg} = \text{velocity of snow relative to ground,}$$
$$v_{cg} = \text{velocity of car relative to ground.}$$

The snow falls vertically relative to the ground and the car moves horizontally relative to the ground; thus the vectors $v_{sg}$ and $v_{cg}$ are at 90°. Hence, the relation between the vectors, as given in Eq. 43, $v_{sc} + v_{cg} = v_{sg}$, is as shown on the sketch. Since 55 km/h = 15.3 m/s, the angle sought is given from $\tan\theta = 15.3/7.8$ so that $\theta = 63°$.

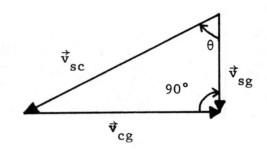

4-78

(a) We use the notation

$$v_{pg} = \text{velocity of plane relative to ground,}$$
$$v_{pa} = \text{velocity of plane relative to the air,}$$
$$v_{ag} = \text{velocity of the air relative to the ground.}$$

Hence, we must have $v_{pa} + v_{ag} = v_{pg}$. The speeds are $v_{pa} = v_{pg} = 135$ mi/h and $v_{ag} = 70$ mi/h, so that the lengths of all the vectors are

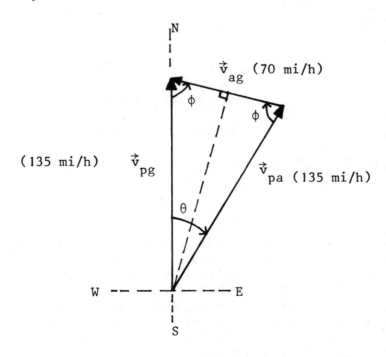

known. As far as the directions are concerned, it is given that $v_{pg}$ is directed to the north, so draw this in first. The other two vectors are arranged to obey the rule for addition of vectors by the graphical method (see Section 3-2 in RHK). The triangle formed is an isosceles, not a right, triangle. (A solution is also possible with the angle $\theta$ drawn counterclockwise from north.) The wind direction is given by the angle $\phi$. To calculate $\phi$, draw the perpendicular bisector (dashed line); this completes a right triangle and bisects the opposite side, so that

$$\phi = \tan^{-1}(35/135) = 75.0°,$$

east of south.

(b) The heading of the plane is the angle $\theta$; since the angles of a triangle must total 180°, we have $\theta = 180° - 2\phi = 30°$, east of north.


## 4-83

In Eq. 46, we identify $v_{PS'} = v_{PB} = 0.42c$, $v_{S'S} = v_{BA} = 0.63c$. Therefore $v_{PS} = v_{PA}$ is given by

$$v_{PA} = \frac{0.42c + 0.63c}{1 + (0.42c)(0.63c)/c^2} = 0.83c.$$

5-4

The force is given by $F = ma$, Newton's second law, so we must find the acceleration $a$ first. Use Eq. 20 of Chapter 2; since the neutron is brought to rest, $v = 0$, so we find

$$0^2 = (1.4 \times 10^7 \text{ m/s})^2 + 2a(1.0 \times 10^{-14} \text{ m}),$$
$$a = -9.8 \times 10^{27} \text{ m/s}^2.$$

We can drop the minus sign since we are asked for the magnitude of the force:

$$F = ma = (1.67 \times 10^{-27} \text{ kg})(9.8 \times 10^{27} \text{ m/s}^2) = 16.4 \text{ N}.$$

5-5

Since the people pull in opposite directions, the net horizontal force is $F = 92 - 90 = 2$ N. The resulting acceleration is given by $a = F/m = (2 \text{ N})/(25 \text{ kg}) = 0.080 \text{ m/s}^2 = 8.0 \text{ cm/s}^2$.

5-7

The initial speed of the car (and passenger) is $v_0 = 53$ km/h = 14.7 m/s and the final speed $v = 0$; the passenger moves a distance $x = 0.65$ m in being brought to rest. With this data now all in SI base units, we can apply Eq. 20 of Chapter 2 to find the acceleration of the passenger in being brought to rest: $a = - (14.7 \text{ m/s})^2/2(0.65 \text{ m}) = -166 \text{ m/s}^2$. We can discard the minus sign (which indicates that the passenger is undergoing deceleration) since in problems of this sort the magnitude of the force is being sought (this is often not stated explicitly). Therefore we have $F = ma = (39 \text{ kg})(166 \text{ m/s}^2) = 6500$ N.

5-15

(a) The acceleration of the sled is $a = F/m = (5.2 \text{ N})/(8.4 \text{ kg}) = 0.62 \text{ m/s}^2$.
(b) We use upper case letters for the girl. Newton's third law requires that the force exerted by the sled on the girl equals, in magnitude, the force exerted by the girl on the sled. Hence, her acceleration is $A = F/M = (5.2 \text{ N})/(40 \text{ kg}) = 0.13 \text{ m/s}^2$.
(c) Since they start from rest, the distances traveled by the sled and the girl in the time $t$ required for them to meet are given by $x = \frac{1}{2}at^2$ and $X = \frac{1}{2}At^2$, each measured from their starting points. Since they were separated by 15 m to begin with,

$$x + X = 15,$$

$$\tfrac{1}{2}(0.62)t^2 + \tfrac{1}{2}(0.13)t^2 = 15 \quad \rightarrow \quad t = 6.32 \text{ s.}$$

Therefore, the distance from the girl's original position to the meeting place is $X = \tfrac{1}{2}At^2 = \tfrac{1}{2}(0.13 \text{ m/s}^2)(6.32 \text{ s})^2 = 2.6 \text{ m.}$

## 5-18

(a) We use Eq. 4: $W = mg = (75 \text{ kg})(9.8 \text{ m/s}^2) = 735 \text{ N.}$
(b) For the weight on Mars, it is only necessary to use the value of $g$ on Mars in Eq. 4: $W = mg = (75 \text{ kg})(3.72 \text{ m/s}^2) = 279 \text{ N.}$
(c) We assume that $g = 0$ in interplanetary space (which is not strictly true); under this assumption, $W = 0$, by Eq. 4.
(d) At each of these locations, Earth, Mars, interplanetary space, the mass of the space traveler is 75 kg, mass being an intrinsic property of the object.

## 5-21

We need the mass of the automobile: $m = W/g = (3900 \text{ lb})/(32 \text{ ft/s}^2) = 121.9$ slug. The net force is $F = ma = (121.9 \text{ slug})(13 \text{ ft/s}^2) = 1580$ lb.

## 5-22

Find the acceleration from Eq. 15 in Chapter 2; noting that 1620 km/h = 450.0 m/s we obtain, using SI base units,

$$v = v_0 + at,$$

$$450 = 0 + a(1.82) \quad \rightarrow \quad a = 247.3 \text{ m/s}^2.$$

The net force is $F = ma = (523 \text{ kg})(247.3 \text{ m/s}^2) = 1.29 \times 10^5$ N. We can write this more compactly as $F = 129$ kN.

## 5-25

(a) Draw a free-body diagram of the sphere showing the forces acting on the sphere: $F_e$, the electric force, $W = mg$, the weight of the sphere (this acts vertically down as always) and $T$, the tension in the string. The sphere remains at rest, so its acceleration $a = 0$. By Newton's second law, this means that the sums of the horizontal and vertical force components must each equal zero. From the diagram, we see that this condition requires that

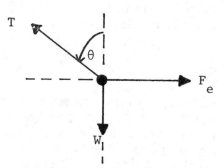

$$Tsin\theta - F_e = 0,$$
$$Tcos\theta - mg = 0.$$

Eliminating $T$ between these equations gives

$$F_e = mgtan\theta = (2.8 \times 10^{-4})(9.8)tan33° = 1.78 \times 10^{-3} \text{ N},$$

or $F_e = 1.78$ mN.
(b) With this result, the tension $T$ can be found from the second equation:

$$T = mgsec\theta = (2.8 \times 10^{-4})(9.8)sec33° = 3.27 \text{ mN}.$$

5-27

We choose up as positive, so that the acceleration and the weight, being directed downward, point in the negative direction. Let $f$ be the retarding force. Apply Newton's second law, in SI base units,

$$F = ma,$$
$$-mg + f = ma,$$
$$-(0.25)(9.8) + f = (0.25)(-9.2) \rightarrow f = 0.150 \text{ N}.$$

5-34

The upward directed tension $T$ in the cord will be less than the weight $W$ of the object if the object is lowered with a downward acceleration $a$ (if $a = 0$, then $T = W$). If we choose down as positive, Newton's second law becomes

$$F = ma,$$
$$W - T = (W/g)a.$$

To find the minimum acceleration $a$ needed, set the tension equal to the breaking strength. We find, using British units,

$$100 - 87 = (100/32)a \rightarrow a = 4.16 \text{ ft/s}^2.$$

In short, to avoid breaking the cord, lower the object with an acceleration $a \geq 4.16$ ft/s²; the greater the acceleration, the smaller the tension.

5-37

First, find the constant acceleration $a$ needed to bring the elevator to rest. Up is positive. Use Eq. 20 of Chapter 2 but write $y$ instead of $x$ to indicate vertical motion. We have $y_0 = 0$ (as usual), $y = -42$ m, $v_0 = -12$ m/s (moving down), $v = 0$ (brought to rest). From the cited equation, and using SI base units throughout, we obtain $a = (-12)^2/2(42) = 1.714$ m/s². This turns out to be positive so the acceleration, as expected, is directed upward. Now sketch a free-body diagram of the elevator; we still take up as positive. Newton's second law takes the form

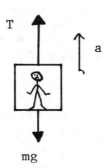

$$T - mg = ma.$$

Substituting data in SI base units:

$$T - (1600)(9.8) = (1600)(1.714),$$

$$T = 18,400 \text{ N}.$$

## 5-45

(a) Descending vertically with constant speed means that the acceleration is zero. In this case, the net force on the landing craft must be zero. Hence, the upward directed rocket thrust $R$ must just equal the downward directed weight $W$, so that $W = R = 3260$ N. (b) With the thrust reduced, the acceleration is no longer zero, but is directed downward. Up is positive. The rocket thrust is still exerted upward. Therefore, Newton's second law, using SI base units, gives

$$R - W = ma,$$
$$2200 - 3260 = m(-0.390) \quad \rightarrow \quad m = 2720 \text{ kg}.$$

(c) Combining the results of (a) and (b) yields $g = W/m = 3260/2720 = 1.20$ m/s$^2$.

## 5-49

Let $L$ be the upward lift of the air on the balloon and $m$ the mass of ballast discarded. Up is positive. Newton's second law, written for conditions before and after dropping the ballast are, with $a$ representing the magnitude of the acceleration,

$$L - Mg = M(-a),$$
$$L - (M - m)g = (M - m)a.$$

Solve for $L$ in the first equation and substitute into the second equation, and then solve for $m$ to get $m = 2Ma/(a + g)$.

## 5-56

(a) For the two blocks treated as a single entity, Newton's second law gives

$$F = (m_1 + m_2)a.$$

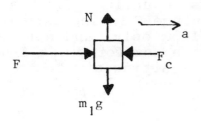

29

Now focus on block $m_1$ and draw a free-body diagram for it (see p. 29). $F_c$ is the contact force that is exerted by block $m_2$; this force retards the motion of $m_1$. Apply Newton's second law to $m_1$ for the horizontal forces to get

$$F - F_c = m_1 a.$$

Now solve for the acceleration $a$ in the first equation and substitute the result into the second equation; then solve the resulting equation for $F_c$ to obtain

$$F_c = \frac{m_2}{m_1 + m_2} F = \frac{1.2}{2.3 + 1.2} (3.2) = 1.10 \text{ N}.$$

(b) Convince yourself, perhaps by drawing a new figure, that this new situation can be described by simply switching the labels on the blocks; that is, interchange $m_1$ and $m_2$ in the formula for $F_c$ above to get for the new contact force

$$F_c = \frac{m_1}{m_2 + m_1} F = \frac{2.3}{1.2 + 2.3} (3.2) = 2.10 \text{ N}.$$

5-58

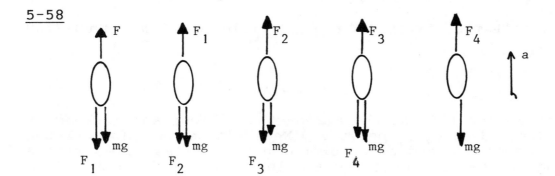

(a) In applying Newton's second law to each link, the following quantities are needed:

$$mg = (0.10 \text{ kg})(9.8 \text{ m/s}^2) = 0.98 \text{ N},$$

$$ma = (0.10 \text{ kg})(2.5 \text{ m/s}^2) = 0.25 \text{ N}.$$

Let $F_i$ (various values of i) represent the forces between adjacent links, counting from the top pair. Then, using Newton's third law, and since $F$ is the external force lifting the chain, exerted on the top link only, and noting that, in this problem, $F$ does not stand for the net force, we find by applying Newton's second law to each link starting at the top (up is positive),

$$F - F_1 - mg = ma,$$
$$F_1 - F_2 - mg = ma,$$
$$F_2 - F_3 - mg = ma,$$
$$F_3 - F_4 - mg = ma,$$
$$F_4 - mg = ma.$$

The last equation gives

$$F_4 = mg + ma = 0.98 + 0.25 = 1.23 \text{ N.}$$

Substituting this into the next to last equation above gives the value of $F_3$, which can then be put into the preceding equation to give $F_2$, and so on. The results obtained are: $F_3 = 2.46$ N, $F_2 = 3.69$ N, and $F_1 = 4.92$ N.
(b) With $F_1$ known, the first equation above gives $F = 6.15$ N. (This result can also be obtained by considering the entire chain as a single object of mass $5m$.)
(c) The net force on each link, by Newton's second law, is $F_{net} = ma = 0.25$ N.

<u>5-59</u>

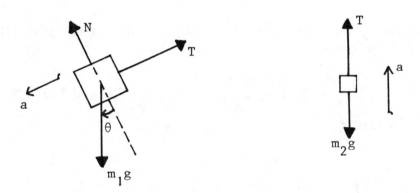

(a) Let $T$ be the tension in the cord. A choice must be made for the direction of the acceleration $a$ as it is not given; we assume that $m_1$ slides down and therefore that $m_2$ accelerates up. Newton's second law for $m_1$ (force components parallel to the plane) and for $m_2$ yields

$$m_1 g \sin\theta - T = m_1 a,$$
$$T - m_2 g = m_2 a.$$

Solve for $T$ in the second equation and substitute the resulting expression into the first equation and then solve for $a$:

$$T = m_2(a + g),$$
$$m_1 g \sin\theta - m_2(a + g) = m_1 a,$$

$$a = \frac{m_1 \sin\theta - m_2}{m_1 + m_2} g = \frac{(3.7)\sin 28° - 1.86}{3.7 + 1.86}(9.8) = -0.217 \text{ m/s}^2.$$

31

As this is negative, the acceleration actually is in the direction opposite to what was assumed; i.e., $m_1$ accelerates up the incline. The value (magnitude) of the acceleration is as found above.
(b) From the third equation above, using the result for the acceleration found in (a) we find

$$T = m_2(a + g) = (1.86)(-0.217 + 9.8) = 17.8 \text{ N}.$$

5-61

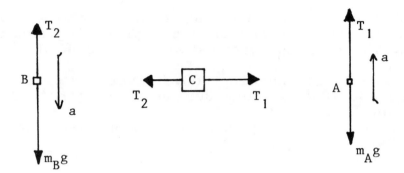

(a) Draw free-body diagrams of the cage (A), counterweight (B) and mechanism (C). For the cage,

$$T_1 - m_A g = m_A a,$$
$$T_1 = m_A(g + a) = (1000)(9.8 + 2.3) = 12,100 \text{ N} = 12.1 \text{ kN}.$$

(b) For the counterweight,

$$T_2 - m_B g = m_B a,$$

$$T_2 = m_B(g + a) = (1400)(9.8 - 2.3) = 10.5 \text{ kN}.$$

Note that with the elevator accelerating upward, the counterweight is accelerating downward; i.e., in the negative direction, so we put $a = -2.3 \text{ m/s}^2$ for the counterweight.
(c) The net force $\mathbf{F} = \mathbf{T}_1 + \mathbf{T}_2$ exerted by the cable on the mechanism equals $F = T_1 - T_2 = 12.1 - 10.5 = 1.6 \text{ kN}$, to the right as pictured. By Newton's third law, the force exerted by the mechanism on the cable is 1.6 kN to the left.


5-66

If we assume that the support cable exerts no retarding force (analogous to a frictionless inclined plane) then we have

$$\Delta T - mg\sin\theta = ma,$$

$$\Delta T - (2800)(9.8)\sin35° = (2800)(0.81) \quad \rightarrow \quad \Delta T = 18.0 \text{ kN},$$

noting that SI base units are used in the calculation.

6-4

The only vertical forces acting are the player's weight $W$ and the normal force $N$. Hence, $N = W = mg = (79 \text{ kg})(9.8 \text{ m/s}^2) = 774.2$ N. Since $f_k = \mu_k N$, we have $\mu_k = f_k/N = 470/774.2 = 0.61$.

6-7

(a) The maximum braking force (assuming no sliding) is the maximum force of static friction between tires and road (see Sample Problem 2). On a level road,

$$f_s = \mu_s N = \mu_s mg = (0.62)(1500)(9.8) = 9100 \text{ N}.$$

(b) On the downgrade the normal force is $N = mg\cos\theta$ (see Sample Problem 1), so we have

$$f_s = \mu_s N = \mu_s mg\cos\theta = (0.62)(1500)(9.8)\cos 8.6° = 9000 \text{ N}.$$

6-9

(a) If the block slips, it will slip down; hence the force $f_s$ of static friction points up. The block will not move in the horizontal direction, so that $N = F = 12$ lb. The maximum available force of static friction is $\mu_s N = (0.6)(12 \text{ lb}) = 7.2$ lb. As this is greater than the weight of the block (which is the force it opposes), the block will not move.

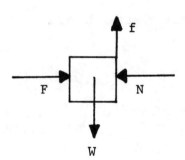

(b) The actual force of friction can be no greater than the net force it opposes, in this case the weight of the block, for if it was greater the block would accelerate up the wall and this does not happen. Thus, the forces exerted by the wall on the block are a force $N = 12$ lb horizontally to the left and the force of static friction equal to 5 lb and directed up.

6-13

The sand, if dry so that there is no cohesion, will form a cone with a slant angle (with respect to the horizontal) given by (see Sample Problem 1) $\theta = \tan^{-1}\mu_s$ (the "angle of friction"). Hence, we have $h = R\tan\theta = \mu_s R$ and the volume $V = \frac{1}{3}\pi R^2 h = \frac{1}{3}\pi R^2(\mu_s R) = \frac{1}{3}\pi\mu_s R^3$.

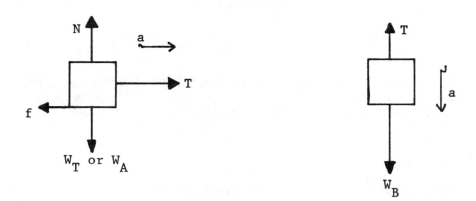

(a) With block C placed on block A, provided there is no slipping, the two blocks form, in essence, a single object with mass $m_T = m_A + m_C$. With C removed, only object A with mass $m_A$ remains on the horizontal surface. The force of friction is $f$. The free-body diagrams for the two objects are shown above. For this part use $m_T$, $f_s = \mu_s N$, and $a = 0$. In this case $N = m_T g$, so that Newton's second law applied to the two objects yields

$$T - \mu_s m_T g = 0,$$
$$T - m_B g = 0.$$

Eliminate the tension $T$ between these two equations to obtain

$$m_B - \mu_s(m_A + m_C) = 0,$$

$$m_C = (m_B/\mu_s) - m_A = (2.6/0.18) - 4.4 = 10 \text{ kg}.$$

(b) With block C removed use $m_A$ in place of $m_T$; $f_k = \mu_k N = \mu_k m_A g$; also $a \neq 0$ ($a$ = magnitude of the acceleration). Newton's second law gives

$$T - \mu_k m_A g = m_A a,$$

$$T - m_B g = m_B(-a).$$

Removing $T$ between these equations yields

$$a = \frac{m_B - \mu_k m_A}{m_B + m_A} g = \frac{2.6 - (0.15)(4.4)}{2.6 + 4.4}(9.8) = 2.72 \text{ m/s}^2.$$

(a) Let $T$ be the tension in the rope; $f$ is the force of friction. With the crate on the verge of moving, $f = f_s = \mu_s N$. Note that $N \neq W$, for part of the weight $W$ of the

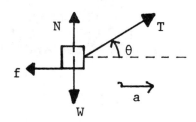

34

crate is balanced by the vertical component of $T$. Applying Newton's second law to the horizontal and vertical force components gives

$$Tcos\theta = f_s = \mu_s N,$$

$$N + Tsin\theta = W.$$

Eliminate the normal force $N$ and solve for the tension $T$:

$$Tcos\theta = \mu_s(W - Tsin\theta),$$

$$T = \frac{\mu_s W}{cos\theta + \mu_s sin\theta} = \frac{150(0.52)}{cos17° + 0.52sin17°} = 70 \text{ lb}.$$

(b) With the crate moving the force of friction becomes $f_k = \mu_k N$; Newton's second law now reads

$$Tcos\theta - \mu_k N = ma = (W/g)a,$$

$$N + Tsin\theta = W.$$

(Note that in problems involving British units, typically the weight is given, rather than the mass.) Eliminate the normal force and solve for the acceleration to get

$$a = g[\frac{T}{W}(cos\theta + \mu_k sin\theta) - \mu_k],$$

$$a = (32)[\frac{70}{150}(cos17° + 0.35sin17°) - 0.35] = 4.6 \text{ ft/s}^2.$$

We use the value of the tension found in part (a), since the problem statement implies that this does not change.


6-23

(a) Gravity tries to pull the block down the inclined surface with a force $Wsin\theta$, where $W$ = weight of the block. The resisting force is static friction, of which the maximum value is $f_s = \mu_s Wcos\theta$. If $f_s > Wsin\theta$ the block does not slide; i.e., if

$$\mu_s mgcos\theta > mgsin\theta \rightarrow \mu_s > tan\theta.$$

Since $\mu_s = 0.63$ and $tan24° = 0.45$ the condition is satisfied. (b) Now the force tending to move the block down the incline is $mgsin\theta + F$. The maximum resisting force is still $\mu_s mgcos\theta$. The block will be on the verge of sliding if

$$mgsin\theta + F = \mu_s mgcos\theta,$$

$$F = mg[\mu_s \cos\theta - \sin\theta],$$

$$F = (1.8 \times 10^7)(9.8)[0.63\cos24° - \sin24°] = 30 \text{ MN},$$

since $1 \text{ MN} = 1 \times 10^6$ N.

6-31

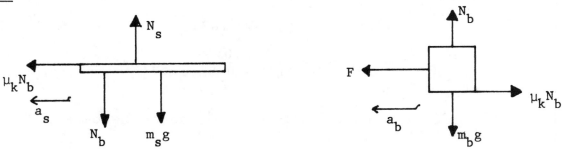

(a) Assume, for the moment, that the block slips on the slab. It is essential to note that, by Newton's third law, forces of friction of equal magnitude act on each of the objects; in fact, it is solely a force of friction that causes the slab to accelerate. Let $F$ be the 110 N force acting on the block. As usual, apply Newton's second law using the free-body diagrams as a guide. Since $N_b = m_b g$, the equations are

$$F - \mu_k m_b g = m_b a_b,$$

$$110 - (0.38)(9.7)(9.8) = 9.7 a_b \;\rightarrow\; a_b = 7.62 \text{ m/s}^2.$$

(b) For the slab,

$$\mu_k m_b g = m_s a_s,$$

$$a_s = \mu_k (m_b/m_s) g = (0.38)(9.7/42)(9.8) = 0.860 \text{ m/s}^2.$$

It remains to be determined whether the block does, in fact, slip on the slab. (If it does not, then the block and slab move together as a single object.) The maximum force of static friction "available" is

$$f_{s,\text{max}} = \mu_s m_b g = (0.53)(9.7)(9.8) = 50.4 \text{ N} < 110 \text{ N},$$

indicating that static friction cannot hold the block in place and the motion of the system is, indeed, as outlined above.


6-33

(a) We choose the SI unit data. Noting that 1 kN = 1000 N, we find for the mass of the car $m = W/g = (10.7 \times 10^3)/(9.8) = 1092$ kg. The required friction force is

$$f = mv^2/r = (1092)(13.4)^2/(61) = 3210 \text{ N.}$$

(b) If the wheels do not slide, then $f = f_s = \mu_s N = \mu_s mg$ and we have

$$\mu_s = f_s/mg = v^2/rg = 0.300$$

after substituting data in SI base units.

## 6-40

Consider the hanging cylinder: since it remains at rest, the tension in the cord must equal its weight, i.e., $T = Mg$. Now examine the disk. The only centripetal force acting on the disk is the tension $T$. Therefore, combining these results gives

$$T = mv^2/r = Mg \rightarrow v = \sqrt{(Mgr/m)}.$$

## 6-42

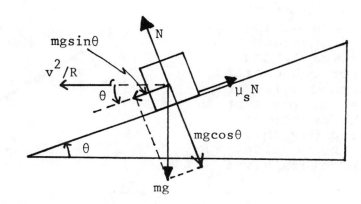

(a) First, convert the speeds to SI base units; we find $v_1 = 95$ km/h = 26.4 m/s and $v_2 = 52$ km/h = 14.4 m/s. By Eq. 8,

$$\tan\theta = v_1^2/Rg,$$

since $v_1$ is the speed for which the curve was designed. This relation is used in the algebra that follows; be alert. At speed $v_2 < v_1$, Newton's second law, applied to force components parallel to the incline, yields

$$mg\sin\theta - \mu_s(mg\cos\theta) = ma = m\left[\left(\frac{v_2^2}{R}\right)\cos\theta\right],$$

$$\mu_s = \tan\theta - \frac{v_2^2}{Rg} = \left(\frac{1}{Rg}\right)[v_1^2 - v_2^2],$$

$$\mu_s = \left(\frac{1}{(210)(9.8)}\right)[(26.4)^2 - (14.4)^2] = 0.238.$$

(b) If the car slides while traveling with a speed $v_2 > v_1$, it will slide up the incline; this reverses the direction of the friction force from that in the diagram above, so we have

$$mg\sin\theta + \mu_s mg\cos\theta = m[(v_2^2/R)\cos\theta],$$

$$v_2^2 = Rg\mu_s + v_1^2 = (210)(9.8)(0.24) + (26.4)^2,$$

$$v_2 = 34.5 \text{ m/s} = 124 \text{ km/h}.$$

## 6-47

Let $r$ be the radius of the circle. The forces acting on the plane are its weight $mg$ which acts vertically down and the lift force $L$ exerted by the air perpendicular to the wings. The acceleration is directed to the center of the circle. Applying Newton's second law in the vertical and horizontal directions gives

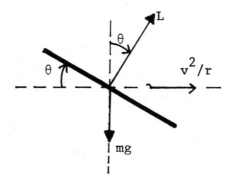

$$L\cos\theta = W = mg,$$
$$L\sin\theta = mv^2/r.$$

Eliminating $L$, and recalling that $\sin\theta/\cos\theta = \tan\theta$, we find that

$$r = v^2/g\tan\theta = (133.9)^2/(9.8\tan38.2°) = 2.32 \text{ km},$$

since $v = 482$ km/h $= 133.9$ m/s.

## 6-52

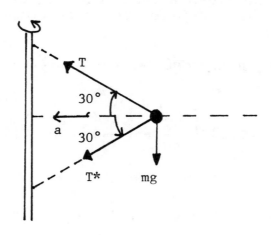

38

(a) The acceleration of the ball is directed horizontally toward the rod, that is, to the center of the ball's circular path. Hence, the vertical force components must sum to zero. Since the angle between the cords and the horizontal is 30°, we have

$$T\sin30° - T*\sin30° - mg = ma_v = 0,$$

$$35\sin30° - T*\sin30° - (1.34)(9.8) = 0,$$

which gives $T* = 8.74$ N.

(b) The net force $F$ acts in the direction of the acceleration, and therefore toward the rod. Thus,

$$F = T\cos30° + T*\cos30° = (T + T*)\cos30°,$$

$$F = (35 + 8.74)\cos30° = 37.9 \text{ N}.$$

(c) $F = ma = mv^2/r$, where $r$ is the perpendicular distance of the ball to the rod (i.e., the radius of the circular path). This distance is $r = 1.7\cos30° = 1.472$ m and therefore

$$v^2 = rF/m = (1.472)(37.9)/(1.34) \rightarrow v = 6.45 \text{ m/s}.$$

6-54

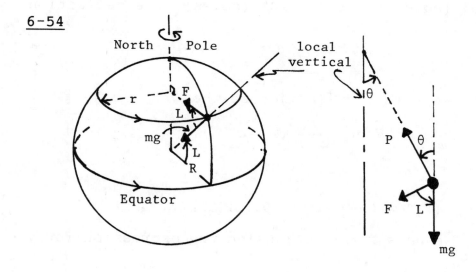

(a) Due to the rotation of the Earth, the plumb bob is moving at constant speed in a circle of radius $r = R\cos L$. As a result, there must be a net force on the bob directed at 90° to the axis of rotation of the Earth. This force makes an angle $L$ with the local vertical and has a magnitude

$$F = \frac{mv^2}{r} = \frac{m(2\pi R\cos L/T)^2}{R\cos L},$$

where $m$ is the mass of the bob. The local situation is shown in the figure at the right above. Since $F = mg + P$, where $P$ is the tension in the cord, resolving this equation into horizontal and vertical components gives

$$mg - P\cos\theta = F\cos L,$$
$$P\sin\theta = F\sin L.$$

Eliminating the tension $P$ between these equations yields

$$\tan\theta = \frac{F\sin L}{mg - F\cos L}.$$

Now $\theta$ is a small angle so that $\tan\theta = \theta$ if $\theta$ is expressed in radians. Also, $mg \gg F\cos L$, because the Earth does not rotate very rapidly. To a good degree of accuracy, then,

$$\theta = \frac{F\sin L}{mg} = \frac{2\pi^2 R\sin 2L}{gT^2}.$$

(b) The maximum value of $\theta$ occurs where $\sin 2L = 1$, or $L = 45°$. With $R = 6.37 \times 10^6$ m, $g = 9.8$ m/s$^2$, and $T = 86,400$ s, we find for the maximum deflection $\theta = 0.00172$ rad = 5'55".
(c) At either $L = 0°$ (equator) or $L = 90°$ (poles), the deflection is $\theta = 0$.


6-57

(a) Since $a = dv/dt$, we have $v = \int (F/m)dt$ so that

$$v = \frac{F_0}{m} \int_0^t e^{-t/T}dt = \frac{F_0 T}{m}(1 - e^{-t/T}),$$

so that

$$v(T) = (F_0 T/m)(1 - e^{-1}) = 0.632 F_0 T/m.$$

(b) Since $v = dx/dt$, we have $x = \int v\, dt$; using the expression for $v$ found above, we find

$$x = \frac{F_0 T}{m} \int_0^t (1 - e^{-t/T})\, dt = \frac{F_0 T}{m}[t + Te^{-t/T} - T],$$

so that

$$x(T) = (F_0 T^2/m)e^{-1} = 0.368 F_0 T^2/m.$$


6-58

The acceleration is zero when the pebble falls at the terminal speed; hence the net force on the pebble is zero. Therefore, the upward force of the water must equal the downward force of gravity, in magnitude, which is $mg = (0.150 \text{ kg})(9.8 \text{ m/s}^2) = 1.47$ N. Thus, the drag force due to the water is 1.47 N, directed upward.

## 6-59

The forces on the falling object are the weight $mg$ down and the drag force $D$; the drag force is directed vertically upward since the object, dropped from rest, falls vertically down. Hence, Newton's second law gives

$$mg - bv^2 = ma.$$

When $v$ reaches the value $v_T$, the drag force equals the weight, so that the acceleration vanishes (becomes zero). The equation above then reads

$$mg - bv_T^2 = 0 \quad \rightarrow \quad v_T = \sqrt{(mg/b)}.$$

## 6-63

The net force on a balloon descending at constant speed is zero. Hence the weight of the balloon must equal, in magnitude, the sum of the upward directed buoyant force (10.3 kN) and upward directed drag force ($bv^2$). Before the dropping of ballast, this condition requires that (using SI base units)

$$10,800 = 10,300 + b(1.88)^2 \quad \rightarrow \quad b = 141.5 \text{ N} \cdot \text{s}^2/\text{m}^2.$$

After dropping ballast, the weight of the balloon is reduced by the weight of dropped ballast: i.e., weight = $10,800 - (26.5)(9.8) = 10,540$ N. The constant $b$ still has the value found above. When the balloon again reaches a constant downward speed, the condition of net upward force = net downward force (in magnitude) requires

$$10,540 = 10,300 + (141.5)v^2 \quad \rightarrow \quad v = 1.30 \text{ m/s}.$$

41

7-3

(a) The worker exerts the 120 N force in the same direction as the crate is sliding so that, by Eq. 1, the work done by the worker is

$$W = Fs\cos\phi = (120 \text{ N})(3.6 \text{ m})\cos 0° = 432 \text{ J}.$$

(b) The work done by gravity is $\pm mgh$, where $h$ is the vertical displacement of the crate. Since the weight (force of gravity) is directed vertically down, we must use the negative sign if the crate moves up the incline. In our case, the crate does move up. The vertical displacement is $h = (3.6 \text{ m})\sin 27° = 1.63 \text{ m}$, so that the work done by gravity is

$$W = -mgh = -(25 \text{ kg})(9.8 \text{ m/s}^2)(1.63 \text{ m}) = -399 \text{ J}.$$

(Note that RHK gives the answer to 2 sig fig, to match the data.)
(c) The normal force, as its name implies, acts at 90° to the surface of the incline. As long as the crate stays on the surface of the incline, no matter how it moves the angle between the normal force and the displacement is 90°. Since $\cos 90° = 0$, Eq. 2 tells us that the work done by the normal force is zero.

7-5

(a) The force exerted by the cord is the tension $T$, which can be found by applying Newton's second law:

$$Mg - T = M(\tfrac{1}{4}g),$$

$$T = \tfrac{3}{4}Mg.$$

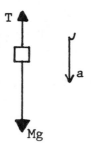

The tension pulls up, while the displacement is down; hence $\phi = 0°$, so that the work done by $T$ is

$$W = Td\cos\phi = (\tfrac{3}{4}Mg)(d)\cos 180° = -\tfrac{3}{4}Mgd.$$

(b) The weight acts vertically down, which is parallel and in the same direction as the displacement. Thus, the work done by gravity is $W = (Mg)(d)\cos 0° = Mgd$.

7-8

(a) Since the block slides down at constant velocity, the net force parallel to the incline must be zero. With the block sliding down, the friction force points up the incline and if this direction is taken as positive, we have

$$f_k + P - mg\sin\theta = 0,$$

where $P$ is the force exerted by the worker. But $f_k = \mu_k N = \mu_k mg\cos\theta$ so we have

$$P = mg(\sin\theta - \mu_k\cos\theta).$$

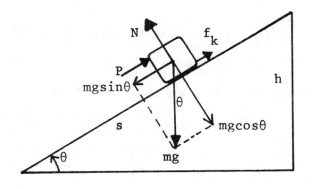

We need the value of $\theta$: writing $h$ = height and $s$ = length of the incline,

$$\sin\theta = h/s = (0.902 \text{ m})/(1.62 \text{ m}) \quad \rightarrow \quad \theta = 33.83°.$$

Using this in the previous equation, we have

$$P = (47.2 \text{ kg})(9.8 \text{ m/s}^2)(\sin 33.83° - 0.11\cos 33.83°) = 215 \text{ N}.$$

(b) With $P$ found, we can calculate the work done by the worker; since the worker pushes up the incline as the ice slides down, $\phi = 180°$ (do not confuse $\phi$ with $\theta$, the angle of the incline); we find

$$W = Ps\cos\phi = (215 \text{ N})(1.62 \text{ m})\cos 180° = -348 \text{ J}.$$

(c) The work done by gravity is $W = \pm mgh$. Since the block slides down, we use the + sign and obtain

$$W = mgh = (47.2 \text{ kg})(9.8 \text{ m/s}^2)(0.902 \text{ m}) = 417 \text{ J}.$$

## 7-12

(a) Converting the data to SI base units, we have for the force constant $k = (15 \text{ N})/(0.01 \text{ m}) = 1500 \text{ N/m}$, and $x = 7.6 \text{ mm} = 0.0076 \text{ m}$. The work done by the force extending the spring is

$$W = \tfrac{1}{2}kx^2 = \tfrac{1}{2}(1500 \text{ N/m})(0.0076 \text{ m})^2 = 0.0433 \text{ J} = 43.3 \text{ mJ}.$$

(b) Eq. 9 gives the work done by the spring. The work done by the agent extending the spring is the negative of this. Furthermore, we have $x_i = 7.6 \text{ mm}$ and $x_f = 7.6 + 7.6 = 15.2 \text{ mm}$. The work done in extending the spring this additional 7.6 mm is

$$W = \tfrac{1}{2}kx_f^2 - \tfrac{1}{2}kx_i^2 = \tfrac{1}{2}(1500)[(0.0152)^2 - (0.0076)^2] = 130 \text{ mJ}.$$

## 7-19

Use Eq. 18, $K = \tfrac{1}{2}mv^2$. The mass of the electron can be found on the inside front cover of RHK (SOME PHYSICAL CONSTANTS). This mass is in kg, the SI base unit. The SI base unit of energy is the joule J, so we must convert the quoted kinetic energy from eV to J; see the inside front cover of RHK (SOME CONVERSION FACTORS); we find

$$4.2 \text{ eV} = (4.2 \text{ eV})(1.6 \text{ X } 10^{-19} \text{ J/eV}) = 6.72 \text{ X } 10^{-19} \text{ J}.$$

Now we can use Eq. 18 to find the speed $v$:

$$v = \sqrt{\frac{2K}{m_e}} = \sqrt{\frac{2(6.72 \text{ X } 10^{-19} \text{ J})}{9.11 \text{ X } 10^{-31} \text{ kg}}} = 1.2 \text{ X } 10^6 \text{ m/s}.$$

## 7-24

(a) The initial speed of the car is $v_i$ = 46 km/h = 12.8 m/s, so that its initial kinetic energy is $K_i = \frac{1}{2}mv_i^2 = \frac{1}{2}(1100)(12.8)^2$ = 90.1 kJ. After applying the brakes, the kinetic energy of the car is $K_f$ = 90.1 kJ - 51 kJ = 39.1 kJ = 39,100 J. Since $K_f = \frac{1}{2}mv_f^2$, we find $v_f$ = 8.43 m/s = 30 km/h.
(b) When the car is at rest its kinetic energy is zero; evidently, then, an additional 39.1 kJ of kinetic energy must be removed.

## 7-29

(a) Calculate the kinetic energy $K$ directly from Eq. 18:

$$K = \frac{1}{2}mv^2 = \frac{1}{2}(8.38 \text{ X } 10^{11} \text{ kg})(3 \text{ X } 10^4 \text{ m/s})^2 = 3.771 \text{ X } 10^{20} \text{ J},$$

$$K = (3.771 \text{ X } 10^{20} \text{ J})/(4.2 \text{ X } 10^{15} \text{ J/Mton}) = 89,800 \text{ Mton},$$

where Mton = "Megatons of TNT".
(b) The energy $E$ and the crater diameter $D$ are related by $D = AE^{1/3}$, where $A$ is a constant that must be evaluated. We assume that all of the kinetic energy is set free in the explosion, so identify $E$ with $K$. We are told that $D$ = 1 km if $E$ = 1 Mton. Therefore

$$A = D/E^{1/3} = (1 \text{ km})/(1 \text{ Mton})^{1/3} = 1 \text{ km/Mton}^{1/3}.$$

Hence, for the comet,

$$D = AE^{1/3} = (1 \text{ km/Mton}^{1/3})(89,800 \text{ Mton})^{1/3} = 44.8 \text{ km}.$$

## 7-31

If $v_f$ is the speed with which the ball leaves the floor, then since, by Eq. 25 of Chapter 2, $v_f^2 = 2gy$, we have

$$\frac{1}{2}mv_f^2 = mgy.$$

Similarly, if $v_i$ is the speed with which the ball hit the floor after being thrown down from height $y$ with speed $v_0$, we have from the same equation quoted above,

$$v_i{}^2 = v_0{}^2 + 2gy \quad \rightarrow \quad \tfrac{1}{2}mv_i{}^2 = \tfrac{1}{2}mv_0{}^2 + mgy.$$

Converting percent (%) to decimals, we are told that

$$\tfrac{1}{2}mv_f{}^2 = (1.00 - 0.15)(\tfrac{1}{2}mv_i{}^2).$$

Using our previous results, this last equation can be recast as

$$mgy = (0.85)(\tfrac{1}{2}mv_0{}^2 + mgy) \quad \rightarrow \quad v_0{}^2 = 2(0.15)gy/(0.85),$$

$$v_0 = (0.594)\sqrt{[gy]} = (0.594)\sqrt{[(9.8)(12.4)]} = 6.55 \text{ m/s}.$$

## 7-35

The woman does work $W$ against gravity in climbing the stairs. Use Eq. 22 and substitute the data in the SI base units given:

$$P = \frac{W}{t} = \frac{mgh}{t} = \frac{(57)(9.8)(4.5)}{3.5} = 720 \text{ W}.$$

## 7-37

Use Eq. 24, $P = Fv$. It is implied that the speed is constant so that, if the swimmer moves in a straight line, the velocity is constant also. In this case of zero acceleration, the force $F$ exerted by the swimmer on the water must be equal in magnitude and opposite in direction to the drag force exerted by the water on the swimmer, so that $F_{net} = 0$. Hence

$$P = Fv = (110 \text{ N})(0.22 \text{ m/s}) = 24 \text{ W}.$$

## 7-40

By Eq. 22, the time $t$ is given by $t = W/P$. If we can ignore all resistive forces, then the work $W$ done by the car equals the net work $W_{net}$. By Eq. 19, $W_{net} = \Delta K = K_f - K_i = K_f$, since $K_i = 0$ with the car starting from rest. Therefore,

$$W = K_f = \tfrac{1}{2}mv_f{}^2 = \tfrac{1}{2}(1230)(29.1)^2 = 5.208 \times 10^5 \text{ J}$$

using SI base units. Recalling that 1 kW = 1000 W = 1000 J/s, we find for the time

$$t = \frac{W}{P} = \frac{5.208 \times 10^5 \text{ J}}{92.4 \times 10^3 \text{ J/s}} = 5.64 \text{ s}.$$

7-45

(a) 1 kW•h = (1000 J/s)(3600 s) = 3.6 X $10^6$ J. Gasoline delivers 140 MJ/gal, so the number of gallons required to obtain 1 kW•h is (3.6 X $10^6$ J)/(140 X $10^6$ J/gal) = 0.0257 gal. At a mileage of 30 mi/gal, this quantity of gasoline will carry the car a distance given by (0.0257 gal)(30 mi/gal) = 0.77 mile.
(b) The time rate of consumption of gasoline while moving at 55 mi/h is (55 mi/h)/(30 mi/gal) = 1.83 gal/h. Hence, the rate of expenditure of chemical energy from the gasoline is

$$P = (1.83 \text{ gal/h})(140 \text{ MJ/gal}) = 256 \text{ MJ/h},$$

$$P = (256 \text{ X } 10^3 \text{ kJ})/(3600 \text{ s}) = 71 \text{ kW}.$$


7-50

(a) It is implied that the force acting on the particle is the net force so we can use Eq. 17, but we need the values of $v_i$ and $v_f$. At any time, the velocity is $v = dx/dt = 3 - 8t + 3t^2$. Therefore, $v_i$ = $v(t = 0)$ = 3 m/s, and $v_f = v(t = 4 \text{ s})$ = 19 m/s. Since the mass is 2.8 kg, we find $W = \frac{1}{2}(2.8)[19^2 - 3^2]$ = 493 J.
(b) By Eq. 24, the rate of doing work, or the power, is $P = Fv$. Use the expression for $v$ found in (a) to find $v(t = 3 \text{ s})$ = 6 m/s. As for the force, we have $F = ma$. But the acceleration $a = dv/dt$; use the expression for $v$ found in (a) and evaluate the derivative to get $a = -8 + 6t$. Therefore $a(t = 3 \text{ s})$ = 10 m/s$^2$ and $F = (2.8)(10)$ = 28 N. Finally, $P(t = 3 \text{ s}) = (28 \text{ N})(6 \text{ m/s})$ = 168 W.


7-54

The power developed is given by $P = Fv$, where $F$ is the force resisting the rotation of the wheel and $v$ is the speed of a point on the rim of the wheel (where the force is applied). Now $F$ is not the force $P$ with which the tool is held against the wheel; rather, we identify $F$ with the force of kinetic friction: $F = f_k = \mu_k N = \mu_k P$;

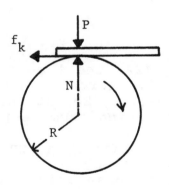

we set $P = N$ by assuming that $P$ is directed to the axle of the wheel. The wheel makes one revolution in time $t = 1/2.53$ = 0.3953 s, so the speed of a point on the rim is given by $v = 2\pi R/t$ = $2\pi(0.207 \text{ m})/(0.3953 \text{ s})$ = 3.290 m/s. Therefore

$$P = Fv = [\mu_k N]v = [(0.32)(180 \text{ N})](3.29 \text{ m/s}) = 190 \text{ W}.$$

The power required from the motor is $P = F_{motor}v$. By Newton's second law,

$$F_{motor} - F = ma$$

where $F$ is the resistive force. At a speed $v = 80$ km/h $= 22.2$ m/s, we find from the given formula,

$$F = 300 + (1.8)(22.2)^2 = 1190 \text{ N}.$$

Also $ma = (W/g)a = (12{,}000/9.8)(0.92) = 1130$ N. Therefore $F_{motor} = 1190 + 1130 = 2320$ N, so that

$$P = F_{motor}v = (2320 \text{ N})(22.2 \text{ m/s}) = 51.50 \text{ kW} = 69.0 \text{ hp},$$

since 1 hp = 746 W.

7-62

(a) Since $v/c = 0.999$, which is close to 1, we must use Eq. 25, not Eq. 18, for the kinetic energy. We first evaluate $mc^2$, and use eV for the energy units:

$$mc^2 = (9.11 \times 10^{-31} \text{ kg})(3 \times 10^8 \text{ m/s})^2/(1.6 \times 10^{-13} \text{ J/MeV}),$$

which gives $mc^2 = 0.512$ MeV. From Eq. 25, then,

$$K = (0.512 \text{ MeV}) \left[ \frac{1}{\sqrt{1 - 0.999^2}} - 1 \right] = 10.92 \text{ MeV}.$$

(b) The new speed is $v = (1 + 0.0005)(0.999c) = 0.9995c$. In this case, we find from Eq. 25 that $K = 15.68$ MeV, an increase of $\Delta K = 15.68 - 10.92 = 4.76$ MeV. The percent increase over the original kinetic energy is $(4.76 \text{ MeV}/10.92 \text{ MeV})(100\%) = 44\%$.

# CHAPTER 8

## 8-1

We assume that the muzzle speed is $v_0$, the speed of the projectile as it passes through the point ($x_0 = 0$) when the spring is, for an instant, in its relaxed state. Eq. 17 applies and can be rearranged to solve for the force constant $k$. Noting that $v_0 = 10$ km/s $= 10$ X $10^3$ m/s, we find

$$k = m[v_0/x_m]^2 = (2.38 \text{ kg})[(10 \text{ X } 10^3 \text{ m/s})/(1.47 \text{ m})]^2 = 110 \text{ MN/m},$$

where 1 MN = 1 X $10^6$ N.

## 8-4

We assume that the man leaves the window with initial speed equal to zero, so that his initial kinetic energy is zero. Also, the stretched net brings him momentarily to rest so his final kinetic energy is zero. Hence, the potential energy stored in the stretched net will equal the gravitational potential energy lost in falling the distance $\Delta y = 36 + 4.4 = 40.4$ ft, and

$$U_{net} = \Delta U_{grav} = (mg)(\Delta y) = (220 \text{ lb})(40.4 \text{ ft}) = 8900 \text{ ft}\bullet\text{lb},$$

noting that weight = $mg$ = 220 lb.

## 8-8

Apply conservation of mechanical energy $E_i = E_f$, or

$$\tfrac{1}{2}mv_i^2 + mgy_i = \tfrac{1}{2}mv_f^2 + mgy_f.$$

We can set $y = 0$ at the initial position of the ball, so that $y_i = 0$ and $y_f = L$. Since the ball "just reaches" the vertically upward position we have $v_f = 0$. The preceding equation can now be solved for $v_i$:

$$\tfrac{1}{2}mv_i^2 + 0 = 0 + mgL \quad \rightarrow \quad v_i = \sqrt{[2gL]}.$$

## 8-11

At the bottom of the ramp put $y_i = 0$, so that $U_i = mgy_i = 0$ there. The truck's energy is initially (at the bottom of the ramp) all kinetic energy. At the point on the escape ramp where the truck is brought to rest $v_f = 0$ and therefore $K_f = 0$. The potential energy there is $U_f = mgy_f$. In terms of the distance $L$ along the ramp, we have $y_f = L\sin 15°$. The mass of the truck is not given; call it $m$ (and hope that it will cancel out). Equating the mechanical energy $E = mgy + \tfrac{1}{2}mv^2$ at the bottom of the ramp to the energy where the truck is brought to rest gives

$$0 + \tfrac{1}{2}mv_i{}^2 = mg(L\sin 15°) + 0.$$

But $v_i$ = 80 mi/h = 80(1.467 ft/s) = 117 ft/s. Note that $m$ does drop out. We then have

$$\tfrac{1}{2}(117 \text{ ft/s})^2 = (32 \text{ ft/s}^2)(L\sin 15°) \quad \rightarrow \quad L = 830 \text{ ft}.$$

## 8-14

(a) With the stone in equilibrium, the upward force of the spring on the stone must equal, in magnitude, the weight of the stone:

$$kx = mg,$$
$$k(0.102 \text{ m}) = (7.94 \text{ kg})(9.8 \text{ m/s}^2) \quad \rightarrow \quad k = 763 \text{ N/m} = 7.63 \text{ N/cm}.$$

(b) With the stone pushed down the additional 28.6 cm, the total compression of the spring is $x$ = 28.6 + 10.2 = 38.8 cm = 0.388 m. The potential energy stored in the spring is now

$$U = \tfrac{1}{2}kx^2 = \tfrac{1}{2}(763 \text{ N/m})(0.388 \text{ m})^2 = 57.4 \text{ J},$$

using the result for the force constant from (a).

(c) The stone is at rest when the spring is released, and the stone is also at rest at the highest point of its path after release; hence $K_i = K_f = 0$. Therefore, the energy stored in the spring will be converted to gravitational potential energy of the stone at the top of the path, so that

$$U_f = mgy_f,$$
$$57.4 \text{ J} = (7.94 \text{ kg})(9.8 \text{ m/s}^2)y_f \quad \rightarrow \quad y_f = 0.738 \text{ m} = 73.8 \text{ cm}.$$

## 8-18

Assume that the work $mgy$ that Ms Meyfarth can do against gravity is the same at both locations. Her mass is the same on the Moon as on Earth, as also is the distance of her center of gravity above the ground. Recalling that "records" are heights above the ground, we have then, using SI base units,

$$\Delta U_{\text{Earth}} = \Delta U_{\text{Moon}},$$
$$m(9.8)(2.02 - 1.10) = m(1.67)(y - 1.10) \quad \rightarrow \quad y = 6.50 \text{ m}.$$

## 8-20

(a) First calculate the force constant $k$ of the spring: $k = F/x$ = (268 N)/(0.0233 m) = 1.150 X $10^4$ N/m. We choose to put $y$ = 0 at the "final" position of the block, when it comes to rest momentarily with the spring compressed; i.e., $y_f$ = 0. In the initial position

49

at the top of the incline $y_i = L\sin 32°$, where $L$ is the distance sought. The block is at rest in both the initial and final positions, so $K_i = K_f = 0$. Conservation of mechanical energy requires that

$$\tfrac{1}{2}mv_i^2 + mgy_i + \tfrac{1}{2}kx_i^2 = \tfrac{1}{2}mv_f^2 + mgy_f + \tfrac{1}{2}kx_f^2.$$

Initially the spring is relaxed (neither stretched nor compressed) so that $x_i = 0$. Inserting data in SI base units into the equation above yields

$$0 + (3.18)(9.8)(L\sin 32°) + 0 = 0 + 0 + \tfrac{1}{2}(1.150 \times 10^4)(0.0548)^2,$$

$$L = 1.05 \text{ m} = 105 \text{ cm}.$$

(b) As the block hits the spring $y_f = (5.48 \text{ cm})\sin 32° = 2.904 \text{ cm} = 0.02904$ m. The spring is not yet compressed, so $x_f = 0$. The initial conditions (block at the top of the incline) are unchanged. Thus, the conservation of mechanical energy equation, reads

$$(3.18)(9.8)(1.05\sin 32°) = \tfrac{1}{2}(3.18)v_f^2 + (3.18)(9.8)(0.02904),$$

$$v_f = 3.22 \text{ m/s}.$$

Note that we left the $y = 0$ position where it was in (a).


8-25

Choose $y = 0$ at the "final" position of the block, momentarily at rest, with the spring compressed a length $x$. At the initial position of the block, just as it is dropped, $y_i = x + 0.436$ (we use SI base units throughout). The initial and final kinetic energies both are zero. The force constant of the spring is $k = 18.6$ N/cm = 1860 N/m. By conservation of mechanical energy,

$$K_i + mgy_i + \tfrac{1}{2}kx_i^2 = K_f + mgy_f + \tfrac{1}{2}kx_f^2,$$

$$0 + (2.14)(9.8)(x + 0.436) + 0 = 0 + 0 + \tfrac{1}{2}(1860)x^2,$$

$$x^2 - 0.02255x - 0.009832 = 0 \quad \rightarrow \quad x = 0.111 \text{ m} = 11.1 \text{ cm}.$$

The method of solution of the last equation was by the quadratic formula (see Appendix H); the negative root was ingored.


8-30

(a) By Eq. 10, with $U(0) = 0$,

$$U(x) - U(0) = -\int_0^x (-3x - 5x^2)\,dx,$$

$$U(x) = \frac{3}{2}x^2 + \frac{5}{3}x^3.$$

From this last expression we can evaluate $U(x = 2.26 \text{ m}) = 26.9$ J. (b) Apply conservation of energy:

$$\tfrac{1}{2}mv_1^2 + U(x_1) = \tfrac{1}{2}mv_2^2 + U(x_2).$$

If we put $x_1 = 4.91$ m and $x_2 = 1.77$ m, then using the expression for $U(x)$ found in (a), we can obtain $U(4.91 \text{ m}) = 233.4$ J and $U(1.77 \text{ m}) = 13.94$ J. The energy conservation equation above now yields

$$\tfrac{1}{2}(1.18)(-4.13)^2 + 233.4 = \tfrac{1}{2}(1.18)v_2^2 + 13.94 \quad \rightarrow \quad v_2 = 19.7 \text{ m/s}.$$

## 8-33

If the ball just barely swings around the nail, then the tension in the string must vanish as the ball passes through the top of its circular path with speed $v$, leaving its weight as the only force acting. Noting that the radius of the circular path is $r = L - d$, Newton's second law for circular motion (see Eq. 5 in Chapter 6), and the conservation of energy require that

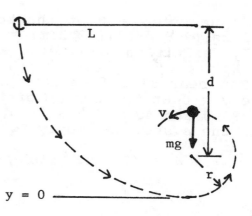

$$mv^2/(L - d) = mg,$$

$$mgL + 0 = mg[2(L - d)] + \tfrac{1}{2}mv^2.$$

In the energy equation, the second equation above, we set $y = 0$ at the bottom of the swing. Eliminating $v^2$ between these equations gives $d = 3L/5$.

## 8-36

Let $v$ be the speed with which the boy leaves the ice. Since he is given a very small push at the top, we can safely assume that his initial speed is zero. Energy conservation during the slide on the ice (no friction) gives

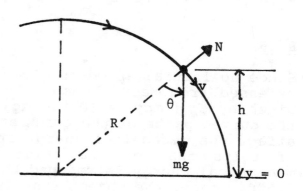

$$0 + mgR = \tfrac{1}{2}mv^2 + mgR\cos\theta,$$

we can be rearranged to read

$$mv^2/R = 2mg(1 - \cos\theta).$$

Now apply Newton's second law, in the radial direction, to get

$$mg\cos\theta - N = mv^2/R.$$

Combining the last two equations to eliminate the speed $v$ yields

$$mg\cos\theta - N = 2mg(1 - \cos\theta).$$

As the boy leaves the ice mound, the normal force $N$, a force of contact, goes to zero. Putting $N = 0$ in the last equation gives $\cos\theta = ⅔$. But $\cos\theta = h/R$ and therefore, at this point, $h = ⅔R$.


## 8-47

(a) If we put $y = 0$ at the bottom of the tree, then $U_i = mgy_i = (25.3\ kg)(9.8\ m/s^2)(12.2\ m) = 3020$ J.
(b) By definition of kinetic energy, at the bottom of the tree we have $K_f = \frac{1}{2}mv_f^2 = \frac{1}{2}(25.3\ kg)(5.56\ m/s)^2 = 391$ J.
(c) The bear slid from rest at the top, so $K_i = 0$. Also, $U_f = 0$ since $y = 0$ at the bottom of the tree. Thus, $\Delta K = K_f - K_i = 391 - 0 = 391$ J and $\Delta U = U_f - U_i = 0 - 3020 = -3020$ J. Consider the bear to be the system and apply Eq. 30 with $\Delta E_{int} = 0$:

$$W_f = \Delta U + \Delta K + \Delta E_{int} = -3020 + 391 = -2630\ J.$$


## 8-52

If $h$ is the height reached with friction acting and $H$ the height reached with the streamlined projectile, then Eq. 30 implies

$$K_i = mgh + W_f,$$
$$K_i = mgH,$$

where $W_f = 68$ kJ and $K_i$ is the kinetic energy of the projectile upon firing. Eliminating $K_i$ between these equations gives $H - h = W_f/mg = (68,000\ J)/(9.4\ kg)(9.8\ m/s^2) = 740$ m.


## 8-56

Since no initial speed is given, we can assume that the object was released from rest; hence, its mechanical energy upon release is given by $E_A = mgy_A = (0.234\ kg)(9.8\ m/s^2)(1.05\ m) = 2.408$ J. Now, the curved parts of the track are frictionless and therefore do not affect the mechanical energy of the object. Therefore, the number of trips that can be taken across the flat part before the mechanical energy is reduced to zero is (2.408 J)/(0.688 J/trip) = 3.500. Hence, the object finally comes to rest in the center of the flat part.

<u>8-57</u>

(a) With no friction, mechanical energy is conserved, so we can apply Eq. 12. Using SI base units and writing $m$ for the unspecified mass of the skier, we have

$$0 + m(9.8)(862) = \tfrac{1}{2}mv^2 + m(9.8)(741) \quad \rightarrow \quad v = 48.7 \text{ m/s}.$$

(b) To obtain $\Delta E_{int}$ of skies and snow we must consider the system consisting of the skier and the snow over which she moves. Eq. 31 applies here; also, it is implied that $v = v_0 = 0$; we obtain

$$\Delta E_{int} = -\Delta U - \Delta K,$$

$$\Delta E_{int} = -(54.4)(9.8)(741 - 862) - 0 = 64.5 \text{ kJ}.$$

<u>8-61</u>

(a) By Eq. 35, $E_0 = mc^2 = (0.120 \text{ kg})(3 \times 10^8 \text{ m/s})^2 = 1.08 \times 10^{16}$ J.
(b) Noting that 1.30 kW = 1300 J/s and that 1 year = $3.16 \times 10^7$ s, the time is

$$t = (1.08 \times 10^{16} \text{ J})/[(1300 \text{ J/s})(3.16 \times 10^7 \text{ s/y})] = 263{,}000 \text{ y}.$$

<u>8-66</u>

The energy needed is $E_0 = K = \tfrac{1}{2}Mv^2$ using, as we are instructed, the nonrelativistic formula for kinetic energy. The mass $M$ of the spaceship is (1820 ton)(907.2 kg/ton) = $1.651 \times 10^6$ kg. The final speed desired is $v = 0.1c = 3 \times 10^7$ m/s. Using these numbers, we find $E_0 = 7.430 \times 10^{20}$ J. The total mass of matter+antimatter that must be annihilated to yield this energy is given by $m = E_0/c^2 = (7.430 \times 10^{20} \text{ J})/(3 \times 10^8 \text{ m/s})^2 = 8260$ kg. [If the relativistic formula for kinetic energy is used, then $m = 8320$ kg.]

<u>8-70</u>

(a) We use Eq. 38; since the energies are in eV, use the value of Planck's constant in eV•s. Since $E_i - E_f = (-3.4 \text{ eV}) - (-13.6 \text{ eV}) = 10.2$ eV, $v = (10.2 \text{ eV})/(4.14 \times 10^{-15} \text{ eV•s}) = 2.46 \times 10^{15}$ s$^{-1}$.
(b) With $E_i > E_f$, the system gives up energy, so that the light is emitted.

# CHAPTER 9

## 9-2

See Eqs. 11a and 11b; read the x and y coordinates of each particle from the axes of the diagram. The total mass of the system is $M = 3$ kg + 4 kg + 8 kg = 15 kg. We find

$$Mx_{cm} = m_1x_1 + m_2x_2 + m_3x_3,$$

$$(15 \text{ kg})x_{cm} = (3 \text{ kg})(0) + (8 \text{ kg})(1 \text{ m}) + (4 \text{ kg})(2 \text{ m}),$$

$$x_{cm} = 1.07 \text{ m}.$$

Similarly,

$$My_{cm} = m_1y_1 + m_2y_2 + m_3y_3,$$

$$(15 \text{ kg})y_{cm} = (3 \text{ kg})(0) + (8 \text{ kg})(2 \text{ m}) + (4 \text{ kg})(1 \text{ m}),$$

$$y_{cm} = 1.33 \text{ m}.$$

## 9-3

Since we want the distance of the center of mass from the center of the Earth, put the center of the Earth at the origin; i.e. $x_E = 0$. As instructed, take needed data from Appendix C and apply Eq. 4; we find $M = 5.98 \times 10^{24}$ kg + $7.36 \times 10^{22}$ kg = $6.0536 \times 10^{24}$ kg so that

$$x_{cm} = \frac{(5.98 \times 10^{24} \text{ kg})(0) + (7.36 \times 10^{22} \text{ kg})(382{,}000 \text{ km})}{6.0536 \times 10^{24} \text{ kg}},$$

$$x_{cm} = 4640 \text{ km}.$$

The radius of the Earth is 6370 km, so the center of mass of the Earth-Moon system lies within the Earth.

## 9-11

(a) By the definition of center of mass, Eq. 4, with one of the bodies at the origin,

$$(2m)x_{cm} = m(0) + m(x) \quad \rightarrow \quad x_{cm} = \tfrac{1}{2}x = \tfrac{1}{2}(56 \text{ mm}) = 28 \text{ mm};$$

i.e., the center of mass lies midway between the bodies, as expected.
(b) Again apply Eq. 4; we put $x = 0$ at the heavier body; then, using grams and millimeters as units, we have

$$(1700)x_{cm} = (884)(0) + (816)(56) \quad \rightarrow \quad x_{cm} = 26.88 \text{ mm};$$

i.e., the center of mass has moved 28 - 26.88 = 1.12 mm toward the heavier body.

(c) In terms of the accelerations of the two bodies, the acceleration of the center of mass is given by Eq. 6:

$$a_{cm} = \frac{1}{M}[m_1a_1 + m_2a_2] = \frac{1}{M}[m_1a + m_2(-a)] = (\frac{\Delta m}{M})a,$$

where $M = m_1 + m_2$, $m_1 = 884$ g and $a$ is the acceleration of this body (directed down as this is the heavier of the two bodies). From Sample Problem 9 in Chapter 5 we have, in our present notation, $a = (\Delta m/M)g$, so that

$$a_{cm} = (\frac{\Delta m}{M})g = (\frac{68}{1700})g = 0.00160g,$$

directed down. In this last equation, $g$ = acceleration due to gravity (not grams).


9-14

Let the shore be at the origin of the $x$ axis, $x$ increasing outward from the shore. Also let

$x_{cm}$ = distance of center of mass of dog+boat from shore,
$x_d$ = distance of dog from shore,
$x_b$ = distance of center of mass of boat from shore.

Then, if $W$ = weight of boat, $w$ = weight of dog, and the star (*) denotes quantities after the dog takes its walk, then by the definition of center of mass, with masses "converted" to weights after multiplication by $g$, we have

$$(W + w)x_{cm} = Wx_b + wx_d,$$

$$(W + w)x_{cm}* = Wx_b* + wx_d*.$$

Since no net external force acts on the system, the conservation of linear momentum requires that $x_{cm} = x_{cm}*$; the equations above then require that

$$W(x_b* - x_b) = w(x_d - x_d*).$$

From Eq. 42 in Chapter 4, (displacement of dog relative to boat) = (displacement of dog relative to shore) + (displacement of shore relative to boat). Since $x$ increases away from the shore and the dog walks towards the shore (i.e., in the direction of negative $x$), the equation described in words above can be written

$$-8.50 = (x_d* - 21.4) - (x_b* - x_b),$$

$$x_b* - x_b = x_d* - 12.9.$$

Substitute this into the third equation above to get

$$W(x_d* - 12.9) = w(x_d - x_d*).$$

But $x_d$ = 21.4 ft, $W$ = 46.4 lb, and $w$ = 10.8 lb. The equation above then yields $x_d*$ = 14.5 ft.


9-16

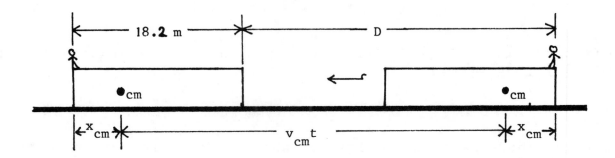

With the ice frictionless, the center of mass of iceboat+man must move with constant velocity even as the man changes his position on the iceboat by walking on it. The center of mass of iceboat+man is at a distance $x_{cm}$ from the end the man is standing at, where

$$(84.4 + 425)x_{cm} = 84.4(0) + (425)(9.1) \rightarrow x_{cm} = 7.592 \text{ m}.$$

The man takes a time $t$ = (18.2 m)/(2.08 m/s) = 8.750 s to walk from one end of the iceboat to the other. In this time, the center of mass of the system moves across the ice a distance

$$v_{cm}t = (4.16 \text{ m/s})(8.75 \text{ s}) = 36.40 \text{ m}.$$

The distance $D$ the iceboat moved in this time measured, say, by the motion of the end of the iceboat (see sketch above) is

$$D = 36.40 + 2(7.592) - 18.2 = 33.4 \text{ m}.$$


9-23

(a) Convert the speeds to SI base units: $v_i$ = 40 km/h = 11.11 m/s and $v_f$ = 50 km/h = 13.89 m/s. Kinetic energy is a scalar quantity so we have,

$$\Delta K = K_f - K_i = \tfrac{1}{2}(2000 \text{ kg})[(13.89 \text{ m/s})^2 - (11.11 \text{ m/s})^2] = 69.5 \text{ kJ}.$$

56

(b)

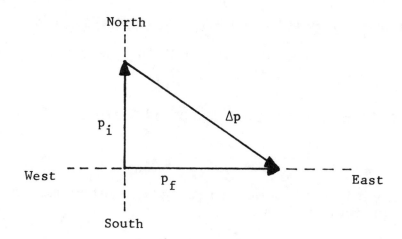

Momentum is a vector. If we choose east as the +x direction and north as the +y direction, we can write

$$\mathbf{p}_i = m\mathbf{v}_i = (2000 \text{ kg})(11.11 \text{ m/s})\mathbf{i} = (22{,}220 \text{ kg}\bullet\text{m/s})\mathbf{i},$$

$$\mathbf{p}_f = m\mathbf{v}_f = (2000 \text{ kg})(13.89 \text{ m/s})\mathbf{j} = (27{,}780 \text{ kg}\bullet\text{m/s})\mathbf{j}.$$

Therefore, in SI base units,

$$\Delta\mathbf{p} = \mathbf{p}_f - \mathbf{p}_i = -22{,}220\mathbf{i} + 27{,}780\mathbf{j}.$$

This vector has magnitude

$$\Delta p = [(22{,}220)^2 + (27{,}780)^2]^{\frac{1}{2}} = 3.56 \times 10^4 \text{ kg}\bullet\text{m/s}.$$

The direction is

$$\theta = \tan^{-1}(p_i/p_f) = \tan^{-1}(-22{,}220/27{,}780) = -38.7°,$$

i.e., clockwise from +x axis, or 38.7° south of east.


9-28

(a) Apply Eq. 22:

$$p = \frac{(9.11 \times 10^{-31} \text{ kg})[(0.99)(3 \times 10^8 \text{ m/s})]}{\sqrt{1 - 0.99^2}} = 1.92 \times 10^{-21} \text{ kg}\bullet\text{m/s}.$$

(b) The needed conversion factor is 1 MeV = $1.60 \times 10^{-13}$ J. We have

$$\text{kg}\bullet\text{m/s} = (\text{kg}\bullet\text{m/s})(c/c) = (\text{kg}\bullet\text{m/s})(3 \times 10^8 \text{ m/s})/c.$$

But $\text{kg}\bullet\text{m}^2/\text{s}^2 = \text{J} = 1/(1.6 \times 10^{-13})$ MeV, so that

$$kg \cdot m/s = (3 \times 10^8 \text{ J})/c = (3 \times 10^8 \text{ MeV})/(1.6 \times 10^{-13})c,$$

$$kg \cdot m/s = 1.875 \times 10^{21} \text{ MeV}/c.$$

Therefore,

$$1.92 \times 10^{-21} \text{ kg} \cdot m/s = (1.92 \times 10^{-21})(1.875 \times 10^{21}) = 3.60 \text{ MeV}/c.$$

## 9-30

Apply conservation of momentum to the system man+cart before and after the man jumps:

$$p_{before} = p_{after},$$

$$(75.2 \text{ kg} + 38.6 \text{ kg})(2.33 \text{ m/s}) = (75.2 \text{ kg})(0) + (38.6 \text{ kg})v,$$

$$v = 6.87 \text{ m/s}.$$

Hence, the speed of the cart increases by $6.87 - 2.33 = 4.54$ m/s.

## 9-35

(a) Let $V$, $v$ be the velocities of the rocket case, mass $M$, and payload, mass $m$, after the separation. The velocity of the center of mass of the rocket (which before separation is the same as the velocity of the rocket) is unchanged by the separation; therefore, if $u = 7600$ m/s (the velocity of the rocket before separation), this velocity can be written in terms of the velocities after separation according to Eq. 13 (taking account of our different notation) as

$$(M + m)u = MV + mv,$$

the initial direction taken as positive. The relative velocity after separation is $v - V = 910$ m/s (the less massive component will have the greater speed), and therefore, in SI base units:

$$(290 + 150)(7600) = 290(v - 910) + 150v \quad \rightarrow \quad v = 8200 \text{ m/s},$$

$$V = 8200 - 910 = 7290 \text{ m/s}.$$

(b) The kinetic energies before and after separation are

$$K_b = \tfrac{1}{2}(M + m)u^2 = \tfrac{1}{2}(440)(7600)^2 = 12.71 \text{ GJ},$$

$$K_a = \tfrac{1}{2}mv^2 + \tfrac{1}{2}MV^2 = \tfrac{1}{2}(150)(8200)^2 + \tfrac{1}{2}(290)(7290)^2 = 12.75 \text{ GJ}.$$

It seems that $12.75 - 12.71 = 0.04 \text{ GJ} = 0.04 \times 10^9 \text{ J} = 4 \times 10^7 \text{ J} = 40$ MJ of energy was stored in the compressed spring.

9-36

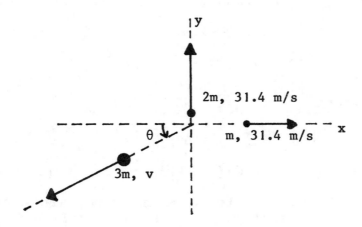

We are not told the masses of the pieces in kg, but can call them *m*, *2m*, and *3m*. The first two pieces go off in perpendicular directions, so we can put their momenta, one along the *x* axis and one along the *y* axis. Now, the vessel was at rest when in a single piece, so the initial momentum was zero. The momentum of the third piece, mass *3m* and speed *v*, must be in the third quadrant as shown, so that these three momenta can add up to zero. The speeds of *m* and *2m* are each 31.4 m/s. Conservation of momentum in the *x* and *y* directions yields

$$0 = m(31.4) - (3m)(v\cos\theta),$$
$$0 = (2m)(31.4) - (3m)(v\sin\theta);$$

we have left off the m/s units to avoid cluttering the equations. The unknown mass *m* cancels; eliminate *v* between the equations to find

$$\sin\theta/\cos\theta = \tan\theta = 2 \quad \rightarrow \quad \theta = 63.4°.$$

Putting this into the first equation above gives

$$v = (31.4)/3\cos63.4° = 23.4 \text{ m/s}.$$

9-39

(*a*) Apply conservation of momentum to the collision between the bullet after emerging from the first block with speed *v* with the second block. We convert the masses to kg (although this is not necessary) to write in SI base units

$$(0.00354)v = (1.78 + 0.00354)(1.48) \quad \rightarrow \quad v = 746 \text{ m/s}.$$

(*b*) Now apply conservation of momentum between the situations as pictured in Fig. 38(*a*) and Fig. 38(*b*); we write *V* for the speed of the bullet when first fired [i.e., as shown in Fig. 38(*a*)]:

$$(0.00354)V = (1.22)(0.63) + (1.78 + 0.00354)(1.48),$$

$$V = 963 \text{ m/s}.$$

9-47

(a) Applying Eq. 36 using SI base units, we have

$$F_{ext}S_{cm} = \Delta K_{cm},$$

$$F_{ext}(0.34) = 0 - \tfrac{1}{2}(116)(3.24)^2 \quad \rightarrow \quad F_{ext} = 1790 \text{ N}.$$

(b) We turn to Eq. 30. The railing does no work, since the point of application is fixed, so that $W = 0$. Also $\Delta U = mg\Delta h = 0$ (horizontal surface). Therefore, we find

$$\Delta U + \Delta K_{cm} + \Delta E_{int} = W,$$

$$\Delta E_{int} = -\Delta K_{cm} = + \tfrac{1}{2}(116)(3.24)^2 = 609 \text{ J}.$$

9-50

(a) The thrust $F$ is the last term in Eq. 43:

$$F = v_{rel}dM/dt = (3270 \text{ m/s})(480 \text{ kg/s}) = 1.57 \text{ X } 10^6 \text{ N}.$$

(b) Since the fuel burns at the constant rate $dM/dt$ we have

$$M_f = M_0 - (dM/dt)t,$$

$$M_f = 2.55 \text{ X } 10^5 \text{ kg} - (480 \text{ kg/s})(250 \text{ s}) = 1.35 \text{ X } 10^5 \text{ kg}.$$

(c) By Eq. 46,

$$v_f = -v_{rel}\ln(M_f/M_0),$$

$$v_f = -(3.27 \text{ km/s})\ln(1.35 \text{ X } 10^5 \text{ kg}/2.55 \text{ X } 10^5 \text{ kg}) = 2.08 \text{ km/s}.$$

9-53

For liftoff, the thrust $F$ must at least equal the weight $Mg$ of the rocket. The thrust is the last term in Eq. 43, so we must have

$$v_{rel}dM/dt = Mg,$$

$$v_{rel}(820 \text{ kg/s}) = (1.11 \text{ X } 10^5 \text{ kg})(9.8 \text{ m/s}^2),$$

$$v_{rel} = 1.33 \text{ km/s}.$$

Note that the mass of fuel does not enter; of course, the mass of fuel will determine for how long the rocket engine can operate.

Assume that the barges pass very close to each other, so that the coal is transported across at virtually zero speed in m/s. That is, the coal is ejected from the slow barge with a velocity relative to the slow barge of virtually zero. Hence, the force on the slow barge $F = v_{rel}dM/dt = 0$. But, although the coal arrives at the fast barge with a cross-stream speed of nearly zero, it arrives at a relative velocity of 21.2 - 9.65 = 11.55 km/h = 3.208 m/s, directed toward the stern of the barge. The force that is applied to the barge as the coal is brought to rest on the barge is

$$F = v_{rel}dM/dt = (3.208 \text{ m/s})(15.42 \text{ kg/s}) = 49.5 \text{ N}.$$

## 9-59

(*a*) We choose the SI data. In injesting the air the plane experiences a resisting force

$$F_1 = v_{rel}dM/dt = (184 \text{ m/s})(70.2 \text{ kg/s}) = 12,920 \text{ N}.$$

The mass ejected each second is the 70.2 kg of air plus 2.92 kg of fuel = 73.12 kg of exhaust. The thrust imparted to the plane in ejecting this is

$$F_2 = v_{rel}dM/dt = (497 \text{ m/s})(73.12 \text{ kg/s}) = 36,340 \text{ N}.$$

Therefore, the net thrust is 36,340 - 12,920 = 23,400 N.
(*b*) The delivered power, by Eq. 24 of Chapter 7, is

$$P = F_{engine}v = (23,400 \text{ N})(184 \text{ m/s}) = 4.31 \text{ MW}.$$

# CHAPTER 10

## 10-1

The initial momentum is $p_i = mv_i = (2300 \text{ kg})(15 \text{ m/s}) = 34,500 \text{ kg·m/s} = 34,500 \text{ N·s}$. The final momentum $p_f = 0$ because the car is brought to rest. Hence, by Eq. 3, the impulse on the car is $J = p_f - p_i = 0 - 34,500 = -34,500 \text{ N·s}$. With $\Delta t = 0.54$ s, Eq. 4 gives

$$\overline{F} = \frac{J}{\Delta t} = \frac{-34,500 \text{ N·s}}{0.54 \text{ s}} = -6.4 \times 10^4 \text{ N} = -64 \text{ kN}.$$

The negative sign indicates that the force is directed opposite to the initial velocity, which was taken to be in the positive direction.

## 10-7

We choose to work in SI base units, so that $m = 150$ g $= 0.150$ kg and $\Delta t = 4.70$ ms $= 0.0047$ s. Suppose that we take the direction from plate to pitcher as positive; then $v_i = -41.6$ m/s and $v_f = 61.5$ m/s. Combining Eqs. 3 and 4 yields

$$\overline{F} = \frac{J}{\Delta t} = \frac{m[v_f - v_i]}{\Delta t} = \frac{(0.150)[61.5 - (-41.6)]}{0.0047} = 3290 \text{ N}.$$

## 10-8

Eqs. 3 and 4, taken together for motion in one dimension, tell us that

$$\overline{F}(\Delta t) = mv_f - mv_i = m(v_f - v_i).$$

If we take the direction of $v_i$ as positive, we then get

$$(-984 \text{ N})(27 \times 10^{-3} \text{ s}) = (0.420 \text{ kg})(v_f - 13.8 \text{ m/s}),$$
$$v_f = -49.5 \text{ m/s}.$$

The ball is now moving in the direction opposite to its initial velocity, and its speed is 49.5 m/s.

## 10-13

Work in the reference frame of the spacecraft before separation. Then $v_i = 0$ for each of the parts into which the spacecraft separates. Their final speeds are

$$v_{1f} = J_1/m_1 = 300 \text{ N·s}/1200 \text{ kg} = 0.250 \text{ m/s},$$

$$v_{2f} = J_2/m_2 = 300 \text{ N} \bullet \text{s}/1800 \text{ kg} = 0.167 \text{ m/s}.$$

Now, by momentum conservation, the two parts move off in opposite directions. Hence, their relative speed is 0.250 + 0.167 = 0.417 m/s = 41.7 cm/s.


## 10-18

(a) The radius of a hailstone is $r = 0.5$ cm and therefore the volume of a hailstone is $V = 4\pi r^3/3 = 4\pi(0.5 \text{ cm})^3/3 = 0.524 \text{ cm}^3$. Since 1 $\text{cm}^3$ of ice has a mass of 0.92 g, the mass $m$ of a hailstone is $(0.524 \text{ cm}^3)(0.92 \text{ g/cm}^3) = 0.48$ g.
(b) The initial momentum (i.e., before hitting the roof) of a hailstone is

$$p_i = mv_i = (0.48 \text{ X } 10^{-3} \text{ kg})(25 \text{ m/s}) = 0.012 \text{ N} \bullet \text{s}.$$

The final momentum is zero since it is assumed that the hailstones come to rest on the roof immediately upon impact. Since the hail falls at 25 m/s, it follows that in a time $t = 1/25$ s = 0.04 s, all the hail within a distance $= v_i t = (25 \text{ m/s})(0.04 \text{ s}) = 1$ m from the roof will strike it during the next 0.04 s. The volume of air occupied by these hailstones is (10 m)(20 m)(1 m) = 200 $\text{m}^3$. With 120 hailstones/$\text{m}^3$, this means that $(120 \text{ /m}^3)(200 \text{ m}^3) = 24{,}000$ hailstones will strike the roof in a time of 0.04 s. The force that is exerted on the roof by these impacts is

$$F = \Delta p/\Delta t,$$

$$F = (24{,}000)[(0.48 \text{ X } 10^{-3} \text{ kg})(25 \text{ m/s})]/(0.04 \text{ s}) = 7200 \text{ N}.$$


## 10-22

(a) Suppose we choose directions to the right as positive. Momentum conservation dictates that, in SI base units,

$$(1.6)(5.5) + (2.4)(-2.5) = (1.6)v + (2.4)(4.9) \quad \rightarrow \quad v = -5.6 \text{ m/s},$$

i.e, a speed of 5.6 m/s to the left.
(b) Calculate the total kinetic energies before and after collision:

$$K_b = \tfrac{1}{2}(1.6)(5.5)^2 + \tfrac{1}{2}(2.4)(2.5)^2 = 31.7 \text{ J};$$

$$K_a = \tfrac{1}{2}(1.6)(5.6)^2 + \tfrac{1}{2}(2.4)(4.9)^2 = 53.9 \text{ J}.$$

Kinetic energy is not conserved (is not the same after collision as it was before collision); therefore the collision is not elastic. In fact, kinetic energy was added during collision; perhaps one of the blocks had an explosive cap attached to the face that suffered the impact.

10-27

With $m$, $v$ = mass and speed of the meteor before impact and $M$, $V$ the mass and speed of the Earth after impact, we have by conservation of momentum,

$$mv = (m + M)V \approx MV,$$

$$(5 \times 10^{10} \text{ kg})(7200 \text{ m/s}) = (5.98 \times 10^{24} \text{ kg})V,$$

$$V = (6.02 \times 10^{-11} \text{ m/s})(3.16 \times 10^7 \text{ s/y}) = 0.0019 \text{ m/y} = 1.9 \text{ mm/y}.$$

10-28

Let $m$, $M$ be the masses of bullet and block and $v$, $V$ their velocities. Conservation of momentum requires that

$$mv_i + MV_i = mv_f + MV_f.$$

But $V_i = 0$ since the block is at rest before the bullet strikes. Inserting the rest of the data in SI base units gives

$$(0.00518)(672) + 0 = (0.00518)(428) + (0.715)V_f,$$
$$V_f = 1.77 \text{ m/s}.$$

Note that we could have retained grams for the mass unit, since the conversion factor of $1 \times 10^{-3}$ from g to kg would have canceled out.

10-34

The speed $v$ with which the weight, mass $m$, strikes the pile is given from Eq. 25 in Chapter 2. Recalling that in British units $g$ = 32 ft/s² and assuming that the weight is dropped from rest, we have

$$v^2 = 0^2 - 2(32)(-6.5 - 0) \quad \rightarrow \quad v = 20.4 \text{ ft/s}.$$

The speed $V$ of the weight+pile combination, total mass $M$, can be found from Eq. 3. Adjusting for the different notation, we find

$$V = mv/M = wv/W = (2.9 \text{ ton})(20.4 \text{ ft/s})/(3.4 \text{ ton}) = 17.4 \text{ ft/s}.$$

(The $g$ in the relations $w = mg$ and $W = Mg$ cancel out.) Now we need the numerical value of $M$:

$$M = W/g = [(3.4 \text{ ton})(2000 \text{ lb/ton})]/(32 \text{ ft/s}^2) = 213 \text{ slug}.$$

This object (weight+pile) is driven a distance $x$ = 1.5 in. = 0.125 ft into the ground, being brought to rest by the force $F$ exerted by the ground, opposite to the displacement of the weight+pile. By the work-energy theorem,

$$Fx\cos\phi = 0 - \tfrac{1}{2}MV^2,$$

$$F(0.125)\cos180° = 0 - \tfrac{1}{2}(213)(17.4)^2 \;\rightarrow\; F = 2.58 \times 10^5 \text{ lb},$$

or about 130 tons.


## 10-35

Let $f = 0.27$ represent the fraction of initial kinetic energy that is dissipated; i.e.,

$$f(\tfrac{1}{2}mv^2) = \tfrac{1}{2}mv^2 - \tfrac{1}{2}(m + M)u^2.$$

Here $m$ = mass of the freight car, $M$ = mass of the caboose; $v$ is the speed of the freight car before collision and $u$ is the speed of the coupled caboose+freight car after impact. The momentum conservation equation is

$$mv = (m + M)u.$$

These two equations can be rewritten, respectively, in the forms

$$(1 - f)mv^2 = (m + M)u^2,$$

$$m^2v^2 = (m + M)^2u^2.$$

Dividing these equations, to eliminate the unknown speeds, gives

$$\frac{1 - f}{m} = \frac{1}{m + M},$$

$$M = \frac{f}{1 - f}\, m,$$

$$W = \frac{f}{1 - f}\, w = \frac{0.27}{0.73}(35 \text{ ton}) = 12.9 \text{ ton}.$$


## 10-39

Let $v$ be the speed of the marbles just before impact; since the collisions are completely inelastic, the marbles comes to rest upon striking the box. Hence, the momentum given to the box and contents by each marble upon striking is $p = m(v_f - v_i) = m(0 - v) = -mv$, or $p = mv$, in magnitude. Since the rate at which marbles strike is $R$, a time interval $t = 1/R$ elapses between impacts. Thus, the average force exerted on the box by the colliding marbles is

$$\overline{F} = p/t = mv/(1/R) = mRv = mR\sqrt{2gh}.$$

(Note that this can also be obtained from the $v_{rel}dM/dt$ term of the rocket equation, Eq. 43 in Chapter 9; in our case the "exhaust" enters, rather than leaves, the "rocket".) After time $t$, $Rt$ marbles with total weight $(mg)Rt$ already reside in the box. Hence, the scale reading $SC$ after time $t$ is

$$SC = mR\sqrt{2gh} + mgRt = mgR[\sqrt{2h/g} + t],$$

$$SC = (0.0046)(9.8)(115)[\sqrt{2(9.62)/(9.8)} + 6.5] = 41.0 \text{ N}.$$

Metric scales generally display mass, not weight. In this case, the scale reading will be $(41.0)/(9.8) = 4.18$ kg.

## 10-46

The total momentum before the collision must equal the momentum of the wreckage after collision:

$$m_A\mathbf{v}_A + m_B\mathbf{v}_B = (m_A + m_B)\mathbf{V}.$$

We can multiply through by $g$ to convert mass to weight:

$$w_A\mathbf{v}_A + w_B\mathbf{v}_B = (w_A + w_B)\mathbf{V}.$$

Now, in units of lb•mi/h,

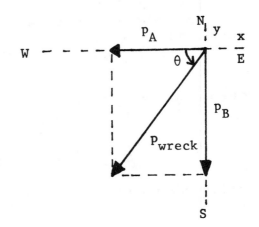

$$w_A\mathbf{v}_A = (2720)(-38.5\mathbf{i}) = -1.047 \times 10^5\mathbf{i},$$

$$w_B\mathbf{v}_B = (3640)(-58.0\mathbf{j}) = -2.111 \times 10^5\mathbf{j}.$$

We also have $w_A + w_B = 6360$ lb. Substituting these last three results into the "weight-momentum" equation gives us

$$\mathbf{V} = -16.46\mathbf{i} - 33.19\mathbf{j},$$

in mi/h. Hence, the speed of the wreckage is

$$V^2 = (16.46)^2 + (33.19)^2 \rightarrow V = 37.0 \text{ mi/h},$$

and this is directed at an angle $\theta$ south of west given by

$$\theta = \tan^{-1}(p_B/p_A) = \tan^{-1}(2.111/1.047) = 63.6°.$$

## 10-49

We call the mass of the neutron $m$ and the mass of the deuteron $M$. The initial speed of the neutron is $v$ and its speed after the collision is $u$. The deuteron, at rest before the collision, has a speed $V$ afterwards. The conservation of momentum, two components, and of kinetic energy become

$$mv = MV\cos\theta,$$

$$0 = mu - MV\sin\theta,$$

$$\tfrac{1}{2}mv^2 = \tfrac{1}{2}mu^2 + \tfrac{1}{2}MV^2.$$

Square the first two equations and add them, using the relation $\sin^2\theta + \cos^2\theta = 1$ to get

$$M^2V^2 = m^2v^2 + m^2u^2.$$

From the third equation above we can write

$$m^2u^2 = m^2v^2 - MmV^2.$$

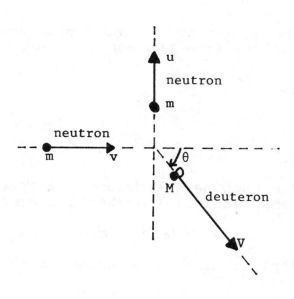

Now combine the last two equations above by eliminating $m^2u^2$ to get the kinetic energy of the deuteron; we find

$$\frac{1}{2}MV^2 = \frac{m^2}{m + M}\,v^2.$$

But $M = 2m$ (almost), and therefore

$$\frac{1}{2}MV^2 = \frac{m}{m + M}(mv^2) = \frac{1}{3}mv^2 = \frac{2}{3}(\frac{1}{2}mv^2),$$

establishing the result quoted in the problem.


## 10-51

(a) See Sample Problem 4, where it is asserted that in an elastic collision between two particles of equal mass, one of which is initially at rest, the recoiling particles move off at 90° to one another. Therefore, the target proton moves off at $\phi_2 = 90° - 64° = 26°$ from the incoming proton's direction.
(b) For the incoming proton, we find from Sample Problem 4,

$$v_{1f} = v_{1i}\cos\phi_1 = (518)\cos 64° = 227 \text{ m/s},$$

$$v_{2f}^2 = v_{1i}^2 - v_{1f}^2 = (518)^2 - (227)^2 \rightarrow v_{2f} = 466 \text{ m/s}.$$


## 10-54

By conservation of mechanical energy, the first pendulum strikes the second with a speed $v$ given from

$$m_1gd = \tfrac{1}{2}m_1v^2 \rightarrow v^2 = 2gd.$$

The speed of the combined pendulum immediately after collision follows from the conservation of momentum:

$$m_1 v + 0 = (m_1 + m_2) V,$$

$$V = (\frac{m_1}{m_1 + m_2}) v.$$

The combined pendulum will reach a height $h$ again given from the conservation of mechanical energy:

$$\tfrac{1}{2}(m_1 + m_2)V^2 = (m_1 + m_2)gh \rightarrow V^2 = 2gh.$$

Apply to this last equation our previous results to find

$$(\frac{m_1}{m_1 + m_2})^2 (v^2) = (\frac{m_1}{m_1 + m_2})^2 (2gd) = 2gh,$$

$$h = (\frac{m_1}{m_1 + m_2})^2 d.$$

10-57

(a) Pool balls all have the same mass; in the absence of a numerical value, call the ball mass $m$. We choose the original direction of motion of the cue ball the $x$ axis. After the collision the second ball moves off at an angle $\phi$ with respect to this $+x$ direction. Conservation of momentum (see Eqs. 24 and 25) requires that

$$mv = m(3.5\cos65°) + m(6.75\cos\phi),$$

$$0 = m(3.5\sin65°) + m(6.75\sin\phi).$$

Cancel the mass $m$. The second equation gives

$$\phi = \sin^{-1}[(-3.5\sin65°)/6.75] = 332°, 208°.$$

If we pick $\phi = 208°$, then the first equation above yields $v < 0$. But $v$ must be positive, for we chose the $+x$ axis in the direction of v. Hence $\phi = 332°$ (= $-28°$).
(b) Put $\phi = 332°$ into the first equation above for $v$ to find that $v = 7.44$ m/s.

## 10-61

The velocity of the center of mass after the collision equals the velocity of the center of mass before the collision. Choosing motion to the right as positive, we use the "before the collision" data to find the velocity of the center of mass; from Eq. 5 in Chapter 9,

$$(m_1 + m_2)v_{cm} = m_1 v_{1i} + m_2 v_{2i},$$

$$(3.16 + 2.84)v_{cm} = (3.16)(-15.6) + (2.84)(12.2),$$

$$v_{cm} = -2.44 \text{ m/s},$$

i.e., at 2.44 m/s to the left.

## 10-66

(a) We apply Eq. 23 of Chapter 9 to the muon, for which $m_\mu c^2 = 105.7$ MeV and $K_\mu = 4.100$ MeV. Therefore,

$$4.100 = \sqrt{(p_\mu c)^2 + (105.7)^2} - 105.7,$$

which gives $p_\mu = 29.72$ MeV/$c$. Since the pion was at rest before the decay, we have $p_\nu = 29.72$ MeV/$c$ also.
(b) Since the kinetic energy of the pion before the decay was zero, we have $Q = K_\mu + K_\nu$. The neutrino has rest energy equal to zero. Hence, Eq. 23 gives $K_\nu = p_\nu c = 29.72$ MeV. Therefore $Q = 33.82$ MeV. By Eq. 43,

$$Q = m_\pi c^2 - m_\mu c^2 - m_\nu c^2,$$

$$33.82 = m_\pi c^2 - 105.7 - 0,$$

$$m_\pi c^2 = 139.5 \text{ MeV}.$$

## 11-3

(a) From Appendix G, we extract the length conversion factor that 1 ly = 9.460 X $10^{12}$ km. The radius $r$ of the Sun's circular path is

$$r = (2.3 \text{ X } 10^4 \text{ ly})(9.46 \text{ X } 10^{12} \text{ km/ly}) = 2.176 \text{ X } 10^{17} \text{ km}.$$

The path itself is a circle with circumference $2\pi r$. The speed $v$ is given in km/s and is constant, so we have for the time $t$ for one revolution,

$$t = 2\pi r/v = 2\pi(2.176 \text{ X } 10^{17} \text{ km})/(250 \text{ km/s}) = 5.47 \text{ X } 10^{15} \text{ s}.$$

(b) In years, $t = (5.47 \text{ X } 10^{15} \text{ s})/(3.16 \text{ X } 10^7 \text{ s/y}) = 1.73 \text{ X } 10^8 \text{ y}$. Therefore, the number $n$ of revolutions completed in 4.5 X $10^9$ y is

$$n = (4.5 \text{ X } 10^9 \text{ y})/(1.73 \text{ X } 10^8 \text{ y/rev}) = 26 \text{ rev}.$$

## 11-7

By Eq. 2, we need both $\Delta\phi$, the total angle turned through during the fall, and the time $\Delta t$ in falling. Since 1 revolution = $2\pi$ rad, the angle $\Delta\phi = (2.5 \text{ rev})(2\pi \text{ rad/rev}) = 5\pi$ rad. The time $\Delta t$ follows from Eq. 24 in Chapter 2 (called simply $t$ there). With $v_0 = 0$, we have

$$-10 = -\tfrac{1}{2}(9.8)t^2 \quad \rightarrow \quad t = \Delta t = 1.43 \text{ s}.$$

Therefore,

$$\omega = \frac{\Delta\phi}{\Delta t} = \frac{5\pi \text{ rad}}{1.43 \text{ s}} = 11 \text{ rad/s}.$$

## 11-12

(a) We are instructed to use revolution as the angle unit and minute as the time unit. Hence $\omega_0 = 78$ rev/min (initial value) and $\omega = 0$ (final value). Also $t = 32$ s = 32/60 = 0.533 min. By Eq. 6,

$$\omega = \omega_0 + \alpha t,$$

$$0 = 78 \text{ rev/min} + \alpha(0.533 \text{ min}) \quad \rightarrow \quad \alpha = -146 \text{ rev/min}^2.$$

The minus sign indicates that the turntable is slowing down.
(b) The number of revolutions is $\phi$ itself in these units and can be found, using only the original data, from Eq. 9:

$$\phi = 0 + \tfrac{1}{2}(78 \text{ rev/min} + 0)(0.533 \text{ min}) = 20.8 \text{ rev}.$$

<u>11-15</u>

(a) We use radians and seconds as the units of angle and time. We are told that $\omega_0$ = 25.2 rad/s and that $\omega$ = 0 when $t$ = 19.7 s. By Eq. 6 we have

$$\omega = \omega_0 + \alpha t,$$

$$0 = 25.2 \text{ rad/s} + \alpha(19.7 \text{ s}) \rightarrow \alpha = -1.28 \text{ rad/s}^2.$$

(b) With the value of the angular acceleration $\alpha$ determined, we can find the angle $\phi$ using one of several equations; we choose Eq. 7:

$$\phi = \phi_0 + \omega_0 t + \tfrac{1}{2}\alpha t^2,$$

$$\phi = 0 + (25.2)(19.7) + \tfrac{1}{2}(-1.28)(19.7)^2 = 248 \text{ rad}.$$

(c) Since we have used radians as the angle unit we must use the fact that 1 rev = $2\pi$ rad to find the number of revolutions:

$$\# \text{ rev} = (\phi \text{ in rad})/(2\pi \text{ rad/rev}) = 248/2\pi = 39.5 \text{ rev}.$$

<u>11-18</u>

(a) If the cord does not slip on the pulley, the distance $s$ traveled by any point on the rim of the pulley must equal the length of the cord. The radius $r$ of the pulley is given as 4.07 cm = 0.0407 m. From Eq. 1 we find the angle $\phi$ through which the pulley turns as the cord unwinds:

$$\phi = s/r = (5.63 \text{ m})/(0.0407 \text{ m}) = 138 \text{ rad}.$$

(b) The pulley starts from rest so that $\omega_0$ = 0. We can use Eq. 7 to find the required time $t$:

$$\phi = \phi_0 + \omega_0 t + \tfrac{1}{2}\alpha t^2,$$

$$138 \text{ rad} = 0 + 0 + \tfrac{1}{2}(1.47 \text{ rad/s}^2)t^2 \rightarrow t = 13.7 \text{ s}.$$

<u>11-21</u>

(a) Use Eq. 7. The rate of change of the rotation speed is very small, so put $\alpha$ = 0. Then, after time $T$ (one period), $\phi$ increases by $2\pi$ rad, so that

$$\phi = \phi_0 + \omega_0 t + \tfrac{1}{2}\alpha t^2,$$

$$2\pi = 0 + \omega_0 T + 0,$$

$$\omega_0 = 2\pi/T = \omega.$$

The last step ($\omega_0 = \omega$) is allowed if $\alpha$ = 0.

(b) In what follows, the angular acceleration is not going to be ignored (i.e., $\alpha \neq 0$). It is given that, at present, $T = 0.033$ s and $dT/dt = 1.26 \times 10^{-5}$ s/y. By definition (see Eq. 5), $\alpha = d\omega/dt$ and, from (a), $\omega = 2\pi/T$. Therefore,

$$\alpha = \frac{d}{dt}\left(\frac{2\pi}{T}\right) = -\frac{2\pi}{T^2}\frac{dT}{dt},$$

$$\alpha = -\frac{2\pi}{(0.033 \text{ s})^2}(1.26 \times 10^{-5} \text{ s/y}) = -0.0727 \text{ (s·y)}^{-1},$$

$$\alpha = -\frac{0.0727 \text{ s}^{-1}}{3.16 \times 10^7 \text{ s}} = -2.30 \times 10^{-9} \text{ rad/s}^2.$$

(c) Call the present value of the angular speed $\omega_0$; its numerical value, from (a) is $\omega_0 = 2\pi/T_0 = 2\pi/(0.033 \text{ s}) = 190.4$ rad/s. When the pulsar stops rotating $\omega = 0$. Hence, by Eq. 6,

$$\omega = \omega_0 + \alpha t,$$

$$0 = 190.4 \text{ s}^{-1} + (-2.30 \times 10\text{-}9 \text{ s}^{-2})t,$$

$$t = 8.278 \times 10^{10} \text{ s} = 2620 \text{ y}.$$

If the present is 1992, then the pulsar will stop in the year 1992 + 2620 = 4612.

(d) Now call the value of the angular velocity when the pulsar was formed $\omega_0$, and the present value $\omega$. The pulsar's age is 1992 - 1054 = 938 y = $2.964 \times 10^{10}$ s. Thus,

$$\omega_0 = \omega - \alpha t = 190.4 - (-2.30 \times 10^{-9})(2.964 \times 10^{10}) = 258.6 \text{ rad/s}.$$

This corresponds to a period of rotation

$$T_0 = 2\pi/\omega_0 = 2\pi/(258.6 \text{ s}^{-1}) = 0.0243 \text{ s} = 24.3 \text{ ms}.$$

## 11-24

The speed of the car is $v = 52.4$ km/h = (52.4)(0.2778 m/s) = 14.56 m/s. We find the angular speed $\omega$ from Eq. 12:

$$v = \omega r,$$

$$14.56 \text{ m/s} = \omega(110 \text{ m}) \quad \rightarrow \quad \omega = 0.132 \text{ rad/s}.$$

## 11-27

(a) Note that in the problem text angular position is given the symbol $\theta$, rather than the $\phi$ used in the chapter text. By Eq. 3, the angular speed at any time $t$ is

$$\omega = d\theta/dt = 0.652t.$$

When $t = 5.60$ s, $\omega = (0.652)(5.60) = 3.65$ rad/s.
(b) The tangential speed is $v = \omega r$. To calculate the speed $v$ at $t = 5.60$ s, we use the value of $\omega$ at $t = 5.60$ s, as found in (a). With $r = 10.4$ m, we get

$$v = (3.65 \text{ rad/s})(10.4 \text{ m}) = 38.0 \text{ m/s}.$$

(c) Find the tangential acceleration from Eq. 13, $a_T = \alpha r$. First we need the angular acceleration $\alpha$. By Eq. 5,

$$\alpha = d\omega/dt = d(0.652t)/dt = 0.652 \text{ rad/s}^2.$$

Hence,

$$a_T = (0.652 \text{ rad/s}^2)(10.4 \text{ m}) = 6.78 \text{ m/s}^2.$$

(d) The radial acceleration $a_R$ at $t = 5.60$ s follows directly from Eq. 14, using the value of $\omega$ at $t = 5.60$ s:

$$a_R = \omega^2 r = (3.65 \text{ rad/s})^2(10.4 \text{ m}) = 139 \text{ m/s}^2.$$

## 11-31

(a) With $\omega_0 = 156$ rev/min, we convert the time to minutes: $t = 2.2(60) = 132$ min. Also $\omega = 0$ (flywheel brought to rest). By Eq. 6,

$$\omega = \omega_0 + \alpha t,$$

$$0 = 156 \text{ rev/min} + \alpha(132 \text{ min}) \rightarrow \alpha = -1.18 \text{ rev/min}^2.$$

(b) The number of rotations (i.e., revolutions) is just $\phi$ (since we are using rev as the angle unit). Use Eq. 9 to find this from the original data:

$$\phi = \phi_0 + \tfrac{1}{2}(\omega + \omega_0)t,$$

$$\phi = 0 + \tfrac{1}{2}(0 + 156 \text{ rev/min})(132 \text{ min}) = 10,300 \text{ rev}.$$

(c) The tangential linear acceleration $a_T$ is given by Eq. 13. To use this equation, we must have the angular acceleration in rad/s² to get m/s² for the linear acceleration. Converting $\alpha$ we find

$$\alpha = -1.18 \ (2\pi \text{ rad})/(60 \text{ s})^2 = -2.059 \times 10^{-3} \text{ rad/s}^2.$$

Hence,

$$a_T = (-2.059 \times 10^{-3} \text{ rad/s}^2)(0.524 \text{ m}) = -1.08 \times 10^{-3} \text{ m/s}^2.$$

(d) The magnitude $a$ of the total linear acceleration follows from

$$a^2 = a_R^2 + a_T^2.$$

We have $a_T$. To find $a_R$ use Eq. 14. Again, we must express the rotational variables in terms of radians, and we also convert minutes to seconds to get SI base units; hence,

$$\omega = 72.5(2\pi \text{ rad})/(60 \text{ s}) = 7.592 \text{ rad/s},$$

$$a_R = \omega^2 r = (7.592 \text{ rad/s})^2(0.524 \text{ m}) = 30.20 \text{ m/s}^2.$$

Evidently $a_R \succ a_T$ in this case. We find that $a = 30.2$ m/s$^2$, the value of $a_T$ being negligible to a 3 significant figure calculation.

## 11-35

Since the belt does not slip, the linear speeds at the rims of the two wheels must be equal; that is,

$$r_A\omega_A = r_C\omega_C.$$

Differentiating with respect to time gives

$$\alpha_C = (r_A/r_C)\alpha_A.$$

Now $\omega_C = \alpha_C t + \omega_{0C}$ and $\omega_{0C} = 0$; therefore

$$\omega_C = (r_A/r_C)\alpha_A t.$$

Since $\omega_C = 100$ rev/min $= 100(2\pi \text{ rad})/(60 \text{ s}) = 10.47$ rad/s, the equation above yields

$$10.47 = (10/25)(1.6)t \rightarrow t = 16.4 \text{ s}.$$

## 11-37

The centripetal acceleration is $a_R = \omega^2 r$. Since the windmill started from rest we have $\omega = \alpha t$, so that $a_R = r\alpha^2 t^2$. The tangential acceleration is $a_T = \alpha r$. If these are equal, we have

$$\alpha r = r\alpha^2 t^2,$$

$$1 = (0.236 \text{ rad/s}^2)t^2 \rightarrow t = 2.06 \text{ s}.$$

## 11-42

(a) Find the velocity from Eq. 16; since k X j = -i and k X k = 0 we find

$$v = \omega \text{ X } r = 14.3k \text{ X } (1.83j + 1.26k) = -26.2i, \text{ m/s}.$$

(*b*) Use Eq. 18. The two terms on the right are

$$\mathbf{a}_{T} = \alpha \text{ X } \mathbf{r} = -2.66\mathbf{k} \text{ X } (1.83\mathbf{j} + 1.26\mathbf{k}) = 4.87\mathbf{i},$$

$$\mathbf{a}_{R} = \omega \text{ X } \mathbf{v} = 14.3\mathbf{k} \text{ X } (-26.2\mathbf{i}) = -375\mathbf{j},$$

using the result from (*a*) for **v**; we also used the cross product relations among the cartesian unit vectors. We have, then,

$$\mathbf{a} = \mathbf{a}_{R} + \mathbf{a}_{T} = 4.87\mathbf{i} - 375\mathbf{j}, \text{ m/s}^{2}.$$

(*c*) The radius $r_{cir}$ of the circle is $r\sin\theta$; by Eq. 15 we find

$$r_{cir} = v/\omega = (26.2 \text{ m/s})/(14.3 \text{ rad/s}) = 1.83 \text{ m},$$

again using our result from (*a*).

## 12-4

Before we can use Eq. 3 to calculate the rotational kinetic energy, we must express the angular speed in rad/s and compute the rotational inertia. The angular speed is

$$\omega = 1.46(2\pi \text{ rad})/s = 9.173 \text{ rad/s}.$$

From Fig. 9(c), we have for the rotational inertia

$$I = \tfrac{1}{2}MR^2 = \tfrac{1}{2}(1220 \text{ kg})(0.59 \text{ m})^2 = 212.3 \text{ kg} \bullet m^2.$$

Therefore, the rotational kinetic energy is

$$K = \tfrac{1}{2}I\omega^2 = \tfrac{1}{2}(212.3 \text{ kg} \bullet m^2)(9.173 \text{ rad/s})^2 = 8930 \text{ J}.$$

## 12-10

(a) The rotational inertia of a homogenous (i.e., solid) cylinder is, from Fig. 9(c), $I = \tfrac{1}{2}MR^2$; numerically, this is

$$I = \tfrac{1}{2}(512 \text{ kg})(0.976 \text{ m})^2 = 243.9 \text{ kg} \bullet m^2.$$

Hence,

$$K = \tfrac{1}{2}I\omega^2 = \tfrac{1}{2}(243.9 \text{ kg} \bullet m^2)(624 \text{ rad/s})^2 = 4.75 \text{ X } 10^7 \text{ J}.$$

(b) Recalling that 1 kW = 1000 W = 1000 J/s, the desired time $t$ is

$$t = \frac{K}{P} = \frac{4.75 \text{ X } 10^7 \text{ J}}{(8130 \text{ J/s})(60 \text{ s/min})} = 97.4 \text{ min}.$$

## 12-15

(a) The resultant torque $\tau$ is given by

$$\tau = \tau_1 + \tau_2 = r_1 \text{ X } F_1 + r_2 \text{ X } F_2,$$

and its magnitude is the magnitude of this expression.
(b) By the right-hand rule, $\tau_1$ is out of the page and $\tau_2$ is directed in. In magnitude,

$$\tau_1 = r_1 F_1 \sin\theta_1 = (1.30 \text{ m})(4.20 \text{ N})\sin 75° = 5.274 \text{ N} \bullet m,$$

$$\tau_2 = r_2 F_2 \sin\theta_2 = (2.15 \text{ m})(4.90 \text{ N})\sin 58° = 8.934 \text{ N} \bullet m.$$

Evidently, the resultant torque is into the page (since $\tau_2 > \tau_1$)

and has magnitude $\tau = 8.934 - 5.274 = 3.66$ N•m.

## 12-19

Each moment arm from the axis of the cylinder is at 90° to the line of action of the respective force, and sin90° = 1. Hence, the total clockwise torque (torques tending, by themselves, to make the cylinder rotate clockwise) is $F_3R_1 + F_2R_2$ and the counterclockwise torque is $F_1R_2$. The resultant torque has magnitude

$$\tau = (F_1 - F_2)R_2 - F_3R_1,$$

$$\tau = (5.88 - 4.13)(0.118) - (2.12)(0.0493) = 0.1020 \text{ N•m.}$$

(We converted cm to m to stay in SI base units.) The rotational inertia of the cylinder is

$$I = \tfrac{1}{2}MR_2{}^2 = \tfrac{1}{2}(1.92 \text{ kg})(0.118 \text{ m})^2 = 0.01337 \text{ kg•m}^2.$$

Finally, by Eq. 18,

$$\alpha = \tau/I = (0.1020 \text{ N•m})/(0.01337 \text{ kg•m}^2) = 7.63 \text{ rad/s}^2,$$

tending to make the cylinder rotate clockwise. (The 30° angle shown in Fig. 43 in the text is irrelevant.)

## 12-23

From Fig. 9(a) the rotational inertia of the hoop is

$$I = MR^2 = (31.4 \text{ kg})(1.21 \text{ m})^2 = 45.97 \text{ kg•m}^2.$$

We need the angular speed in rad/s:

$$\omega = (283)(2\pi \text{ rad})/(60 \text{ s}) = 29.64 \text{ rad/s.}$$

The work required to stop the hoop follows from the work-energy theorem $W = \Delta K = K_f - K_i$, so that

$$W = 0 - \tfrac{1}{2}I\omega^2 = 0 - \tfrac{1}{2}(45.97 \text{ kg•m}^2)(29.64 \text{ rad/s})^2 = 20.19 \text{ kJ.}$$

Therefore, the average power $P$, in absolute value, is

$$P = W/t = (20.19 \text{ kJ})/(14.8 \text{ s}) = 1.36 \text{ kW.}$$

## 12-30

(a) A horizontal friction force $f_k = \mu_k N$ acts on the wheel at the top of the axle. The normal force $N$ equals the weight $Mg$ of the wheel, so that $f_k = \mu_k Mg$. This exerts a torque $\tau = rF\sin\theta$; the angle $\theta = 90°$ and sin90° = 1, so that $\tau = a\mu_k Mg$. The rotational inertia of the wheel is $I = Mk^2$ (see Problem 11) so that, by Eq. 18, the angular acceleration follows from

$$\tau = I\alpha,$$

$$-\mu_k a Mg = (Mk^2)\alpha \ \rightarrow \ \alpha = -\mu_k a g/k^2.$$

(We attached a minus sign to the torque since it acts to slow the wheel.) With the wheel being brought to rest ($\omega = 0$), Eq. 6 of Chapter 11 gives

$$t = -\omega_0/\alpha = \omega_0 k^2/\mu_k ga.$$

(b) The number $n$ of revolutions is $(\phi - \phi_0)/2\pi$; by Eq. 8 of Chapter 11 we find

$$n = -\omega_0^2/4\pi\alpha = \omega_0^2 k^2/4\pi\mu_k ga.$$

## 12-33

Apply conservation of mechanical energy:

$$U_i + K_i = U_f + K_f.$$

In the initial position $K_i = 0$ since it is implied that the stick falls from rest. The initial potential energy is $U_i = Mgy_i$, where $y_i$ is the position of the center of mass of the stick when upright. If we take the floor as $y = 0$, then clearly $y_i = \frac{1}{2}L$, $L$ the length of the stick. Hence, $U_i = \frac{1}{2}MgL$. As the stick hits the floor $U_f = Mgy_f = 0$ since $y_f = 0$. The kinetic energy is one of rotation about the lower end of the stick, so we have

$$K_f = \frac{1}{2}I_{end}\omega^2 = \frac{1}{2}I_{end}(v_f/L)^2.$$

Here $I_{end}$ is the rotational inertia of the stick about the end in contact with the floor; from Fig. 9($f$) we see that $I_{end} = \frac{1}{3}ML^2$. With these results, the conservation of energy equation becomes

$$\tfrac{1}{2}MgL + 0 = 0 + \tfrac{1}{2}(\tfrac{1}{3}ML^2)(v_f/L)^2,$$

$$v_f = \sqrt{3gL} = \sqrt{3(9.8)(1.27)} = 6.11 \text{ m/s}.$$

## 12-38

(a) The radial acceleration of a point at the top of a chimney of height $h$ is, by Eq. 14 of Chapter 11, $a_R = \omega^2 h$. By conservation of mechanical energy,

$$mg(\tfrac{1}{2}h) = \tfrac{1}{2}I\omega^2 + mg(\tfrac{1}{2}h\cos\theta).$$

Here $I = \frac{1}{3}mh^2$ since the chimney is, in effect, rotating about an

axis through its base; see Fig. 9(f).
Substituting this expression for $I$
into the energy equation gives

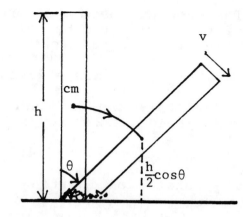

$$mgh(1 - \cos\theta) = (\tfrac{1}{3}mh^2)\omega^2,$$

and therefore

$$a_R = \omega^2 h = 3g(1 - \cos\theta).$$

(b) By Eq. 13 of Chapter 11, the
tangential acceleration $a_T = h\alpha$.
Use the result from (a)

$$\omega^2 = \frac{3g}{h}(1 - \cos\theta),$$

to find that

$$2\omega\frac{d\omega}{dt} = (\frac{3g}{h})\sin\theta(\frac{d\theta}{dt}),$$

$$\frac{d\omega}{dt} = \alpha = (\frac{3g}{2h})\sin\theta.$$

(Note that $\omega = d\theta/dt$.) Thus, the tangential acceleration is

$$a_T = (3g/2)\sin\theta.$$

(c) Using the results from (a) and (b) we find for the total linear
acceleration

$$a = \sqrt{a_R^2 + a_T^2} = \frac{3g}{2}\sqrt{(1 - \cos\theta)(5 - 3\cos\theta)}.$$

This is zero at $\theta = 0°$, and increases as $\theta$ increases, passing $g$
near 34.5°. If the chimney does not break up, then as the top hits
the ground we have $\theta = 90°$ and $a = 3.35g > g$.
(d) As the chimney tips over, the weight of the upper part acts
transverse to (across) the column. Chimneys standing upright are
not subject to this force (although they are subject to the lesser
transverse force due to the wind), and therefore are not designed
to withstand it.


12-41

(a) Apply conservation of energy:

$$Mgh = \tfrac{1}{2}I\omega^2 + \tfrac{1}{2}Mv^2.$$

79

The angular speed $\omega$ of the flywheel is proportional to the speed $v$ of the car; let $c$ be the proportionality constant, so that $\omega = cv$. Using this to eliminate $\omega$ from the energy equation gives

$$Mgh = \tfrac{1}{2}I(cv)^2 + \tfrac{1}{2}Mv^2 = \tfrac{1}{2}v^2[c^2I + M].$$

To evaluate the constant $c$, use the data given that $v = 86.5$ km/h must yield $\omega = 237$ rev/s. Converting to SI base units, we find that

$$c = \omega/v = (1489 \text{ rad/s})/(24.03 \text{ m/s}) = 61.96 \text{ m}^{-1}.$$

The rotational inertia $I$ of the flywheel, from Fig. 9(c), is

$$I = \tfrac{1}{2}mR^2 = \tfrac{1}{2}(W/g)R^2 = \tfrac{1}{2}(194/9.8)(0.54)^2 = 2.886 \text{ kg}\bullet\text{m}^2.$$

The vertical distance the car descends is $h = L\sin\theta = 1500\sin5° = 130.7$ m. The mass $M$ of the car is 822 kg. Putting all of these data, which are in SI base units, into the energy equation above yields

$$(822)(9.8)(130.7) = \tfrac{1}{2}v^2[(61.96)^2(2.886) + 822],$$

$$v = 13.30 \text{ m/s} = 47.9 \text{ km/h}.$$

(b) Return to the first energy equation; this time replace $v$ with $\omega/c$ to obtain

$$Mgh = \tfrac{1}{2}\omega^2[I + M/c^2].$$

Differentiate this with respect to time $t$, noting that

$$d(\omega^2)/dt = 2\omega(d\omega/dt) = 2\omega\alpha.$$

Hence,

$$Mg(dh/dt) = (\omega\alpha)[I + M/c^2].$$

But $dh/dt = d(L\sin\theta)/dt = v\sin\theta = (\omega/c)\sin\theta$, so that

$$(Mg\omega/c)\sin\theta = (\omega\alpha)[I + M/c^2].$$

Now solve this for the angular acceleration $\alpha$, and then apply the second formula above for $Mgh$ to obtain

$$\alpha = cv^2/2L = (61.96)(13.30)^2/2(1500) = 3.65 \text{ rad/s}^2.$$

(c) The power $P$ is, by Eq. 19,

$$P = \tau\omega = (I\alpha)\omega = (I\alpha)(cv),$$

$$P = (2.886)(3.65)(61.96)(13.30) = 8680 \text{ W} = 8.68 \text{ kW}.$$

It is a useful exercise to substitute the proper SI units in this equation to verify the units of power obtained.

12-44

We assume that the work can be computed from the work-energy theorem; i.e., the work equals the change in kinetic energy of the hoop. This kinetic energy can be taken as the sum of the translational kinetic energy and the kinetic energy of rotation about the center of mass. Alternatively, the kinetic energy can be considered solely as one of rotation about the point of contact of hoop and floor. We adopt the second view, and therefore have

$$W = \Delta K = 0 - \tfrac{1}{2}I_B\omega^2,$$

where $I_B$ is the rotational inertia of the hoop about the point of contact with the floor (see Fig. 24). By the parallel-axis theorem, Eq. 4, we have

$$I_B = I_{cm} + Mh^2 = MR^2 + MR^2 = 2MR^2.$$

Also $\omega = v/R$. Put all this together and we get

$$W = -\tfrac{1}{2}(2MR^2)(v^2/R^2) = -Mv^2 = -(137 \text{ kg})(0.153 \text{ m/s})^2 = -3.21 \text{ J}.$$

Often, the minus sign is omitted in giving such a result.


12-48

(a) See Sample Problem 9, where it is shown that for a sphere

$$a_{cm} = (5/7)g\sin\theta.$$

Therefore, to obtain the desired acceleration of $0.133g$, the inclination angle $\theta$ must be given from

$$0.133g = (5/7)g\sin\theta,$$

$$\theta = \sin^{-1}[(7/5)(0.133)] = 10.7°.$$

(b) For a block, $a_{cm} = a = g\sin\theta$ (see Chapter 5); hence, for the inclination angle found above,

$$a = g\sin10.7° = 0.186g.$$


12-49

If we know the speed $v$ with which the ball leaves the right-hand end of the track, then the desired distance $x$ can be found from Eq. 23 in Chapter 4. Since the track is horizontal at this end we have $\phi_0 = 0$; also put $y = -h$ ($y$ positive up), and we write $v$ for $v_0$. With these considerations, Eq. 23 of Chapter 4 becomes

$$h = gx^2/2v^2.$$

To find $v$, apply conservation of energy to the motion of the ball on the track. The mass and radius of the ball are not given; call them $M$ and $R$. Apply Eq. 24a and pick $I_{cm}$ from Fig. 9(g). Also, we are writing $v$ for $v_{cm}$. We get

$$K = \frac{1}{2}Mv^2 + \frac{1}{2}\left(\frac{2}{5}MR^2\right)\left(\frac{v^2}{R^2}\right) = \frac{7}{10}Mv^2.$$

Thus, conservation of mechanical energy $E = U + K$ gives

$$MgH + 0 = Mgh + \frac{7}{10}Mv^2.$$

$$v = \sqrt{\frac{10}{7}g(H - h)} = \sqrt{\frac{10}{7}(9.8)(60 - 20)} = 23.7 \text{ m/s}.$$

With $v$ determined, we now have from our very first equation,

$$20 = (9.8)x^2/2(23.7)^2 \rightarrow x = 48 \text{ m}.$$

## 12-51

(a) Let $T$ be the tension in each cord. If we take down as positive, Newton's second law gives

$$W - 2T = Ma = (W/g)a.$$

If there is no slipping between the cords and the cylinder, the motion is equivalent to a cylinder rolling down a vertical surface. From Sample Problem 9, we have for $\theta = 90°$,

$$a_{cm} = a = \tfrac{2}{3}g.$$

Substituting this into the preceding equation gives

$$W - 2T = (W/g)(\tfrac{2}{3}g) \rightarrow T = W/6.$$

(b) From (a), $a = \tfrac{2}{3}g$. Note that, in place of using the result from Sample Problem 9, we could have used the dynamical method $\tau = I\alpha$:

$$(2T)R = (\tfrac{1}{2}MR^2)(a/R).$$

This equation together with Newton's second law, the first equation above, now form two equations in the two unknowns, the tension $T$ and the acceleration $a$. The reader should solve these equations and verify that the same solution as given above is obtained.

Let the radius of the body be $R$ and its mass $M$. The kinetic energy (rotation plus translation) is given by Eq. 24a. We will write $v$ instead of $v_{cm}$. Conservation of energy requires that

$$Mgh = K_i,$$

where the kinetic energy is that at the bottom of the hill. We have then

$$Mgh = \tfrac{1}{2}Mv^2 + \tfrac{1}{2}I_{cm}(v^2/R^2).$$

Whatever the body is, presuming a circular cross section, we can write

$$I_{cm} = \beta(MR^2),$$

where $\beta$ is a positive number. Putting this and the given expression for $h$ into the energy equation we obtain

$$Mg(3v^2/4g) = \tfrac{1}{2}Mv^2(1 + \beta) \quad \rightarrow \quad \beta = 1/2.$$

From Fig. 9(c) we see that the body must be either a solid circular cylinder or disk. The same conclusion follows from Fig. 9(b) with $R_1 = 0$ and $R_2 = R$. Fig. 9(i) does not apply as the axis shown does not permit rolling.

Since the impulse is central, the initial motion of the ball is pure sliding. If $v_0$ is to the right (taken as the positive direction), the friction force $f_k$ points to the left. Let clockwise rotations be positive. Then from Newton's second law and the two successive integrations of it:

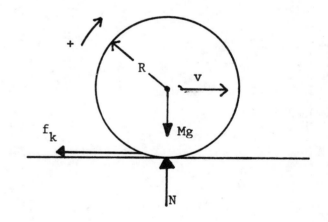

$$-f_k = +Ma = M(dv/dt),$$

$$v = -f_k t/M + v_0,$$

$$x = -\tfrac{1}{2}(f_k/M)t^2 + v_0 t.$$

The rotation equation is

$$\tau = f_k R = I\alpha = (2MR^2/5)d\omega/dt,$$

$$\omega = 5f_k t/2MR,$$

the initial angular speed being zero. Now let $t$ stand for the time at which rolling sets in. At this moment, $v = R\omega$ so that

$$-f_k t/M + v_0 = 5f_k t/2M \quad \rightarrow \quad t = 2Mv_0/7f_k.$$

Substituting this expression for $t$ into the equation for distance traveled $x$ given above yields

$$x = 12Mv_0^2/f_k.$$

Finally,

$$f_k = \mu_k N = \mu_k Mg.$$

Put this into the expression for $x$ directly above to find

$$x = 12v_0^2/\mu_k g.$$

## 13-6

The angular momentum is given from Eq. 2; since $p = mv$, we have

$$l = mvr\sin\theta.$$

The person moves at constant speed $v$ in a circular path of radius $r$ equal to the radius of the Earth; from Appendix C,

$$r = 6370 \text{ km} = 6.37 \times 10^6 \text{ m}.$$

In uniform circular motion, the velocity **v** is tangent to the circle so that $\theta$, the angle between **r** and **v**, is 90°. We need the speed $v$. This is given by

$$v = 2\pi r/t = 2\pi(6.37 \times 10^6 \text{ m})/(86,400 \text{ s}) = 463.2 \text{ m/s}.$$

In this equation $t = 1$ day $= 86,400$ s is the period of rotation of the Earth. (Appendix C gives this period as 0.997 day, but this is the period with respect to the fixed stars; we use the period with respect to the Sun, which is the basis of civil time.) Thus we have

$$l = (84.3 \text{ kg})(463.2 \text{ m/s})(6.37 \times 10^6 \text{ m})\sin90°,$$

$$l = 2.49 \times 10^{11} \text{ kg} \cdot \text{m}^2/\text{s}.$$

## 13-10

(a) Use Eq. 8, which we write in scalar form as

$$\tau = dL/dt.$$

If $\tau$ is constant, as it is in this Problem, then

$$L = \int \tau dt = \tau t + L_0,$$

where $L_0$ is the initial angular momentum (value at $t = 0$). If the disk is at rest when the motor is turned on then $L_0 = 0$ and we get

$$L = \tau t = (15.8 \text{ N} \cdot \text{m})(33 \times 10^{-3} \text{ s}) = 0.521 \text{ kg} \cdot \text{m}^2/\text{s}.$$

(b) By Eq. 11, $L = I\omega$ so that

$$\omega = L/I = (0.521 \text{ kg} \cdot \text{m}^2/\text{s})/(1.22 \times 10^{-3} \text{ kg} \cdot \text{m}^2) = 427 \text{ rad/s},$$

which equals 4080 rev/min.

By Eq. 11, the angular momentum is $L = I\omega$, so we need to find both $I$ and $\omega$ first to find the angular momentum. Assuming, as instructed, that the Earth can be considered a uniform sphere, we have from Fig. 9($g$) of Chapter 12,

$$I = (2/5)MR^2 = (2/5)(5.98 \times 10^{24} \text{ kg})(6.37 \times 10^6 \text{ m})^2,$$

$$I = 9.706 \times 10^{37} \text{ kg} \cdot \text{m}^2.$$

The Earth rotates once (through $2\pi$ radians) in 1 day = 86,400 s. (Ignore the value of the rotation period given in Appendix C.) By Eq. 3 of Chapter 11,

$$\omega = d\phi/dt = \phi/t = (2\pi \text{ rad})/(86,400 \text{ s}) = 7.272 \times 10^{-5} \text{ rad/s}.$$

(The Earth rotates at constant angular speed so we do not need the derivative.) Therefore

$$L = I\omega = (9.706 \times 10^{37} \text{ kg} \cdot \text{m}^2)(7.272 \times 10^{-5} \text{ rad/s}),$$

$$L = 7.06 \times 10^{33} \text{ kg} \cdot \text{m}^2/\text{s}.$$

($a$) By Eq. 8 the average torque is

$$\tau = \frac{\Delta L}{\Delta t} = \frac{0.788 - 3.07}{1.53} = -1.49 \text{ Nm}.$$

($b$) The initial and final angular speeds are (using SI base units),

$$\omega_0 = L_0/I = 3.07/0.142 = 21.62 \text{ rad/s},$$

$$\omega = L/I = 0.788/0.142 = 5.549 \text{ rad/s}.$$

By Eq. 9 of Chapter 11,

$$\phi = \tfrac{1}{2}(\omega + \omega_0)t = \tfrac{1}{2}(5.549 + 21.62)(1.53) = 20.8 \text{ rad}.$$

($c$) The work done is, by Eq. 14 in Chapter 12,

$$W = \tau\phi = (-1.49 \text{ N} \cdot \text{m})(20.8 \text{ rad}) = -31.0 \text{ J}.$$

($d$) The average power is

$$P_{av} = W/t = (-31.0 \text{ J})/(1.53 \text{ s}) = -20.3 \text{ W}.$$

The motion consists of a translation and a rotation. For the motion of translation, use Eq. 3 of Chapter 9:

$$J = p_f - p_i = m(v_f - v_i).$$

Inserting the numerical data, we have

$$12.8 \text{ N}\bullet\text{s} = (4.42 \text{ kg})(v_f - 0) \rightarrow v_f = 2.90 \text{ m/s}.$$

This is the velocity of the center of mass immediately after the collision. For the rotation, we use a result quoted in Problem 9. If we also apply Eq. 4 of Chapter 10, the result becomes

$$Jr = I(\omega_f - \omega_i).$$

Now $r = 46.4$ cm $= 0.464$ m. The rotational inertia $I$ is about the center of mass of the stick. From Fig. 9(e) of Chapter 12,

$$I = (1/12)ML^2 = (1/12)(4.42 \text{ kg})(1.23 \text{ m})^2 = 0.5573 \text{ kg}\bullet\text{m}^2.$$

Therefore, the rotation equation becomes

$$(12.8 \text{ N}\bullet\text{s})(0.464 \text{ m}) = (0.5573 \text{ kg}\bullet\text{m}^2)(\omega_f - 0),$$

$$\omega_f = 10.7 \text{ rad/s}.$$

Let $f_k(t)$ be the horizontal frictional force exerted by the block on the cylinder. By Newton's third law, the cylinder exerts a force of $-f_k(t)$ on the block. Apply the impulse and angular impulse (see Problem 9) equations, integrating over the time interval during which the block slips on the cylinder to obtain

$$R\int f_k(t)dt = I\omega,$$

$$-\int f_k(t)dt = M(v_2 - v_1),$$

$\omega$ being the final angular velocity of the cylinder. When slipping ceases, $v_2 = R\omega$. Invoking this, and eliminating $\int f_k dt$ from the two equations and solving for $v_2$ gives

$$M(v_1 - v_2) = \frac{I}{R}\omega = \frac{I}{R}(v_2/R),$$

$$V_2 = \frac{Mv_1}{M + I/R^2} = \frac{v_1}{1 + I/MR^2}.$$

## 13-23

We call the force of friction $f$. Positive directions for rotation and translation are shown on the sketch. The impulse and angular impulse equations, assuming that $F \gg f$, are

$$\int F dt = mv_0,$$

$$h \int F dt = I\omega_0,$$

where $\omega_0$ is the initial angular velocity. Solving for this yields

$$\omega_0 = hmv_0/I.$$

During the subsequent motion, until slipping ceases, the force $f$ of friction exerts a torque on the ball tending to decrease the angular speed and accelerate the ball. Hence, with $\tau = -fR$, we have

$$-fR = I\alpha = (2mR^2/5)d\omega/dt.$$

Integrating gives

$$ft = -(2mR/5)(\omega - \omega_0) = -(2mR/5)(\omega - hmv_0/I),$$

using the expression for $\omega_0$ derived above. From Newton's second law,

$$f = m(dv/dt),$$

$$v = ft/m + v_0.$$

Substituting for $ft$ from above gives

$$m(v - v_0) = -(2mR/5)(\omega - hmv_0/I).$$

When rolling sets in, $v = R\omega = 9v_0/7$, the last by supposition. Replacing $v$ with this results in

$$2mv_0/7 = -(2mR/5)[(9v_0/7R) - (5hmv_0/2mR^2)] \rightarrow h = 4R/5.$$

## 13-27

We use the conservation of angular momentum, because it is reasonable to assume that no external torques act. Since $L = I\omega$

the conservation law can be written as (see Eq. 16),

$$I_i\omega_i = I_f\omega_f.$$

Under the assumption that the Sun and the white dwarf it eventually becomes both are uniform spheres, we can write $I = (2/5)MR^2$ for each. Also, we wish to work with the rotation period $T$ rather than angular speed $\omega$. But $\omega = 2\pi/T$. Applying these relations, our conservation equation above becomes

$$(2/5)M_iR_i^2(2\pi/T_i) = (2/5)M_fR_f^2(2\pi/T_f),$$

$$M_iR_i^2/T_i = M_fR_f^2/T_f.$$

Now let us associate i with the present Sun and f with the later white dwarf configuration. With no mass loss during the transition we have $M_i = M_f$. We are told to use $R_f = R_{Earth} = 6370$ km. From Appendix C the radius of the present Sun is $R_i = 696,000$ km. Solving the equation above for $T_f$, the rotation period of the white dwarf, gives

$$T_f = [R_f/R_i]^2 T_i,$$

$$T_f = [(6370 \text{ km})/(696,000 \text{ km})]^2(25 \text{ d}) = 0.0021 \text{ d} = 3.0 \text{ min}.$$

## 13-29

The initial angular momentum of the system train+wheel is zero; by the conservation of angular momentum, the angular momentum with the power on must be zero also. Thus, the wheel and the train rotate in opposite directions relative to the Earth. The magnitude of their angular momenta must be equal:

$$L_{wheel} = L_{train}.$$

Since the rotational inertia of the wheel is $MR^2$, we have

$$L_{wheel} = MR^2\omega.$$

The speed $V$ of the train relative to the Earth is $V = v - R\omega$. Thus

$$MR^2\omega = mVR = m(v - R\omega)R,$$

$$\omega = \frac{m}{M + m}\left(\frac{v}{R}\right).$$

## 13-34

Use the definition of radius of gyration given in Problem 11 of Chapter 12 to find the rotational inertia of the merry-go-round (note that we must convert $k$ to meters):

$$I = Mk^2 = (176 \text{ kg})(0.916 \text{ m})^2 = 147.7 \text{ kg} \cdot \text{m}^2.$$

Since the child runs tangent to the rim, we have $r\perp = R$, where $R$ is the radius of the merry-go-round. Hence, the angular momentum of the running child is

$$l = r\perp p = Rmv = (1.22 \text{ m})(44.3 \text{ kg})(2.92 \text{ m/s}) = 157.8 \text{ kg} \cdot \text{m}^2/\text{s}.$$

Now apply conservation of angular momentum. The angular momentum of the merry-go-round is zero initially. The rotational inertia of the child, about the axis of the merry-go-round, after jumping on is $(44.3)(1.22)^2 = 65.94 \text{ kg} \cdot \text{m}^2$. Therefore, we have

$$157.8 \text{ kg} \cdot \text{m}^2 = [(147.7 + 65.94)\text{kg} \cdot \text{m}^2]\omega \quad \rightarrow \quad \omega = 0.739 \text{ rad/s}.$$

## 13-39

(a) The total angular momentum of the two skaters, just before one grabs the pole of length $d$, about a point midway between their paths is

$$L_0 = 2[mv(\tfrac{1}{2}d)] = mvd.$$

The mass of the pole is set equal to zero since we are told that the pole is "light." After the second skater grabs the pole the total angular momentum is

$$L_a = I_a\omega_a = 2[m(\tfrac{1}{2}d)^2]\omega_a = \tfrac{1}{2}md^2\omega_a,$$

where we treated each skater as a point object at the end of the pole. Each skater is a distance $\tfrac{1}{2}d = 1.46$ m from the center of mass of the system, about which point they move in a circular path at angular speed $\omega_a$. No forces act that can change the total angular momentum, so that $L_0 = L_a$, or

$$mvd = \tfrac{1}{2}md^2\omega_a,$$

$$\omega_a = 2v/d = 2(1.38 \text{ m/s})/(2.92 \text{ m}) = 0.945 \text{ rad/s}.$$

(b) The forces which the skaters exert on the pole to pull themselves closer are internal to the system so that the total angular momentum is still conserved. Each skater is now at a distance $\tfrac{1}{2}d_b$ from the axis, and they now revolve at angular speed $\omega_b$. By conservation of angular momentum $L_b = L_0$, so that

$$mvd = 2[m(\tfrac{1}{2}d_b)^2]\omega_b,$$

$$\omega_b = 2vd/d_b^2 = 2(1.38 \text{ m/s})(2.92 \text{ m})/(0.940 \text{ m})^2 = 9.12 \text{ rad/s}.$$

(c) The kinetic energy of two point objects, each of mass $m$ and separated by distance $d$, revolving about their center of mass at an angular speed $\omega$ is

$$K = \tfrac{1}{2}I\omega^2 = \tfrac{1}{2}[2m(\tfrac{1}{2}d)^2]\omega^2 = \tfrac{1}{4}md^2\omega^2.$$

Therefore,

$$K_a = \tfrac{1}{4}md_a^2\omega_a^2 = \tfrac{1}{4}md^2(4v^2/d^2) = mv^2,$$

so that

$$K_a = (51.2 \text{ kg})(1.38 \text{ m/s})^2 = 97.5 \text{ J}.$$

In position (b),

$$K_b = \tfrac{1}{4}md_b^2\omega_b^2 = \tfrac{1}{4}md_b^2(4v^2d^2/d_b^2) = (mv^2)(d/d_b)^2,$$

$$K_b = (97.5 \text{ J})(2.92 \text{ m}/0.940 \text{ m})^2 = 941 \text{ J}.$$

Evidently, the skaters did $941 - 97.5 = 843.5$ J of work in pulling themselves closer; this work is done at the expense of their internal energy.

## 13-42

Apply Eq. 21. We have

$$Mgr = (0.492 \text{ kg})(9.8 \text{ m/s}^2)(0.0388 \text{ m}) = 0.1871 \text{ N}\bullet\text{m}.$$

The angular speed is $\omega = (28.6)(2\pi \text{ rad})/\text{s} = 180.0$ rad/s so that

$$L = I\omega = (5.12 \times 10^{-4} \text{ kg}\bullet\text{m}^2)(180.0 \text{ rad/s}) = 0.09216 \text{ kg}\bullet\text{m}^2/\text{s}.$$

Therefore the angular speed of precession is

$$\omega_P = Mgr/L = (0.1871 \text{ N}\bullet\text{m})/(0.09216 \text{ kg}\bullet\text{m}^2/\text{s}) = 2.030 \text{ rad/s},$$

or $2.030/2\pi = 0.323$ rev/s. The direction is the same as that of the spin.

## 14-4

When the nut cracks, it exerts a force of 46 N on each arm of the nutcracker at the point of contact. Let us assume (in the lack of other data) that these forces act at 90° to the arms. If this 46 N force is the minimum force needed to crack the nut (as is implied), then at the "cracking force", we can still consider each arm of the nutcracker to be in equilibrium. Apply the torque condition (Eq. 7) about the hinge of the nutcracker as origin. The force exerted by the nut acts "outward", opposite to the applied force $F$. Hence,

$$\Sigma \tau = 0,$$

$$F(13 \text{ cm}) - (46 \text{ N})(2.6 \text{ cm}) = 0 \quad \rightarrow \quad F = 9.2 \text{ N}.$$

Note that it is not necessary to convert the cm to m, since the common conversion factor of 0.01 would cancel out.

## 14-6

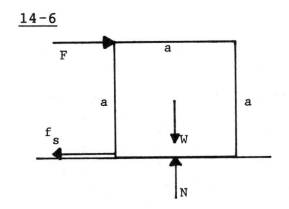 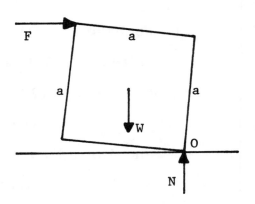

For the cube to slide, the applied force $F$ must at least equal the maximum available force of static friction which, by Eq. 1 in Chapter 6, is $f_s = \mu_s N$. In this situation (cube on a horizontal surface, $F$ acting horizontally), the normal force is $N = W$, the weight of the cube. Thus, for sliding, we must have

$$F \geq \mu_s W = 0.46W.$$

If the cube tips (leaves the floor except at the edge O), the normal force acts at the edge through point O, and therefore exerts no torque about O. The clockwise torque about O must exceed the restoring counterclockwise torque due to gravity (the weight of the cube). That is

$$F(a) \geq W(\tfrac{1}{2}a) \quad \rightarrow \quad F \geq 0.50W.$$

Comparing the conditions for sliding and tipping, we see that the cube will slide before it tips as $F$ increases.

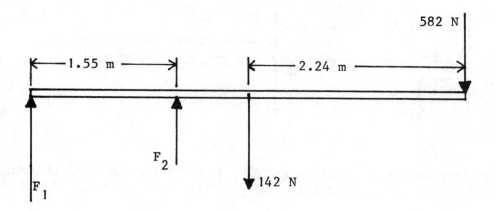

Follow the method of Sample Problem 1. The weight $W$ = 142 N of the board acts at the center of gravity at the midpoint of the board. We have assumed in drawing the force diagram above that the forces $F_1$ and $F_2$ exerted by the pedestals on the board are directed vertically up. Eq. 1 yields

$$F_1 + F_2 - 142 - 582 = 0,$$

$$F_1 + F_2 = 724 \text{ N}.$$

If we choose the left end of the board as the point about which to apply Eq. 7, we find

$$F_2(1.55) - (142)(2.24) - (582)(4.48) = 0,$$

$$F_2 = 1890 \text{ N}.$$

This turns out to be positive so that the force $F_2$ actually is directed vertically up. The force exerted by the board on the pedestal, by Newton's third law, acts opposite to this or vertically down, so that this pedestal is under compression. If we substitute this value of $F_2$ into the force equation, we find that $F_1$ = -1170 N. The negative sign means that $F_1$ acts in the direction opposite to that assumed in drawing the diagram; i.e., $F_1$ acts vertically down. The force exerted by the board on the pedestal acts opposite to this, or vertically up, so that this pedestal is in tension. This means that the board must be attached to this pedestal.

14-14

See our sketch, p. 94. Take torques about the pivot P (to eliminate the unknown normal force $N$ exerted by the pivot); we find

$$-Mg(50 - 45.5) + 2mg(45.5 - 12) = 0;$$

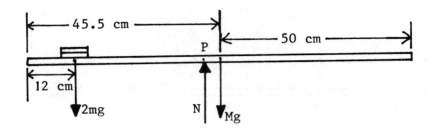

we are using Gaussian units (grams for mass, cm for distances; $g$ = 980 cm/s², but this cancels out). Since $m$ = 5 g, this equation yields

$$M = [2(5\ g)(33.5\ cm)]/(4.5\ cm) = 74.4\ g.$$

## 14-19

The normal force vanishes as the wheel leaves the ground. Taking torques about O (to eliminate the unknown force exerted by the step on the wheel at O) gives

$$Wx = F(r - h).$$

By the Pythagorean theorem,

$$x^2 + (r - h)^2 = r^2.$$

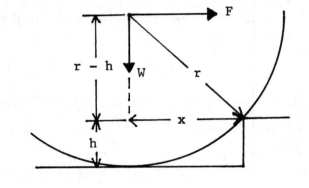

Solving this for $x$ and substituting into the previous equation gives

$$F = W\frac{\sqrt{2rh - h^2}}{r - h}.$$

## 14-24

(a) We assume that the forces are directed as shown on the sketch, p. 95. There are two forces to be found: $F_c$ exerted by the catch and $F_h$ exerted by the hinge. We can eliminate $F_h$, and so find $F_c$ first, by taking torques about the hinge. We select the data in the British system of units, expressing forces in pounds and distances in inches. Eq. 7 yields

$$\Sigma\tau = 0 = -F_c(36) + (25)(18 - 4) - F_h(0),$$

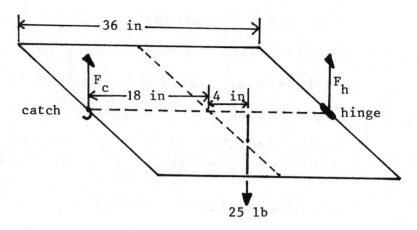

25 lb

which yields $F_c = 9.72$ lb.
(b) We can now find $F_h$ from the force condition, Eq. 1:

$$\Sigma F = 0 = 9.72 - 25 + F_h,$$

and this gives $F_h = 15.3$ lb.


## 14-25

(a) We work the problem algebraically; see the sketch for the symbols used. The angle $\phi$ made by the beam with the horizontal is found by equating two expressions for the angle made by the beam with the vertical:

$$\phi + 90° = 180 - 2\theta,$$

$$\phi = 90° - 2\theta.$$

Taking torques about the hinge gives

$$T(L\sin\theta) - W(\tfrac{1}{2}L\cos\phi) = 0,$$

$$T\sin\theta = \tfrac{1}{2}W\sin2\theta = W\sin\theta\cos\theta,$$

$$T = W\cos\theta = 52.7\cos27° = 47.0 \text{ lb.}$$

(b) Now take torques about the point where the wire is attached to the wall. We find

$$F_h L - W(\tfrac{1}{2}L\cos\phi) = 0.$$

Using the relation between the angles $\phi$ and $\theta$ found above, this equation becomes

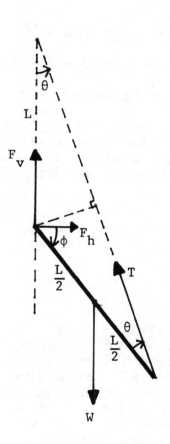

$$F_h L = W(\tfrac{1}{2}L\sin2\theta),$$

$$F_h = \tfrac{1}{2}W\sin2\theta = \tfrac{1}{2}(52.7)\sin54° = 21.3 \text{ lb.}$$

Now apply Eq. 1 ($\Sigma F = 0$); up is positive so that we find

$$F_v + T\sin(\theta + \phi) - W = 0.$$

But $\theta + \phi = 90° - \theta$ and therefore $\sin(\theta + \phi) = \cos\theta$. Using this relation and the relation for $T$ found in ($a$), the equation above becomes

$$F_v + (W\cos\theta)\cos\theta - W = 0,$$

$$T = W\sin^2\theta = (52.7)\sin^2 27° = 10.9 \text{ lb.}$$

## 14-28

Summing the horizontal and vertical force components to zero, and summing the torques about O to zero successively yields

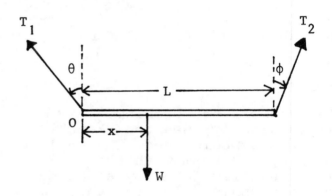

$$T_2\sin\phi - T_1\sin\theta = 0,$$

$$T_2\cos\phi + T_1\cos\theta - W = 0,$$

$$-Wx + (T_2\cos\phi)L = 0.$$

Multiply the first equation by $\cos\theta$, the second equation by $\sin\theta$ and add the resulting equations, thereby eliminating $T_1$. The resulting equation can be solved for $T_2$; we find (see Appendix H for a needed trigonometric identity)

$$T_2 = \frac{W\sin\theta}{\sin(\theta + \phi)}.$$

## 14-32

See the sketch on p. 97. The force $f_s$ of static friction opposes the tendency of the plank to slip back to the left and therefore is directed to the right. The roller, being free of friction, exerts only a normal force (called $R$ to distinguish it from the normal force $N$ exerted by the floor) on the plank. Examine the situation at $\theta = 68°$, where $f_s = \mu_s N$, its limiting value. Summing the forces and torques to zero (the latter about the bottom of the plank to eliminate two forces)

$$R\cos\theta + N - W = 0,$$

96

$$R\sin\theta - \mu_s N = 0,$$

$$Rd - W(\tfrac{1}{2}L\cos\theta) = 0.$$

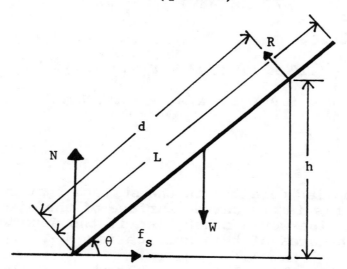

To eliminate $N$, multiply the first equation by $\mu_s$ and add to the second equation to get

$$R(\sin\theta + \mu_s\cos\theta) = \mu_s W.$$

The torque equation gives $R = (WL/2d)\cos\theta$; put this into the equation directly above to find that

$$L(\sin\theta\cos\theta + \mu_s\cos^2\theta) = 2d\mu_s = 2(h/\sin\theta)\mu_s,$$

$$\mu_s = \frac{\sin^2\theta\cos\theta}{2h - L\sin\theta\cos^2\theta}L.$$

Numerically, $L = 6.23$ m, $h = 2.87$ m and $\theta = 68°$; the equation above then gives $\mu_s = 0.407$.


## 14-42

First, calculate the strain; this is

$$\Delta L/L = (2.8\text{ cm})/(1500\text{ cm}) = 0.00187.$$

To find the stress we need first the cross-sectional area of the rope; this is

$$A = \tfrac{1}{4}\pi d^2 = \tfrac{1}{4}\pi(9.6 \times 10^{-3}\text{ m})^2 = 7.24 \times 10^{-5}\text{ m}^2.$$

Hence, the stress is

$$F/A = mg/A = (95 \text{ kg})(9.8 \text{ m/s}^2)/(7.24 \text{ X } 10^{-5} \text{ m}^2),$$

$$F/A = 1.29 \text{ X } 10^7 \text{ N/m}^2.$$

Therefore, by Eq. 28,

$$1.29 \text{ X } 10^7 \text{ N/m}^2 = E(0.00187),$$

$$E = 6.9 \text{ X } 10^6 \text{ N/m}^2 = 6.9 \text{ MN/m}^2.$$

## 14-45

The force $F$ is applied parallel to the surface that ruptures as the hole is formed; this is the curved surface of the ejected slug, the surface to which glue would be applied if the slug was reinserted into the hole. The area of this surface is

$$A = \pi DL = \pi(1.46 \text{ cm})(1.27 \text{ cm}) = 5.825 \text{ cm}^2 = 5.825 \text{ X } 10^{-4} \text{ m}^2.$$

(We need this area in m² since the area unit in the quoted value of the shear strength of steel is m².) Therefore,

$$\text{Shear stress} = F/A,$$

$$345 \text{ X } 10^6 \text{ N/m}^2 = F/(5.825 \text{ X } 10^{-4} \text{ m}^2),$$

$$F = 2.01 \text{ X } 10^5 \text{ N/m}^2 = 201 \text{ kN/m}^2.$$

## 14-48

(a) The volume of material between the roof of the tunnel and the ground surface is $V = (5.77 \text{ m})(61.5 \text{ m})(152 \text{ m}) = 5.394 \text{ X } 10^4 \text{ m}^3$. The weight of this material is $W = mg = \rho Vg$. The ultimate strength of steel is given in Table 1 in units of N/m², i.e., in SI base units. Therefore, we must convert the density from g/cm³ to kg/m³. The conversion factor is 1000, as found in Appendix G. Thus, we have

$$W = \rho Vg = (2830 \text{ kg/m}^3)(5.394 \text{ X } 10^4 \text{ m}^3)(9.8 \text{ m/s}^2),$$

$$W = 1.496 \text{ X } 10^9 \text{ N}.$$

(b) To provide a safety factor, we use as the force $F$ that the columns must support

$$F = (\text{safety factor})W = 2(1.496 \text{ X } 10^9 \text{ N}) = 2.992 \text{ X } 10^9 \text{ N}.$$

We select the ultimate strength of steel from Table 1 because the safety factor is against rupture (collapse) of the tunnel. To find the number of columns needed, find the total area $A$ of columns required:

$$\text{stress} = F/A,$$

$$400 \times 10^6 \text{ N/m}^2 = (2.992 \times 10^9 \text{ N})/A \rightarrow A = 7.48 \text{ m}^2.$$

Therefore, the number # of columns needed is

$$\# = (7.48 \times 10^4 \text{ cm}^2)/(962 \text{ cm}^2) = 77.75 \rightarrow \# = 78.$$

Note that the 7.18 m height of the tunnel does not enter into these calculations.

## 15-2

(a) The period $T$ is the time required for the motion to begin to repeat itself; therefore in this case $T = 0.484$ s.
(b) By Eq. 9, the frequency is $\nu = 1/T = 1/(0.484$ s$) = 2.07$ Hz.
(c) The angular frequency, by Eq. 10, is $\omega = 2\pi\nu = 2\pi(2.07$ s$^{-1}) = 13.0$ rad/s.
(d) The force constant $k$ is given by Eq. 7:

$$k = m\omega^2 = (0.512 \text{ kg})(13.0 \text{ rad/s})^2 = 86.5 \text{ N/m.}$$

(e) From Eq. 11, the maximum speed $v_{max}$ is

$$v_{max} = \omega x_m = (13.0 \text{ rad/s})(0.347 \text{ m}) = 4.51 \text{ m/s.}$$

(f) By Newton's second law, the maximum force is

$$F_{max} = ma_{max} = m(\omega^2 x_m) = (0.512 \text{ kg})(13.0 \text{ rad/s})^2(0.347 \text{ m}),$$

$$F_{max} = 30.0 \text{ N.}$$

## 15-6

(a) See Fig. 2 (top); the amplitude is $x_m = \frac{1}{2}(2$ mm$) = 1$ mm.
(b) By Eqs. 10 and 11,

$$v_{max} = \omega x_m = 2\pi\nu x_m = 2\pi(120 \text{ s}^{-1})(1 \times 10^{-3} \text{ m}) = 0.754 \text{ m/s.}$$

(c) Examine the same equations as cited in (b) above to find

$$a_{max} = \omega^2 x_m = (2\pi\nu)^2 x_m = 4\pi^2\nu^2 x_m,$$

$$a_{max} = 4\pi^2(120 \text{ s}^{-1})^2(1 \times 10^{-3} \text{ m}) = 568 \text{ m/s}^2.$$

## 15-9

The force constant of the spring is $k = F/x = (50$ lb$)/(4$ in$) = 12.5$ lb/in $= 12.5$ lb/$(1/12$ ft$) = 150$ lb/ft. Now use Eq. 9, but replace the mass $m$ with $W/g$, where $W$ is the weight and $g = 32$ ft/s$^2$. Since the frequency is $\nu = 2$ Hz, we have

$$2 = \frac{1}{2\pi}\sqrt{\frac{(150)(32)}{W}},$$

Solving this equation we find $W = 30.4$ lb.

## 15-12

Find the force constant $k$ by applying Eq. 2 to the two situations. If we consider magnitudes only, then the minus sign can be dropped. We work in SI base units and therefore convert g to kg and cm to m. Also, in each case, $F = mg$, with $g = 9.8$ m/s². We have then

$$(2.14)(9.8) = kx,$$

$$(2.14 + 0.325)(9.8) = k(x + 0.018),$$

where $x$ is the distance the 2.14-kg object stretches the spring. If we subtract these equations we get

$$(0.325)(9.8) = (0.018)k \rightarrow k = 176.9 \text{ N/m.}$$

Now apply Eq. 8 for the period $T$; with the 0.325-kg body removed, only the 2.14-kg object is attached to the spring. Hence,

$$T = 2\pi\sqrt{\frac{m}{k}} = 2\pi\sqrt{\frac{2.14}{176.9}} = 0.691 \text{ s.}$$

## 15-15

If it does not slip on the shake table, then the block also undergoes simple harmonic motion. Under this condition, the maximum force on the block in the horizontal direction will be

$$F_{max} = ma_{max} = m(4\pi^2 v^2 x_m).$$

This force is supplied by static friction. If the amplitude $x_m$ increases, $F_{max}$ increases also. But the force of static friction cannot increase beyond $\mu_s N$, where $N = mg$ in this case. Hence, the maximum amplitude can be found from

$$\mu_s mg = 4\pi^2 v^2 x_{m,max},$$

$$x_{m,max} = \mu_s g/4\pi^2 v^2 = (0.63)(9.8 \text{ m/s}^2)/4\pi^2(2.35 \text{ s}^{-1})^2,$$

$$x_{m,max} = 0.0283 \text{ m} = 2.83 \text{ cm.}$$

## 15-19

(a) The amplitude of each motion is $\frac{1}{2}L$ and their periods of motion both are $T = 1.50$ s. Call their displacements $x_1$ and $x_2$. Recalling that $30° = \pi/6$ rad, we can write

$$x_1 = \tfrac{1}{2}L\cos\left(\frac{2\pi t}{T}\right),$$

$$x_2 = \tfrac{1}{2}L\cos\left(\frac{2\pi t}{T} - \frac{\pi}{6}\right).$$

Particle 1 is at one end of the line at time $t = 0$. Particle 2 is at this same point at time

$$t = \frac{\pi/6}{2\pi/T} = 0.125 \text{ s.}$$

Hence, particle 1 leads particle 2. It is required to find their distance apart at time $t = 0.125 + 0.5 = 0.625$ s. Using the equations for their displacements given above, it is found that at this time

$$x_1 = \tfrac{1}{2}L\cos(5\pi/6) = -0.433L,$$

$$x_2 = \tfrac{1}{2}L\cos(4\pi/6) = -0.250L.$$

Hence, their distance apart is $x_2 - x_1 = 0.183L$.
(b) To establish the directions of motion, examine the velocities at $t = 0.625$ s. Since $v = dx/dt$, these velocities involve the sines of the angles encountered in (a). But $\sin(5\pi/6)$ and $\sin(4\pi/6)$ both are positive. This indicates that the particles are moving in the same direction at this time.


15-26

(a) The amplitude is $x_m = 9.84$ cm $= 0.0984$ m. With all the data now in SI base units we can use Eq. 14:

$$E = \tfrac{1}{2}kx_m^2,$$

$$1.18 \text{ J} = \tfrac{1}{2}k(0.0984 \text{ m})^2 \rightarrow k = 244 \text{ N/m.}$$

(b) The maximum speed is

$$v_{\max} = \omega x_m = \sqrt{\frac{k}{m}}\, x_m,$$

$$m = kx_m^2/v_{max}^2,$$

$$m = (244 \text{ N/m})(0.0984 \text{ m})^2/(1.22 \text{ m/s})^2 = 1.59 \text{ kg}.$$

(c) The frequency is

$$v = \omega/2\pi = v_{max}/2\pi x_m,$$

$$v = (1.22 \text{ m/s})/[2\pi(0.0984 \text{ m})] = 1.97 \text{ Hz}.$$

## 15-29

(a) Convert all the data to SI base units, so that $m = 12.3$ kg, $x_m$ = $1.86 \times 10^{-3}$ m and $a_{max} = 7930$ m/s². The period $T$ can be found from

$$a_{max} = \omega^2 x_m = (2\pi/T)^2 x_m,$$

$$7930 \text{ m/s}^2 = (2\pi/T)^2(1.86 \times 10^{-3} \text{ m}),$$

$$T = 3.04 \times 10^{-3} \text{ s} = 3.04 \text{ ms}.$$

(b) The angular frequency is

$$\omega = \sqrt{\frac{a_{max}}{x_m}} = \sqrt{\frac{7930 \text{ m/s}^2}{0.00186 \text{ m}}} = 2065 \text{ rad/s}.$$

Therefore

$$v_{max} = \omega x_m = (2065 \text{ rad/s})(1.86 \times 10^{-3} \text{ m}) = 3.84 \text{ m/s}.$$

(c) At the equilibrium position the potential energy is zero and $v$ = $v_{max}$. Thus,

$$E = \tfrac{1}{2}mv_{max}^2 = \tfrac{1}{2}(12.3 \text{ kg})(3.84 \text{ m/s})^2 = 90.7 \text{ J}.$$

## 15-34

The speed $V$ imparted to the block+bullet assembly immediately after impact will be the maximum speed of the subsequent simple harmonic motion. This speed is found from the conservation of momentum applied to the collision:

$$mv + 0 = (m + M)V,$$

$$V = mv/(m + M) = v_{max} = \omega x_m.$$

The angular frequency is

$$\omega = \sqrt{\frac{k}{m + M}}.$$

Therefore,

$$x_m = \frac{V}{\omega} = \frac{mv}{m + M}\sqrt{\frac{m + M}{k}} = \frac{mv}{\sqrt{k(m + M)}}.$$

15-36

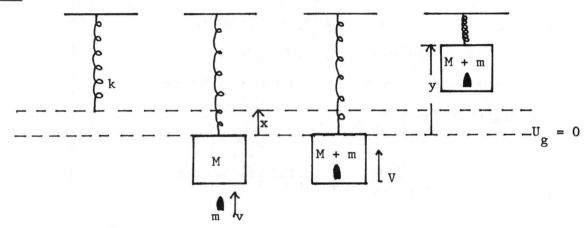

(*a*) Momentum conservation gives

$$mv = (m + M)V \quad \rightarrow \quad V = mv/(m + M).$$

After the bullet has come to rest in the block, energy conservation demands that as the block+bullet rises to maximum height,

$$\tfrac{1}{2}(m + M)V^2 + \tfrac{1}{2}kx^2 = \tfrac{1}{2}k(y - x)^2 + (m + M)gy.$$

The equilibrium position of the block before the bullet is fired is taken as the zero level of gravitational potential energy; $x$ is the distance the block stretches the spring in this position. The difference in equilibrium positions before and after the bullet impacts must be accounted for. The initial equilibrium position is given from

$$Mg = kx.$$

Put this and the result for $V$ into the energy equation to get

$$ky^2 + 2mgy = (mv)^2/(m + M),$$

104

which, with the insertion of the numerical data in proper SI base units (some unit conversions are needed), becomes

$$500y^2 + 0.98y - 13.89 = 0,$$

$$y = 0.166 \text{ m} = 16.6 \text{ cm}.$$

The block+bullet assembly will oscillate about the new equilibrium position $x_e$ found from

$$(m + M)g = kx_e.$$

Hence, the new equilibrium position is lower than the initial equilibrium position by

$$x_e - x = mg/k = (0.05 \text{ kg})(9.8 \text{ m/s}^2)/(500 \text{ N/m}),$$

$$x_e - x = 9.8 \times 10^{-4} \text{ m} = 0.098 \text{ cm}.$$

Therefore, the amplitude of oscillation will be 16.6 + 0.1 = 16.7 cm = $x_m$.

(b) The desired fraction is

$$\tfrac{1}{2}kx_m^2/\tfrac{1}{2}mv^2 = (500 \text{ N/m})(0.167 \text{ m})^2/(0.05 \text{ kg})(150 \text{ m/s})^2 = 0.0124,$$

or 1.24%.


## 15-37

(a) The translational kinetic energy is $K_t = \tfrac{1}{2}Mv^2$. Since the rotational inertia of a cylinder about its axis is $I = \tfrac{1}{2}MR^2$, and the cylinder is rolling without slipping so that $v = R\omega$, the kinetic energy of rotation about its axis is

$$K_r = \tfrac{1}{2}I\omega^2 = \tfrac{1}{2}(\tfrac{1}{2}MR^2)(v/R)^2 = \tfrac{1}{4}Mv^2.$$

The potential energy is $U = \tfrac{1}{2}kx^2$. Hence, the total energy $E$ is

$$E = \tfrac{3}{4}Mv^2 + \tfrac{1}{2}kx^2.$$

Now evaluate $E$ at the moment of release; since the cylinder was released from rest $v = 0$ and

$$E = \tfrac{1}{2}kx_i^2 = \tfrac{1}{2}(294 \text{ N/m})(0.239 \text{ m})^2 = 8.397 \text{ J}.$$

At the equilibrium point $x = 0$ and $v = v_{max}$; energy conservation yields

$$E = \tfrac{3}{4}Mv_{max}^2 = 8.397,$$

$$K_t = \tfrac{1}{2}Mv_{max}^2 = \tfrac{2}{3}(8.397) = 5.60 \text{ J.}$$

(b) As can be seen in (a), we have $K_r = \tfrac{1}{2}K_t$. Therefore, the rotational kinetic energy of the cylinder as it passes through the equilibrium position is $\tfrac{1}{2}(5.60) = 2.80$ J.

(c) As before, we have $E = \tfrac{3}{4}Mv^2 + \tfrac{1}{2}kx^2$. Differentiate this with respect to time $t$; note that $dE/dt = 0$ (conservation of energy). After factoring out $2dx/dt$, we find

$$\frac{d^2x}{dt^2} + \left(\frac{2k}{3M}\right)x = 0.$$

We recognize the first term as the acceleration $a$, so we have

$$F = Ma = -(\tfrac{2}{3}k)x.$$

This meets the requirement for simple harmonic motion (see Eq. 2). By Eqs. 7 and 10,

$$\tfrac{2}{3}k/M = \omega^2,$$

$$T = \frac{2\pi}{\omega} = 2\pi\sqrt{\frac{3M}{2k}}.$$

## 15-40

Evidently the period is $T = 180$ s/72 osc = 2.5 s/osc = 2.5 s. But the period of a simple pendulum is given by the unnumbered equation on p. 325 of RHK:

$$T = 2\pi\sqrt{\frac{L}{g}}.$$

Therefore,

$$g = 4\pi^2L/T^2 = 4\pi^2(1.53 \text{ m})/(2.5 \text{ s})^2 = 9.66 \text{ m/s}^2.$$

## 15-43

(a) The frequency is $\nu = 1/T$ and we can find $T$ from Eq. 28. The rotational inertia $I$ of the hoop about the nail at its rim follows from Fig. 9(a) of Chapter 12 and the parallel-axis theorem: $I = MR^2 + Mh^2 = MR^2 + MR^2 = 2MR^2$. We also have $d = R$ so that Eq. 28 gives

$$T = 2\pi\sqrt{\frac{I}{Mgd}} = 2\pi\sqrt{\frac{2MR^2}{MgR}} = 2\pi\sqrt{\frac{2R}{g}},$$

and therefore

$$\nu = \frac{1}{T} = \frac{1}{2\pi}\sqrt{\frac{g}{2R}} = \frac{1}{2\pi}\sqrt{\frac{9.8 \text{ m/s}^2}{2(0.653 \text{ m})}} = 0.436 \text{ Hz}.$$

(b) If we compare the final formula for $T$ above with the expression for $T$ for a simple pendulum (see Problem 40 above), we find that they will be identical if we put $L = 2R = 2(0.653 \text{ m}) = 1.31 \text{ m}$, and this is the length sought.

## 15-46

This is a torsion oscillator, with a period given by Eq. 21. The sphere oscillates about a (vertical) diameter so that the rotational inertia is given from Fig. 9(g) in Chapter 12:

$$I = (2/5)MR^2 = (2/5)(95.2 \text{ kg})(0.148 \text{ m})^2 = 0.8341 \text{ kg} \cdot \text{m}^2.$$

To find the torsional constant $\kappa$, insert the given data into Eq. 17:

$$\tau = \kappa\theta,$$

$$0.192 \text{ N} \cdot \text{m} = \kappa(0.85 \text{ rad}) \rightarrow \kappa = 0.2259 \text{ N} \cdot \text{m}.$$

Therefore the period is

$$T = 2\pi\sqrt{\frac{I}{\kappa}} = 2\pi\sqrt{\frac{0.8341}{0.2259}} = 12.1 \text{ s}.$$

## 15-48

(a) The rotational inertia of the pendulum is the sum of the rotational inertias of the disk and rod. For the disk, apply the parallel-axis theorem and refer to Fig. 9(c) in Chapter 12. In SI base units we get

$$I_{disk} = \tfrac{1}{2}(0.488 \text{ kg})(0.103 \text{ m})^2 + (0.488 \text{ kg})(0.627 \text{ m})^2,$$

$$I_{disk} = 0.1944 \text{ kg} \cdot \text{m}^2.$$

For the rod, Fig. 9($f$) in Chapter 12 gives directly

$$I_{rod} = \tfrac{1}{3}(0.272 \text{ kg})(0.524 \text{ m})^2 = 0.0249 \text{ kg} \cdot \text{m}^2.$$

Hence, the rotational inertia of the pendulum is $I = 0.1944 + 0.0249 = 0.219 \text{ kg} \cdot \text{m}^2$.

($b$) Apply Eq. 4 of Chapter 9 to the rod, considered localized at its center of mass, and disk, likewise localized at its center of mass. Measuring distances in m from the pivot and masses in kg, we have for the location of the center of mass of the pendulum

$$(0.488 + 0.272)d = (0.488)(0.627) + (0.272)(0.262),$$

$$d = 0.496 \text{ m}.$$

($c$) Use Eq. 28 with the results above. The total mass of the pendulum is $M = 0.488 + 0.272 = 0.760$ kg. We have then

$$T = 2\pi\sqrt{\frac{I}{Mgd}} = 2\pi\sqrt{\frac{0.219}{(0.760)(9.8)(0.496)}} = 1.53 \text{ s}.$$

## 15-51

For a stick swinging about an axis through one end, the appropriate rotational inertia is $I = \tfrac{1}{3}ML^2$. Also the distance $d$ between the center of mass and the axis is $d = \tfrac{1}{2}L$. Therefore,

$$I/Mgd = (\tfrac{1}{3}ML^2)/Mg(\tfrac{1}{2}L) = 2L/3g.$$

Hence, from Eqs. 21 and 1, we have

$$\nu = \frac{1}{2\pi}\sqrt{\frac{3g}{2L}} = \frac{A}{\sqrt{L}},$$

where $A$ is a constant (everything except the $L$). For the original stick of length $L_o$, the frequency is

$$\nu_o = \frac{A}{\sqrt{L_o}}.$$

The new stick has length $L = \frac{2}{3}L_o$, so that its frequency is

$$\nu = \frac{A}{\sqrt{\frac{2}{3}L_o}} = \frac{\nu_o\sqrt{L_o}}{\sqrt{\frac{2}{3}L_o}} = \sqrt{\frac{3}{2}}\,\nu_o = 1.22\nu_o.$$

## 15-62

(a) From Eq. 38 the required condition is

$$\tfrac{1}{3}x_m = x_m e^{-bt/2m}.$$

Solving for $t$ by taking ln of both sides gives

$$t = (2m\ln 3)/b = [2(1.52 \text{ kg})\ln 3]/(0.227 \text{ kg/s}) = 14.7 \text{ s}.$$

(b) By direct substitution of the given parameters, $k = 8.13$ N/m, $m = 1.52$ kg, $b = 0.227$ kg/s, into Eq. 39 we find $\omega' = 2.312$ rad/s. Therefore the period of oscillation is $T = 2\pi/2.312 = 2.718$ s. The number of oscillations in the time found in (a) is $14.7/2.718 = 5.41$.

## 15-67

The speed of 10 mi/h = 14.7 ft/s. With the "bumps" 13 ft apart, the car encounters them at the rate $v = (14.7 \text{ ft/s})/(13 \text{ ft}) = 1.13 \text{ s}^{-1}$. Therefore $\omega = 2\pi v = 7.10$ rad/s. The total weight of the loaded car is $2200 + 4(180) = 2920$ lb for a mass $m = 2920/32 = 91.25$ slug. Hence the spring force constant is

$$k = m\omega^2 = (91.25 \text{ slug})(7.10 \text{ rad/s})^2 = 4600 \text{ lb/ft} = 383 \text{ lb/in}.$$

The reduction in weight when the four people get out is $F = 4(180) = 720$ lb, and therefore the car rises a distance $x$ given from

$$x = F/k = (720 \text{ lb})/(383 \text{ lb/in}) = 1.9 \text{ in}.$$

## 15-69

Put $m_1 = 1.13$ kg and $m_2 = 3.24$ kg into Eq. 46 to find the reduced mass $m = 0.8378$ kg. The period of oscillation follows from Eq. 8 with $k = 252$ N/m; we find $T = 0.362$ s.

16-3

Call the mass of the space probe $m$. Apply Eq. 1 to the force exerted by the Sun on the space probe and to the force exerted by the Earth on the spaceprobe. These forces act in opposite directions. We call the distance between the space probe and the Earth $x$, and the distance between the Earth and the Sun $d$. Since the probe is on the line joining the Earth and Sun, it follows that the distance between the probe and the Sun is $d - x$. Let $M_E$ be the mass of the Earth and $M_S$ be the mass of the Sun. Setting the forces equal gives

$$\frac{GM_E m}{x^2} = \frac{GM_S m}{(d - x)^2}.$$

The mass $m$ of the space probe cancels out. If we take the square root of both sides and solve for $x$, thereby avoiding a quadratic equation, we obtain

$$x = \frac{\sqrt{M_E}}{\sqrt{M_E} + \sqrt{M_S}} (d).$$

Now turn to Appendix C to obtain the needed numerical data: $M_E = 5.98 \times 10^{24}$ kg, $M_S = 1.99 \times 10^{30}$ kg, $d = 1.50 \times 10^8$ km. Putting these quantities into the equation above yields $x = 260,000$ km.


16-7

We ignore the rotation of the Earth, so that $g = g_0$. The location sought is at a distance $r$ from the center of the Earth. By Eq. 3,

$$r = \sqrt{\frac{GM_E}{g}}.$$

Solve for $r$ by substituting $G = 6.67 \times 10^{-11}$ N•m²/kg², $M_E = 5.98 \times 10^{24}$ kg and $g = 7.35$ m/s²; we find $r = 7.37 \times 10^6$ m = 7370 km. To find the altitude above the Earth's surface, subtract the radius of the Earth: altitude = 7370 km - 6370 km = 1000 km.


16-11

We are not told the length of the pendulum, so we can choose a length. Suppose we pick $L = 2$ m. In this case, we find for the period $T$ of the pendulum in Paris

$$T = 2\pi\sqrt{\frac{L}{g}} = 2\pi\sqrt{\frac{2\text{ m}}{9.81\text{ m/s}^2}} = 2.837 \text{ s.}$$

In Cayenne, the period is greater by the fraction 2.5 min/1 day = 2.5 min/1440 min = $1.736 \times 10^{-3}$. That is, the period in Cayenne is greater by the amount $(1.736 \times 10^{-3})(2.837 \text{ s}) = 0.0049$ s, so that the period is now $T = 2.837 + 0.0049 = 2.842$ s. Rearranging the formula above for the period to solve for $g$ we find

$$g = 4\pi^2 L/T^2 = 4\pi^2(2\text{ m})/(2.842\text{ s})^2 = 9.78 \text{ m/s}^2$$

in Cayenne. You should convince yourself that our result is independent of the value of our choice for the length of the pendulum.

16-16

Let the suspended object have a mass $m$, and the spring a force constant $k$. The forces on the object are its true weight $mg$ directed down and the spring force $kx$ directed up. This latter force is equal in magnitude to the force exerted on the object by the spring and is equal to the scale reading. The object is moving on the equator (a circle) with linear speed $V$ relative to space; hence, by Newton's second law,

$$mg - kx = mV^2/R,$$

$R$ the radius of the Earth. The speed $V$ is made up of the speed $R\omega$ due to the Earth's rotation and the speed $v$ of the ship relative to the Earth; that is

$$V = R\omega \pm v.$$

The + sign holds if the ship is sailing in the direction of the Earth's rotation, i.e., to the east, and the - sign if it is sailing in the opposite direction. Substitute this into the first equation and solve for the scale reading to find

$$kx = mg - m(R\omega \pm v)^2,$$

$$kx = mg - mR\omega^2 \pm 2mv\omega - mv^2/R.$$

In the last equation above, the + sign applies to a ship sailing west. Now, $v \ll R\omega$; that is, the speed of the ship relative to the Earth is very much smaller than the speed due to the Earth's rotation (about 1000 mph at the equator). This means that the last term in the last equation above is numerically very much smaller than the other terms in that equation and can be dropped with little error. When the ship is at rest (dead in the water, $v = 0$), the scale reading $W_0$ is given by

$$W_0 = mg - mR\omega^2,$$

so that when the ship is moving,

$$W = W_0 \pm 2mv\omega = W_0(1 \pm 2mv\omega/W_0) \approx W_0(1 \pm 2mv\omega/mg),$$

$$W = W_0(1 \pm 2\omega v/g),$$

the + sign applying to a ship sailing west.


## 16-21

We know that in simple harmonic motion the maximum speed (see Eq. 11 in Chapter 15) is given by $v_{max} = \omega x_m$. In Sample Problem 4, the amplitude $x_m$ equals the radius $R$ of the Earth. Also $\omega = 2\pi/T$, and the numerical value of the period $T$ of oscillation is derived in the Sample Problem. Putting all this together gives

$$v_{max} = (2\pi/T)R = (2\pi/5060 \text{ s})(6370 \text{ km}) = 7.91 \text{ km/s}.$$


## 16-22

The measured value of $g$ at the bottom of the shaft is not affected by the matter that lies in the thin shell between distances $R - D$ and $R$ from the center of the Earth. That is, if we let $M'$ be the mass of that part of the Earth inside a sphere of radius $R - D$,

$$g = \frac{GM'}{(R - D)^2}.$$

Since we are to assume that the Earth is a uniform sphere, we have

$$\frac{M'}{M_E} = \frac{4\pi(R - D)^3/3}{4\pi R^3/3} = \frac{(R - D)^3}{R^3}.$$

Therefore

$$g = \frac{GM_E(R - D)}{R^3} = \frac{GM_E}{R^2} \frac{R - D}{R} = g_s\left(1 - \frac{D}{R}\right).$$


## 16-25

(a) The rotation of the Earth is ignored. We have, by Eq. 3,

$$g = GM_E/R_E^2 = (6.67 \times 10^{-11})(5.98 \times 10^{24})/(6.37 \times 10^6)^2,$$

$$g = 9.83 \text{ m/s}^2.$$

(b) The matter exterior to the Moho has no gravitational effect; only the mass $M_{Moho}$ of that part of the Earth interior to the Moho contributes. Now, from Fig. 41,

$$M_{Moho} = 1.93 \times 10^{24} \text{ kg} + 4.01 \times 10^{24} \text{ kg} = 5.94 \times 10^{24} \text{ kg},$$

so that at the Moho

$$g = GM_{Moho}/R_{Moho}{}^2,$$

$$g = (6.67 \times 10^{-11})(5.94 \times 10^{24})/(6345 \times 10^3)^2 = 9.84 \text{ m/s}^2.$$

(c) For the uniform sphere (uniform density) model, the result of Problem 22 gives

$$g = (9.83 \text{ m/s}^2)(1 - 25/6370) = 9.79 \text{ m/s}^2.$$

This is different from the result of (b), and such measurements of $g$ can rule out the uniform density model of the Earth's interior.


16-33

(a) The escape speed at the Earth's surface, by Eq. 17, is

$$v_{esc} = \sqrt{\frac{2GM_E}{R_E}}.$$

Using the relation

$$g = \frac{GM_E}{R_E^2},$$

this becomes

$$v_{esc} = \sqrt{2gR_E}.$$

Since the initial speed of the rocket is $v = 2\sqrt{(gR_E)} > v_{esc}$, the rocket escapes from the Earth.
(b) By the conservation of energy,

$$K_i + U_i = K_f + U_f,$$

$$\tfrac{1}{2}mv^2 - GmM_E/R_E = \tfrac{1}{2}mV^2 + 0,$$

the potential energy being nearly zero when the rocket is very far from the Earth, its final speed out there being called $V$. Solving the equation above for $V$ yields

$$V^2 = v^2 - 2GM_E/R_E = 4gR_E - 2gR_E = 2gR_E,$$

$$V = \sqrt{2gR_E}.$$

## 16-36

Suppose that the projectile reaches a distance $r$ from the center of the Earth. Since the projectile was fired vertically, its speed is zero at the highest point reached. When fired with speed $v = 9.42$ km/s, the projectile was at the surface of the Earth, a distance $R_E$ from the center of the Earth. Conservation of energy requires that

$$\tfrac{1}{2}mv^2 - GM_Em/R_E = 0 - GM_Em/r.$$

The mass $m$ of the projectile cancels and we find for the altitude $h = r - R_E$,

$$h = \frac{R_E}{x - 1},$$

in which

$$x = \frac{2GM_E}{R_E v^2} = 1.411,$$

the value of $x$ following after substituting the numerical quantities. Therefore $h = (6370 \text{ km})/(1.411 - 1) = 15,500$ km.

## 16-41

(a) Let $M$ be the mass of each star and $r_i$ their initial separation. Also, let $v$ be the desired speed when their separation is $\tfrac{1}{2}r_i$. The conservation of energy principle gives

$$-\frac{GM^2}{r_i} = -\frac{GM^2}{\tfrac{1}{2}r_i} + 2\left(\tfrac{1}{2}Mv^2\right),$$

$$v = \sqrt{\frac{GM}{r_i}} = \sqrt{\frac{(6.67 \times 10^{-11})(1.56 \times 10^{30})}{93400}} = 3.34 \times 10^7 \text{ m/s}.$$

114

(*b*) As they are about to collide, their separation is $r = 2R$. Once again, we use conservation of energy:

$$- \frac{GM^2}{r_i} = - \frac{GM^2}{2R} + 2(\tfrac{1}{2}Mv^2),$$

$$v^2 = GM[\frac{1}{2R} - \frac{1}{r_i}].$$

Upon substituting the numerical data into this last equation, we find $v = 54,900$ km/s.


16-45

See Sample Problem 7. The radius of the orbit of Phobos is $r = 9400$ km $= 9.4 \times 10^6$ m and the period of revolution is $T = 7$ h 39 min $= 27,540$ s. With the data now in SI base units we find for the mass $M$ of Mars

$$M = \frac{4\pi^2 r^3}{GT^2} = \frac{4\pi^2 (9.4 \times 10^6)^3}{(6.67 \times 10^{-11})(27,540)^2} = 6.5 \times 10^{23} \text{ kg.}$$


16-48

The triangle shown in Fig. 46 is a right triangle. The hypotenuse is the radius $r$ of the geosynchronous orbit. Therefore,

$$L = \cos^{-1}(R_E/r) = \cos^{-1}(6370 \text{ km}/42,200 \text{ km}) = 81.3°.$$

We found the numerical value of the radius of the geosynchronous orbit in Sample Problem 9.


16-49

(*a*) Find the mass $M$ and radius $R$ of the Moon in Appendix C. Since the spacecraft orbits "at very low altitude", we take the radius $r$ of the orbit to be the radius $R$ of the Moon itself. By Eq. 28,

$$v = \sqrt{\frac{GM}{r}} = \sqrt{\frac{(6.67 \times 10^{-11})(7.36 \times 10^{22})}{1.74 \times 10^6}} = 1680 \text{ m/s.}$$

(*b*) In one revolution, the spacecraft travels the circumference of the circular path. The circumference of a circle, in terms of the radius, is found in Appendix H. Using this, we have

$$T = 2\pi r/v = 2\pi(1.74 \times 10^6 \text{ m})/(1680 \text{ m/s}) = 6508 \text{ s} = 108 \text{ min.}$$

## 16-51

See Fig. 15. The orbit is symmetrical about the solid vertical line in that figure, so that the distance between $M$ and $F'$ is $2ea = 2(0.0167)(1.50 \times 10^{11} \text{ m}) = 5.01 \times 10^9 \text{ m}$. In terms of the radius of the Sun, this is $(5.01 \times 10^9 \text{ m})/(6.96 \times 10^8 \text{ m}) = 7.20$ solar radii.

## 16-54

(*a*) See Section 16-8, under *The Law of Orbits*. The perigee distance is $R_p = 1180 + 6370 = 7550$ km from the center of the Earth; the apogee distance is $R_a = 2360 + 6370 = 8730$ km. But

$$R_p = a(1 - e), \quad R_a = a(1 + e).$$

Therefore

$$a = \tfrac{1}{2}(R_p + R_a) = \tfrac{1}{2}(7550 + 8730) = 8140 \text{ km.}$$

(*b*) We can find $e$ from either the apogee or perigee distance. If we use the perigee distance, we have

$$e = 1 - R_p/a = 1 - 7550/8140 = 0.0725.$$

(*c*) For elliptical orbits, we replace $r$ with $a$ in Eq. 24. We found the semimajor axis $a$ in (*a*) above, so we can proceed immediately:

$$T^2 = 4\pi^2 a^3/GM,$$

$$T^2 = 4\pi^2(8140 \times 10^3)^3/(6.67 \times 10^{-11})(5.98 \times 10^{24}),$$

$$T = 7306 \text{ s} = 122 \text{ min.}$$

## 16-63

(*a*) We use the notation $R$, $h$, $M$, $m$ for the radius of the Earth, the altitude of the circular orbit of the satellite (so that the radius of the orbit is $r = R + h$), mass of the Earth, and mass of the satellite. For circular orbits the orbital speed is given by Eq. 28. We use this with $M = 5.98 \times 10^{24}$ kg, $R = 6.37 \times 10^6$ m and $h = 640$ km $= 6.4 \times 10^5$ m. We find

$$v = \sqrt{\frac{GM}{R + h}} = 7.54 \text{ km/s.}$$

(b) The period $T$ of revolution can be found from

$$T = \frac{2\pi(R + h)}{v} = \frac{2\pi(7010 \text{ km})}{7.54 \text{ km/s}} = 5840 \text{ s} = 97.4 \text{ min.}$$

(c) The total mechanical energy, by Eq. 27, is

$$E = -\frac{GMm}{2(R + h)}.$$

In the initial orbit $h = 640{,}000$ m; using this we find that $E_i = -6.259$ GJ. After 1500 revolutions the mechanical energy is

$$E_f = -6.259 \text{ GJ} - (1500 \text{ rev})(1.40 \times 10^5 \text{ J/rev}) = -6.469 \text{ GJ.}$$

The new altitude above the Earth's surface can therefore be found from Eq. 27 with the new total energy; we find $h = 412$ km. The equations cited in (a) and (b) now yield $v = 7.67$ km/s and $T = 92.6$ min for the new speed and period of revolution.

(d) Let $F$ be the force. The rate of energy loss (power) is given by Eq. 19 in Chapter 12: $P = \tau\omega$. But the angular speed $\omega = 2\pi/T$ so that we have

$$P = dE/dt = F(R + h)(2\pi/T).$$

The period $T$ is given from Eq. 24; substituting this into the equation above yields

$$F = \sqrt{\frac{R + h}{GM}}\left(\frac{dE}{dt}\right).$$

For the highest (initial) orbit,

$$dE/dt = (140 \text{ kJ/rev})/(5840 \text{ s/rev}) = 24 \text{ J/s,}$$

in absolute value. The previous equation now yields for the force $F = 3.18$ mN.

(e) The resistive force exerts a torque on the satellite and its orbital angular momentum diminishes. If all influences outside of the Earth-satellite system are ignored, then the system is isolated and its total angular momentum must remain constant. This implies that the rotational angular momentum of the Earth increases (the Earth spins faster).

(a) Calculate the weight $W = mg$ of the truck on Eros. The acceleration due to gravity at the surface of Eros is

$$g = GM/R^2 = (6.67 \times 10^{-11})(5 \times 10^{15})/(7000)^2 = 0.00681 \text{ m/s}^2.$$

Therefore $W = (2000 \text{ kg})(0.00681 \text{ m/s}^2) = 13.6$ N. You should be able to lift this easily.
(b) The required orbital speed $v$, for a circular orbit very close to the surface of Eros, is

$$v = \sqrt{\frac{GM}{r}} = \sqrt{\frac{(6.67 \times 10^{-11})(5 \times 10^{15})}{7000}} = 6.9 \text{ m/s}.$$

Not easy, but maybe you could run that fast.


16-69

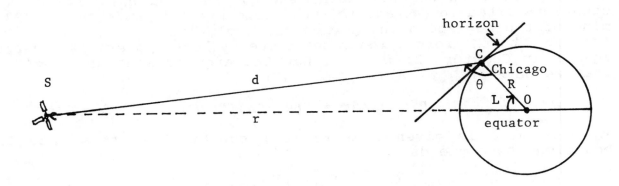

From Sample Problem 9, the radius $r$ of the geosynchronous orbit is 42,200 km. The triangle COS has no right angle, but we can use a trigonometric formula (see Appendix H under TRIANGLES) to find $d$, the straight-line distance between the satellite and Chicago:

$$d^2 = r^2 + R^2 - 2rR\cos L,$$

$$d^2 = (42,200)^2 + (6370)^2 - 2(42,200)(6370)\cos 47.5°,$$

$$d = 38,190 \text{ km}.$$

Now use another of the formulas found in Appendix H:

$$d/\sin L = r/\sin\theta,$$

$$38,190/\sin 47.5° = 42,200/\sin\theta,$$

$$\theta = 54.6°, \ 125.4°.$$

The angle 54.6° cannot apply since it would require aiming the antenna below the horizon. Thus, the antenna should be pointed at an angle 125.4° - 90° = 35.4° above the horizon, toward the south (i.e., toward the equatorial plane of the satellite's orbit).

16-70

The object spends most of its journey under the gravitational influence of the Sun. Only near the Earth does the Earth's field dominate. An approximate answer can be obtained by neglecting the Sun's field when the object is very close to the Earth, and neglecting the Earth's field when the object is far from the Earth ("patched-conic" approximation). This transition occurs at about 260,000 km from the Earth (see Problem 3), this being taken as the radius of the Earth's "sphere of influence." The object must leave the Earth's sphere of influence with a speed $u$, relative to the Sun, sufficient to project it onto a parabolic escape trajectory. This parabolic escape speed is $u = v_0\sqrt{2}$, where $v_0$ is the Earth's orbital speed; see Problem 32. Thus, as the object leaves the Earth's sphere of influence, its speed $V$ relative to the Earth will be

$$V = u - v_0 = (\sqrt{2} - 1)v_0.$$

Apply conservation of energy within the Earth's sphere of influence so that, if $v_{esc}$ is the speed sought,

$$\tfrac{1}{2}mv_{esc}^2 - GMm/R = \tfrac{1}{2}mV^2 - 0,$$

ignoring the potential energy far from the Earth; $M$, $R$ are the mass and radius of the Earth. Substituting for $V$ gives

$$v_{esc}^2 = 2GM/R + [(\sqrt{2} - 1)v_0]^2.$$

By Eq. 17 and Table 2, $2GM/R = (11.2 \text{ km/s})^2$. From Appendix C, $v_0 = 29.8$ km/s. Hence

$$v_{esc}^2 = (11.2)^2 + [(\sqrt{2} - 1)(29.8)]^2 \rightarrow v_{esc} = 16.7 \text{ km/s}.$$

16-73

The motion of the stars is about the center of mass of the system and therefore this must be located. It lies at the intersection of the symmetrical dividing lines; see Fig. 8 in Chapter 9. (Only two of the lines are needed.) The length of one of these lines is $L\sin 60° = \tfrac{1}{2}L\sqrt{3}$. Hence, the center of mass (cm) is at a distance $x$ from each side of the triangle, where $x$ is found from Eq. 4 of Chapter 9:

$$(3M)x = M(\tfrac{1}{2}L\sqrt{3}) + 2M(0) \rightarrow x = L\sqrt{3}/6.$$

The distance $s$ of any of the stars from the center of mass is given by

$$s = \tfrac{1}{2}L\sqrt{3} - x = \tfrac{1}{3}L\sqrt{3}.$$

Now $s$ is the radius of the circular path of each star; for uniform circular motion

$$F_R = Mv^2/s,$$

$F_R$ the resultant force on each star. But

$$F_R = 2F\cos 30° = F\sqrt{3},$$

where $F$ is the gravitational force between any pair of the stars. By the law of gravitation,

$$F = GM^2/L^2.$$

Putting the last four equations together yields

$$\frac{GM^2}{L^2}\sqrt{3} = \frac{Mv^2}{\tfrac{1}{3}L\sqrt{3}},$$

$$v = \sqrt{\frac{GM}{L}}.$$

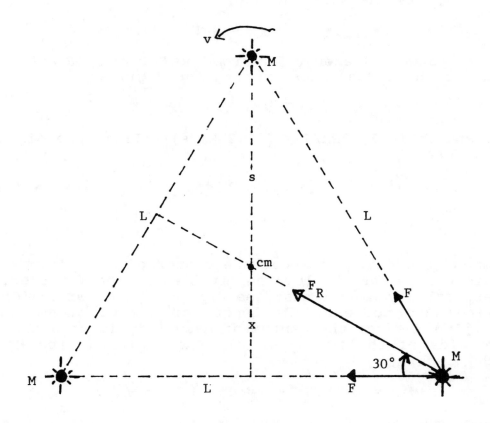

## 17-1

We need the area of the syringe in m² to get the pressure in Pa, i.e., in SI base units; hence, we calculate

$$\Delta A = \tfrac{1}{4}\pi d^2 = \tfrac{1}{4}\pi(0.0112 \text{ m})^2 = 9.852 \times 10^{-5} \text{ m}^2.$$

Therefore, by Eq. 2,

$$\Delta p = \Delta F/\Delta A = (42.3 \text{ N})/(9.852 \times 10^{-5} \text{ m}^2) = 4.29 \times 10^5 \text{ N/m}^2,$$

$$\Delta p = 429 \text{ kPa}.$$

## 17-5

Let $p$ be the pressure inside the box. The air remaining in the box exerts a force $pA$ against the lid from the inside in the same direction as the force $F$ needed to pull off the lid. The force acting to keep the lid on the box is $PA$, where $P$ is the outside air pressure. If $F$ is the smallest force needed to remove the lid, then assume equilibrium conditions, so that

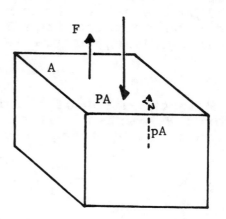

$$pA + F = PA,$$

$$p = P - F/A = 15 - 108/12 = 6 \text{ lb/in}^2.$$

## 17-9

Eq. 9 applies, but we must express the density in kg/m³; since 1 g/cm³ = 1000 kg/m³, we find

$$p = p_0 + \rho gh,$$

$$p = 1.013 \times 10^5 \text{ Pa} + (1024 \text{ kg/m}^3)(9.8 \text{ m/s}^2)(118 \text{ m}),$$

$$p = 1.29 \times 10^6 \text{ Pa} = 1.29 \text{ MPa}.$$

## 17-12

(a) Use Eq. 13 with $a$ = 8.55 km. For the exponent to remain without dimension, the height $y$ must be in km. We have, then,

$$p = p_0 e^{-y/a} = (1 \text{ atm}) e^{-5/8.55} = 0.557 \text{ atm.}$$

(b) By Eq. 13 again, with both pressures in atm, we find

$$0.5 = (1) e^{-y/8.55},$$

$$y = 5.93 \text{ km.}$$

## 17-14

(a) Eq. 2 cannot be applied to the entire dam face (!) since the pressure varies with depth, according to Eq. 9. But, the force $dF$ exerted by the water on the very narrow shaded rectangle of the dam face *can* be computed from Eq. 2 since, with the rectangle being very narrow, the pressure is constant over the rectangle: hence,

$$dF = pdA = (\rho gx)(Wdx),$$

where $W$ is the width of the dam ($W$ is not the weight of anything). The force $F$ exerted against the entire dam face follows from the equation above by integration over the dam face; to wit,

$$F = \int dF = \rho gW \int_0^D xdx = \tfrac{1}{2}\rho gWD^2.$$

(b) The torque due to the force $dF$ is $d\tau = dF(D - x)$, since $D - x$ is the moment arm. Substituting from (a) gives for the total torque

$$\tau = \int_0^D \rho gWx(D - x)\, dx = \frac{1}{6}\rho gWD^3.$$

(c) The line of action of $F$ as given in (a) to yield the torque calculated in (b) must be at a distance $r$ from the bottom of the dam where

$$(\tfrac{1}{2}\rho gWD^2)\, r = \frac{1}{6}\rho gWD^3,$$

so that $r = \tfrac{1}{3}D$.

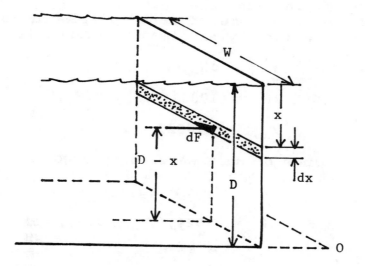

17-17

Air pressure inside the submarine annuls the external atmospheric pressure exerted by the atmosphere on the ocean surface and transmitted by the water to the submarine hull. Combining Eqs. 2 and 9, and working in SI base units, gives

$$\Delta F = p[\Delta A] = \rho g h [\Delta A],$$

$$\Delta F = (1024 \text{ kg/m}^3)(9.8 \text{ m/s}^2)(112 \text{ m})[(1.22 \text{ m})(0.59 \text{ m})],$$

$$\Delta F = 8.09 \times 10^5 \text{ N}.$$

17-23

The net effect is that gravity takes a slab of liquid $\frac{1}{2}(h_2 - h_1)$ in thickness and allows it to fall a vertical distance equal to this thickness. Thus, the work $W$ done by gravity is

$$W = mg[\frac{1}{2}(h_2 - h_1)].$$

But the mass $m$ of the slab equals the density $\rho$ times the volume of the slab, and this volume is the base area $A$ times the thickness; hence,

$$W = [\frac{1}{2}(h_2 - h_1)A\rho]g[\frac{1}{2}(h_2 - h_1)] = \frac{1}{4}A\rho(h_2 - h_1)^2 g.$$

17-31

(a) For the boat to float in vertical equilibrium, the upward buoyant force must equal the weight of the boat, regardless of whether the water is of the fresh or salt variety. By Archimedes' principle, the buoyant force equals the weight of water displaced, and therefore the weight of the water displaced equals the weight

of the boat, 35.6 kN, in each case.
(b) Let's calculate the volume of water displaced in each case.
First, for the fresh water, by Archimedes' principle,

$$W_{boat} = \rho_{fresh}V_{fresh}g,$$

$$35.6 \times 10^3 = (1000)(V_{fresh})(9.8),$$

$$V_{fresh} = 3.6327 \text{ m}^3.$$

Now calculate the volume of salt water displaced:

$$W_{boat} = \rho_{salt}V_{salt}g,$$

$$35.6 \times 10^3 = (1024)(V_{salt})(9.8),$$

$$V_{salt} = 3.5475 \text{ m}^3.$$

It follows that the volume of salt water displaced is less by the
amount 3.6327 - 3.5475 = 0.0852 m³.

17-36

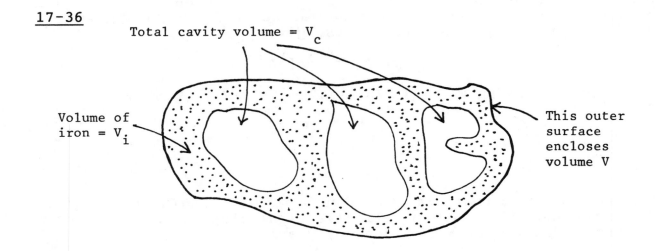

Let $V$ be the enclosed volume of the casting, including the volume
$V_i$ occupied by iron, and the volume $V_c$ of the empty cavities, so
that $V = V_i + V_c$; see the sketch above. The cavities do not
contribute to the weight $W$ of the casting so that we have

$$W = m_i g = (\rho_i V_i)g,$$

$$6130 \text{ N} = (7870 \text{ kg/m}^3)V_i(9.8 \text{ m/s}^2) \rightarrow V_i = 0.0795 \text{ m}^3.$$

The buoyant force is 6130 - 3970 = 2160 N. When the casting is
submerged, water cannot enter the cavities. Hence, by Archimedes'
principle,

$$F_B = \rho_w Vg,$$

$$2160 \text{ N} = (1000 \text{ kg/m}^3)V(9.8 \text{ m/s}^2) \rightarrow V = 0.2204 \text{ m}^3.$$

The volume of the cavities must be $V_c = V - V_i = 0.2204 - 0.0795 = 0.141 \text{ m}^3$.


## 17-39

The weight of the gas carried is

$$W_{gas} = \rho_{gas}Vg = (0.796)(1.17 \times 10^6)(9.8) = 9.13 \text{ MN},$$

where 1 MN = 1 X $10^6$ N. This equals the total weight $W$ of the loaded dirigible since we are to ignore the weight of the airship itself. The buoyant force, by Archimedes' principle with air the displaced fluid, is

$$F_B = \rho_{air}Vg = (1.21)(1.17 \times 10^6)(9.8) = 13.87 \text{ MN}.$$

Note that $F_B > W$, so that when the dirigible is to be tethered it must be "held down". Hence, the tension $T$ in the tie-down rope is given from

$$T + W = F_B,$$

$$T = F_B - W = 13.87 - 9.13 = 4.74 \text{ MN}.$$


## 17-44

(a) For the minimum area $A$ of ice, let the ice sink its full thickness $h$ so that the top surface of the ice is at the water line. For equilibrium,

$$F_B = W_{ice} + W_{car}.$$

By Archimedes' principle, the buoyant force $F_B$ equals the weight of water displaced; this is $F_B = \rho_w g V_{ice}$, since the volume of water displaced equals the volume of the block of ice. Also, the weight of the ice is $W_{ice} = \rho_{ice} g V_{ice}$. Therefore, we have

$$\rho_w g V_{ice} = \rho_{ice} g V_{ice} + M_{car} g.$$

The acceleration due to gravity $g$ cancels. Also, the volume of ice is the area times the thickness: $V_{ice} = Ah$; hence,

$$(\rho_w - \rho_{ice})V_{ice} = M_{car},$$

$$(1000 - 917)A(0.305) = 1120,$$

$$A = 44.2 \text{ m}^2.$$

(*b*) If the car is placed off-center, the ice will tip, with part of its volume outside of the water. This reduces the buoyant force, so that a thicker slab of ice will be needed to support the car.


## 17-46

(*a*) The forces acting on the log are its weight $W$ and the buoyant force $F_B$, which is a function of the displacement $x$ of the log, measured by the vertical displacement from equilibrium. If a length $L$ of the log is submerged when the log is in equilibrium, then by Archimedes' principle

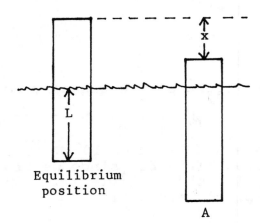

$$W = \rho g A L,$$

$\rho$ the density of water and $A$ the cross-sectional area of the log. When the log is displaced a distance $x$, with $x$ positive downward, Newton's second law gives

$$W - F_B = ma = (W/g)a,$$

$$\rho g A L - \rho g A(L + x) = (\rho g A L/g)a,$$

$$a = -(g/L)x = -\omega^2 x.$$

This last equation describes simple harmonic motion: see Eq. 11 of Chapter 15.
(*b*) The period of oscillation is

$$T = \frac{2\pi}{\omega} = 2\pi\sqrt{\frac{L}{g}} = 2\pi\sqrt{\frac{2.56}{9.8}} = 3.21 \text{ s.}$$


## 17-50

Evidently the pressure at the surface of Venus is

$$p = 90(1.013 \times 10^5 \text{ Pa}) = 9.117 \times 10^6 \text{ Pa.}$$

By Section 5, $p = \rho g h$. From Appendix C, we find that on Venus $g = 8.60 \text{ m/s}^2$. Therefore,

$$p = \rho g h,$$

$$9.117 \times 10^6 \text{ Pa} = (13.6 \times 10^3 \text{ kg/m}^3)(8.60 \text{ m/s}^2)h,$$

and this equation yields $h = 78$ m.

## 17-53

Use Eq. 21; we will need the radius of the bubble in m to give the area in m² (SI base units); we obtain

$$\Delta U = \gamma(\Delta A) = \gamma(4\pi r^2),$$

$$\Delta U = (0.025 \text{ N/m})(4\pi)(0.014 \text{ m})^2 = 6.16 \times 10^{-5} \text{ J} = 61.6 \ \mu\text{J}.$$

## 17-56

The surface tension force must equal the weight of the risen liquid. There are two lines of contact of the liquid with the container on which the surface tension can act: a circle of radius $R$ (outer boundary) and a circle of radius $r$ (inner boundary). The weight of the risen liquid is $W = mg = \rho V g = \rho(Ay)g$. The cross-sectional area of the risen liquid is the difference between the areas enclosed by the outer boundary and the inner boundary. Putting all this together gives

$$F_{\text{tension}} = W,$$

$$(2\pi R + 2\pi r)\gamma = \rho(\pi R^2 - \pi r^2)yg,$$

$$y = 2\gamma/\rho g(R - r).$$

Now substitute the given data being careful to use SI base units: $r = 0.013$ m, $R = 0.017$ m, $\gamma = 0.0728$ N/m, $\rho = 1000$ kg/m³, and, of course, $g = 9.8$ m/s². We find $y = 3.7$ mm.

# CHAPTER 18

## 18-2

The area $A$ of the hose is

$$A = \tfrac{1}{4}\pi D^2 = \tfrac{1}{4}\pi(0.75 \text{ in})^2 = 0.442 \text{ in}^2.$$

The total area of the holes is

$$24a = 24[\tfrac{1}{4}\pi d^2] = 24[\tfrac{1}{4}\pi(0.05 \text{ in})^2] = 0.0471 \text{ in}^2.$$

Now apply the equation of continuity, Eq. 3:

$$A_1 v_1 = A_2 v_2,$$

$$A v_1 = (24a) v_2,$$

$$(0.442 \text{ in}^2)(3.5 \text{ ft/s}) = (0.0471 \text{ in}^2) v_2 \quad \rightarrow \quad v_2 = 33 \text{ ft/s}.$$

## 18-4

The work $W$ done by the pump on a mass $m$ of water is

$$W = mgh + \tfrac{1}{2}mv^2,$$

where $v$ is the ejection speed of the water through the window at height $h$ above the water line. The power $P$ supplied by the pump is

$$P = dW/dt = (dm/dt)(gh + \tfrac{1}{2}v^2).$$

But the mass flow rate, by Eqs. 2 and 4 is

$$dm/dt = \rho R = \rho A v.$$

Therefore we have

$$P = A v \rho [gh + \tfrac{1}{2}v^2],$$

so that, substituting the data in SI base units, we get

$$P = [\pi(0.0097)^2](5.3)(1000)[(9.8)(2.9) + \tfrac{1}{2}(5.3)^2] = 66.5 \text{ W}.$$

## 18-10

(a) Apply the equation of continuity, Eq. 3:

$$A_1 v_1 = A_2 v_2,$$

$$(4.20 \text{ cm}^2)(5.18 \text{ m/s}) = (7.60 \text{ cm}^2) v_2 \quad \rightarrow \quad v_2 = 2.86 \text{ m/s}.$$

(b) Apply Bernoulli's equation, Eq. 9. All terms must be espressed in SI base units. For the upper level ($y = 0$ at the lower level),

$$p + \tfrac{1}{2}\rho v^2 + \rho g y = \text{constant},$$

$$152{,}000 + \tfrac{1}{2}(1000)(5.18)^2 + (1000)(9.8)(9.66) = \text{constant},$$

$$2.601 \times 10^5 \text{ Pa} = \text{constant}.$$

The "constant" has the same value at the lower level $y = 0$ (from whence its name). Also, the speed at the lower level is found from part (a). Therefore, for the lower level,

$$p + \tfrac{1}{2}(1000)(2.86)^2 + 0 = 2.601 \times 10^5,$$

$$p = 2.56 \times 10^5 \text{ Pa} = 256 \text{ kPa}.$$

## 18-13

The air inside the office is at rest. Let $p_{in}$ be the office air pressure and $p_{out}$ be the outside air pressure. Since the heights inside and outside the window are equal, the terms $\rho g y$ in Bernoulli's equation cancel and that equation therefore reduces to

$$p_{in} = p_{out} + \tfrac{1}{2}\rho v^2.$$

The window area is $A = (4.26 \text{ m})(5.26 \text{ m}) = 22.41 \text{ m}^2$. The pressures act in opposite directions on the window, so that the net force on the window is

$$F = (p_{in} - p_{out})A = \tfrac{1}{2}\rho v^2 A,$$

$$F = \tfrac{1}{2}(1.23 \text{ kg/m}^3)(28 \text{ m/s})^2(22.41 \text{ m}^2) = 10.8 \text{ kN}.$$

## 18-17

Use Bernoulli's equation, Eq. 9. First, evaluate the "constant" at the surface of the gasoline, a height $y = 53$ m above the hole. Since the area of the liquid surface is much greater than the area of the bullet hole, the equation of continuity, Eq. 3, predicts that the liquid surface will fall very slowly compared with the speed of efflux. For simplicity, then, we take the speed at which the surface falls as zero. We must use SI base units, so at the surface of the gasoline $p = 3.10(101.3 \text{ kPa}) = 3.131 \times 10^5$ Pa. We have then

$$p + \tfrac{1}{2}\rho v^2 + \rho g y = \text{constant},$$

$$3.131 \times 10^5 + 0 + (660)(9.8)(53) = 6.559 \times 10^5 \text{ Pa} = \text{constant}.$$

The "constant" has the same value at the bullet hole. At the hole the pressure equals atmospheric pressure since at this location the tank is open. Also $y = 0$ there, by choice. Therefore, again applying Bernoulli's equation,

$$1.01 \times 10^5 + \tfrac{1}{2}(660)v^2 + 0 = 6.559 \times 10^5,$$

$$v = 41.0 \text{ m/s.}$$

## 18-22

(a) Since the enlargement is abrupt, the pressure $p_1$ still acts over an area $a_2$ in the wide portion close to the enlargement. By Newton's second law

$$F = m(\Delta v/t) = (m/t)(\Delta v),$$

$$p_2 a_2 - p_1 a_1 = (\rho a_1 v_1)(v_1 - v_2),$$

since $\rho a v$ is the mass flow rate. Use the continuity equation

$$a_1 v_1 = a_2 v_2,$$

and this equation becomes

$$p_2 - p_1 = \rho v_2 (v_1 - v_2).$$

(b) If the pipe widened so gradually that there is no turbulence Bernoulli's equation can be applied; since the pipe is horizontal the $\rho g y$ terms cancel and we find

$$p_1 + \tfrac{1}{2}\rho v_1^2 = p_2 + \tfrac{1}{2}\rho v_2^2,$$

$$p_2 - p_1 = \tfrac{1}{2}\rho(v_1^2 - v_2^2).$$

(c) Comparing the expressions in (a) and (b) above, we can say that the loss in pressure due to the abruptness of the enlargement is

$$\Delta p_2 = \tfrac{1}{2}\rho(v_1^2 - v_2^2) - \rho v_2(v_1 - v_2) = \tfrac{1}{2}\rho(v_1 - v_2)^2.$$

## 18-26

The terms $\rho g y$ in Bernoulli's equation cancel out since $y_t = y_u$ for all practical purposes (t = top, u = under). Also, $v_u = 0$; writing $v$ for $v_t$ gives

$$p_t + \tfrac{1}{2}\rho v^2 = p_u.$$

Hence, if A is the area of the plate,

$$F = (p_u - p_t)A = \tfrac{1}{2}\rho v^2 A,$$

where $F$ is the net upward force exerted by the air on the plate. This must equal the weight $W = mg$ of the plate if it is to be held

130

in position. Therefore,

$$\tfrac{1}{2}\rho v^2 A = mg,$$

$$\tfrac{1}{2}(1.21 \text{ kg/m}^3)v^2(0.091 \text{ m})^2 = (0.488 \text{ kg})(9.8 \text{ m/s}^2),$$

$$v = 30.9 \text{ m/s}.$$

## 18-30

The Venturi tube is presumed horizontal, so the $\rho g y$ terms in Eq. 8 cancel, leaving

$$p_1 + \tfrac{1}{2}\rho v_1^2 = p_2 + \tfrac{1}{2}\rho v_2^2.$$

By Eq. 3, $A_1 v_1 = A_2 v_2$. Combining these two equations to eliminate $v_1$ gives

$$p_1 - p_2 = \tfrac{1}{2}\rho v_2^2 \left[1 - \left(\frac{A_2}{A_1}\right)^2\right].$$

Now each area can be written in terms of its diameter: $A = \tfrac{1}{4}\pi d^2$. With this, the preceding equation gives

$$v_2 = \sqrt{\frac{2(p_1 - p_2)}{\rho[1 - (d_2/d_1)^4]}} = 7.141 \text{ m/s}.$$

(You should insert the numerical data and verify the value of $v_2$ cited above.) We can now use Eq. 4 to find the volume flux:

$$R = A_2 v_2 = \tfrac{1}{4}\pi d_2^2 v_2,$$

$$R = \tfrac{1}{4}\pi(0.113 \text{ m})^2(7.141 \text{ m/s}) = 0.0716 \text{ m}^3/\text{s} = 71.6 \text{ L/s}.$$

## 18-35

(a) Let the outward direction be positive so that Newton's second law $F = ma$ gives,

$$-(p + dp)A + pA = -(\rho A dr)v^2/r,$$

$$dp/dr = -\rho v^2/r.$$

(b) Apply Bernoulli's equation to two nearby streamlines, assuming the same "constant" for each; we find

$$p + \tfrac{1}{2}\rho v^2 = (p + dp) + \tfrac{1}{2}\rho(v + dv)^2.$$

If we ignore the products of small quantities (i.e., the products of differentials), this becomes

$$dp = -\rho v dv.$$

Invoking (a) gives

$$\rho v^2 dr/r = -\rho v dv,$$

$$dr/r = -dv/v.$$

Integrating gives

$$\ln r = -\ln v + C,$$

$$rv = \text{Constant.}$$

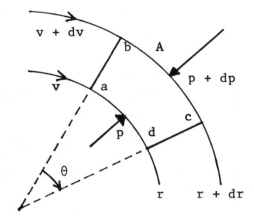

(c) By the rule in Problem 18-34, compute $\oint \mathbf{v} \cdot d\mathbf{s}$, choosing the path abcd in the sketch. Note that $\mathbf{v}$ is at 90° to $d\mathbf{s}$ along ab and cd so the integral vanishes over these sections. The length of the section bc is $(r + dr)\theta$ and the length of the section da is $r\theta$. The section da is traversed opposite to $\mathbf{v}$ giving rise to $\cos 180° = -1$ in the scalar product. Putting all this together yields

$$\oint \mathbf{v} \cdot d\mathbf{s} = (v + dv)[(r + dr)\theta] - v[r\theta].$$

The angle $\theta$ at the center of curvature is a common factor. If we multiply out the remaining terms and use the result from (b) that $rdv = -vdr$, we see that the integral vanishes. (We set the product of differentials $drdv = 0$.)


18-38

We set the Reynolds number equal to 2000 at the transition to turbulent flow. The density of blood can be found in Table 2 in Chapter 17. By Eq. 22, then,

$$v = \eta R/\rho D.$$

Take the viscosity of blood from Table 1. We find

$$v = (0.004 \ \text{N} \cdot \text{s/m}^2)(2000)/(1060 \ \text{kg/m}^3)(0.0038 \ \text{m}) = 2.0 \ \text{m/s.}$$


18-39

(a) We need the speed $v$ in order to calculate the Reynolds number. The volume flow rate is $0.0535 \ \text{L/min} = 0.0535(10^{-3} \ \text{m}^3)/(60 \ \text{s}) = 8.917 \times 10^{-7} \ \text{m}^3\text{/s}$. The cross-sectional area of the pipe is $A =$

$\pi(0.0188 \text{ m})^2 = 1.110 \times 10^{-3} \text{ m}^2$. Thus $v = 8.917 \times 10^{-7}/1.110 \times 10^{-3} = 8.03 \times 10^{-4}$ m/s. By Eq. 22, working in SI base units,

$$R = \rho Dv/\eta = (13{,}600)(0.0376)(8.03 \times 10^{-4})/(1.55 \times 10^{-3}),$$

$$R = 265 < 2000.$$

(b) See Eq. 20; if we divide both sides by the density $\rho$ the left-hand side becomes the volume flow rate; i.e., we have

$$\frac{dV}{dt} = \frac{\pi R^4 \Delta p}{8\eta L}.$$

We evaluated the volume flow rate in m$^3$/s in (a) above. The other quantities needed are all given, and if we put them into the equation above we find that $\Delta p = 0.0355$ Pa.

## 19-2

(a) The frequency is $\nu$ = (number of oscillations)/(elapsed time);
hence $\nu$ = (12 osc)/(30 s) = 0.40 osc/s = 0.40 Hz.
(b) The wave speed is $v = x/t$ = (15 m)/(5 s) = 3 m/s.
(c) We can now use Eq. 3 to find the wavelength:

$$v = \lambda\nu,$$

$$3 \text{ m/s} = \lambda(0.40 \text{ s}^{-1}),$$

$$\lambda = 7.5 \text{ m}.$$

## 19-6

(a) The phase of the wave is $(kx - \omega t)$. Since the units of $\omega t$ are
(rad/s)(s) = radian, we must express the phase difference $\Delta$(phase)
in radians also. Now $55° = (55°/180°)(\pi)$ = 0.9599 rad. We have then

$$\Delta(\text{phase}) = (kx_1 - \omega t_1) - (kx_2 - \omega t_2),$$

$$0.9599 = k(x_1 - x_2),$$

since it is implied that the phase difference occurs at the same
time, so put $t_1 = t_2$. By Eq. 13,

$$k = \omega/v = 2\pi\nu/v = 2\pi(493 \text{ Hz})/(353 \text{ m/s}) = 8.775 \text{ m}^{-1}.$$

The previous equation now yields

$$0.9599 = (8.775 \text{ m}^{-1})\Delta x \quad \rightarrow \quad \Delta x = 0.109 \text{ m}.$$

(b) This time we have $x_1 = x_2$ since the displacements are at the
same place. Our equation for the phase difference in (a) now gives

$$\Delta(\text{phase}) = \omega(t_2 - t_1) = 2\pi\nu(\Delta t),$$

$$\Delta(\text{phase}) = 2\pi(493 \text{ Hz})(1.12 \times 10^{-3} \text{ s}) = 3.469 \text{ rad} = 199°.$$

## 19-10

We calculate the speed from Eq. 18. First, we must compute the
linear mass density $\mu$ (in SI base units, kg/m, since the tension is
in newtons); we find

$$\mu = \text{mass/length} = (0.0625 \text{ kg})/(2.15 \text{ m}) = 0.02907 \text{ kg/m}.$$

By Eq. 18, then,

$$v = \sqrt{\frac{F}{\mu}} = \sqrt{\frac{487 \text{ N}}{0.02907 \text{ kg/m}}} = 129 \text{ m/s}.$$

## 19-13

We will find the linear mass density from Eq. 18. The tension is $F = 16.3$ N, so all we need to do is find the wave speed $v$. This is given by Eq. 13, $v = \omega/k$. Comparing the given equation of the wave with the standard form displayed in Eq. 12, we see that

$$v = \omega/k = (317 \text{ s}^{-1})/(23.8 \text{ m}^{-1}) = 13.32 \text{ m/s}.$$

Therefore, by Eq. 18,

$$\mu = F/v^2 = (16.3 \text{ N})/(13.32 \text{ m/s})^2 = 0.0919 \text{ kg/m} = 91.9 \text{ g/m}.$$

## 19-19

The linear mass density of the wire is

$$\mu = (0.0978 \text{ kg})/(10.3 \text{ m}) = 9.495 \times 10^{-3} \text{ kg/m}.$$

Therefore, by Eq. 18,

$$v = \sqrt{\frac{F}{\mu}} = \sqrt{\frac{248 \text{ N}}{0.009495 \text{ kg/m}}} = 161.6 \text{ m/s}.$$

Let the later disturbance be generated at time $t = 0$. The distance $x'$ traveled by this disturbance in time $t$ is $x' = vt$. But, in this same time, the other disturbance, which was generated earlier, has traveled a distance $x = v(t + 0.0296)$. When the disturbances meet we have $x + x' = 10.3$ m. Hence, at the instant of meeting,

$$v(t + 0.0296) + vt = 10.3,$$

$$161.6(2t + 0.0296) = 10.3 \quad \rightarrow \quad t = 0.01707 \text{ s}.$$

The waves therefore meet at a point $(161.6 \text{ m/s})(0.01707 \text{ s}) = 2.76$ m from that end of the wire from which the later disturbance originated.

## 19-23

(a) If the linear mass density is constant along a wire we have, by definition, $\mu = M/L$. For a variable linear mass density we write $\mu = dm/dx$, so that the mass of the wire is

$$M - \int \mu \, dx - \int_0^L (kx) \, dx - \tfrac{1}{2}kL^2.$$

(b) Since $v = dx/dt$, we see that $dt = dx/v$, so that the desired travel time is given by

$$t - \int \frac{dx}{v} - \int \frac{dx}{\sqrt{F/\mu}} - \int \frac{dx}{\sqrt{F/kx}} - \sqrt{\frac{k}{F}} \int_0^L \sqrt{x} \, dx,$$

$$t - \sqrt{\frac{k}{F}} \, (\tfrac{2}{3}L^{3/2}).$$

But from (a), $k = 2M/L^2$; substituting this into the equation above yields the result quoted in the problem.

### 19-28

See Eq. 29 and Sample Problem 3. The power output $P$ of the source is given by $P = 4\pi r^2 I$. The two locations record the intensity from the same source, so we have

$$4\pi r_1^2 I_1 = 4\pi r_2^2 I_2.$$

If we let location 1 be the more distant position, then $r_2 = r_1 - 5.3$. We are given the numerical values of the intensities. Putting all these data into the previous equation yields

$$r_1^2(1.13) = (r_1 - 5.3)^2(2.41),$$

$$r_1\sqrt{(1.13)} = (r_1 - 5.3)\sqrt{(2.41)},$$

$$r_1 = 16.81 \text{ m}.$$

Hence the power output of the source is

$$P = 4\pi(16.81 \text{ m})^2(1.13 \text{ W/m}^2) = 4.01 \text{ kW}.$$

### 19-32

We presume that the waves are traveling at the same speed so that, with their frequencies equal, their wavelengths are equal; i.e., both $\omega$ and $k$ are the same for the two waves. We can write the equations of the waves in the form

$$y_1 = (3.20)\sin\alpha,$$

$$y_2 = (4.19)\sin(\alpha + \tfrac{1}{2}\pi) = (4.19)\cos\alpha,$$

where $\alpha = kx + \omega t$. Let $y$ represent the resultant wave, so that

$$y = y_1 + y_2 = (3.20)\sin\alpha + (4.19)\cos\alpha.$$

We can define two numbers A and $\delta$ so that

$$A\cos\delta = 3.20,$$

$$A\sin\delta = 4.19.$$

We will then have

$$y = A\cos\delta\sin\alpha + A\sin\delta\cos\alpha = A\sin(\alpha + \delta),$$

using a trigonometric identity from Appendix H. We see that $\delta$ is the phase difference between the resultant wave and wave 1. To find the value of the amplitude A, square the two defining equations above for A and $\delta$ and add to get

$$A^2(\cos^2\delta + \sin^2\delta) = A^2 = (3.20)^2 + (4.19)^2,$$

$$A = 5.27 \text{ cm.}$$

## 19-37

The difference in the lengths of the paths traveled by the direct wave from $S$ and that reflected from height $H$ must equal an integral number of wavelengths for constructive interference:

$$L_H - d = N\lambda.$$

However, waves reflected from the layer when at altitude $H + h$ are out of phase with the direct wave, so that

$$L_{H+h} - d = N\lambda + \tfrac{1}{2}\lambda;$$

that is, the waves are out of phase by one-half of a wavelength. If we subtract these equations, we find

$$L_{H+h} - L_H = \tfrac{1}{2}\lambda.$$

But the path lengths are

$$L_H = 2\sqrt{(\tfrac{1}{2}d)^2 + H^2} = \sqrt{d^2 + 4H^2}.$$

$$L_{H+h} = \sqrt{D^2 + 4(H + h)^2}.$$

Putting these two results into the previous equation, and then multiplying by two, yields

$$\lambda = 2\sqrt{d^2 + 4(H + h)^2} - 2\sqrt{d^2 + 4H^2}.$$

## 19-40

(a) We find the wave speed from Eq. 18. The linear mass density in SI base units is $\mu = 7.16 \times 10^{-3}$ kg/m. Substituting this with $F = 152$ N into Eq. 18 yields $v = 146$ m/s.

(b) There are 3 loops. Each loop is bounded by a node at each end. The distance between adjacent nodes is $\frac{1}{2}\lambda$. Therefore

$$3(\tfrac{1}{2}\lambda) = 89.4 \text{ cm,}$$

$$\lambda = 59.6 \text{ cm.}$$

(c) We can now find the frequency from Eq. 13:

$$v = \lambda \nu,$$

$$146 \text{ m/s} = (0.596 \text{ m})\nu,$$

$$\nu = 245 \text{ Hz.}$$

## 19-44

Associated with a small element of length $dx$ and mass $dm$ of the string is kinetic energy

$$dK = \tfrac{1}{2}(dm)u^2 = \tfrac{1}{2}(\mu dx)u^2,$$

where $u$ is the transverse speed of the element. We find $u$ from Eq. 40; we have

$$u = \frac{\partial y}{\partial t} = -2y_m\omega\sin kx \, \sin\omega t.$$

Using this result, our expression for the kinetic energy of the element becomes

$$dK = 2y_m^2\omega^2\mu\sin^2\omega t \, \sin^2 kx \, dx.$$

The maximum value of $dK$ occurs at times $t$ such that $\sin^2 \omega t = 1$. This maximum value of $dK$ is

$$dK = 2y_m^2 \omega^2 \mu \sin^2 kx \; dx.$$

To find the total maximum kinetic energy in one loop, integrate over the loop. Recalling that adjacent nodes are separated by $\frac{1}{2}\lambda$, and $\frac{1}{2}\lambda = \frac{1}{2}(2\pi/k) = \pi/k$, we have

$$K - \int dK - 2y_m^2 \omega^2 \mu \int_0^{\pi/k} \sin^2 kx \; dx - \pi y_m^2 \omega^2 \mu / k.$$

Finally, introduce $\omega = 2\pi\nu$ and $v = \omega/k$ to get

$$K = \pi y_m^2 (2\pi\nu)(\omega/k)\mu = 2\pi^2 y_m^2 v\nu\mu.$$

## 19-48

(a) The frequency follows from Eq. 44, which applies to the string:

$$\nu_1 = (1)v/2L = (1)(250 \text{ m/s})/2(0.15 \text{ m}) = 833 \text{ Hz}.$$

(b) The frequency of the sound wave in air is the same as that of the wave on the string. Hence, the wavelength in air is

$$\lambda_{air} = v_{air}/\nu_{air} = v_{air}/\nu_1 = (348 \text{ m/s})/(833 \text{ Hz}) = 41.8 \text{ cm}.$$

## 19-53

With 4 loops the length of the string must be $L = 4(\frac{1}{2}\lambda)$, or $\lambda = \frac{1}{2}L = \frac{1}{2}(92.4 \text{ cm}) = 46.2 \text{ cm}$. Hence the wave speed is

$$v = \lambda\nu = (0.462 \text{ m})(60 \text{ Hz}) = 27.72 \text{ m/s}.$$

The linear mass density is

$$\mu = m/L = (0.0442 \text{ kg})/(0.924 \text{ m}) = 0.04784 \text{ kg/m}.$$

The tension $F$ now follows from Eq. 18:

$$F = \mu v^2 = (0.04784 \text{ kg/m})(27.72 \text{ m/s})^2 = 36.8 \text{ N}.$$

## 19-54

(a) The tension in the compound wire (each section) is

$$F = mg = (10 \text{ kg})(9.8 \text{ m/s}^2) = 98 \text{ N}.$$

In terms of the density, the linear mass density of each section

139

is given by

$$\mu = \text{mass/length} = \rho AL/L = \rho A,$$

where $A$ is the cross-sectional area of the wire and $\rho$ is the density of the section. Numerically, we find

$$\mu_1 = (2600 \text{ kg/m}^3)(1 \times 10^{-6} \text{ m}^2) = 0.0026 \text{ kg/m},$$

$$\mu_2 = (7800 \text{ kg/m}^3)(1 \times 10^{-6} \text{ m}^2) = 0.0078 \text{ kg/m}.$$

With these and the value of the common tension, Eq. 18 gives for the wave speeds $v_1 = 194.1$ m/s and $v_2 = 112.1$ m/s. The distance between adjacent nodes is $\frac{1}{2}\lambda$, so that if it is required that the joint be a node, then

$$n_1(\tfrac{1}{2}\lambda_1) = L_1; \quad n_2(\tfrac{1}{2}\lambda_2) = L_2.$$

We are told that $L_1 = 0.6$ m and $L_2 = 0.866$ m. Also $v = \lambda \nu$ for each section. Using these relations the equations directly above can be put in terms of frequency:

$$n_1 = 0.00618\nu_1; \quad n_2 = 0.0155\nu_2.$$

It can be found by trial and error that the smallest integers $n_1$ and $n_2$ giving $\nu_1 = \nu_2$ are $n_1 = 2$ and $n_2 = 5$. The resulting frequency is $2/0.00618 = 5/0.0155 = 323$ Hz.

(b) There are $5 + 2 = 7$ loops or 8 nodes in all. If the two at the ends are not counted, the number of nodes is $8 - 2 = 6$.

# CHAPTER 20

## 20-1

(a) The wavelength is given by Eq. 13 in Chapter 19, which applies to longitudinal waves as well as to transverse waves:

$$\lambda = v/\nu = (343 \text{ m/s})/(4.5 \times 10^6 \text{ s}^{-1}) = 0.0762 \text{ mm}.$$

(b) Using the sound speed in tissue, we find

$$\lambda = (1500 \text{ m/s})/(4.5 \times 10^6 \text{ s}^{-1}) = 0.333 \text{ mm}.$$

## 20-8

Let $D$ be the required distance. The travel times for the P and S waves are

$$t_P = D/v_P, \quad t_S = D/v_S.$$

The seismograph does not tell us these times separately, but it does tell us their difference. Recording the $t$'s in seconds and measuring $D$ in km, we have

$$t_S - t_P = D\left[\frac{1}{v_S} - \frac{1}{v_P}\right],$$

$$180 = D\left[\frac{1}{4.5} - \frac{1}{8.2}\right],$$

$$D = 1800 \text{ km}.$$

## 20-9

The time $t_1$ required for a stone to fall from rest to the bottom of the well of depth $d$ is

$$t_1 = \sqrt{2d/g}.$$

The time $t_2$ needed for sound, traveling at speed $v$, to cover this same distance back up the well is

$$t_2 = d/v.$$

The total time $t$ that elapses between dropping the stone and hearing the splash is, therefore,

$$t = \sqrt{2d/g} + d/v.$$

Transpose the $d/v$ to the other side and then square both sides to obtain a quadratic equation for $d$:

$$gd^2 - [(2v)(gt + v)]d + (vt)^2 g = 0.$$

Solve the equation using the quadratic formula (see Appendix H); we find

$$d = v[(\frac{v}{g} + t) \pm \sqrt{(\frac{v}{g})(\frac{v}{g} + 2t)}\ ].$$

The negative sign is appropriate since $t = 0$ should indicate $d = 0$. With $g = 9.8$ m/s$^2$, $v = 343$ m/s, a time $t = 3$ s gives $d = 40.7$ m.

20-15

By Eq. 18 the power of the wave is $P = IA$. But the definition of power is $P = E/t$ where $E$ is the energy that passes in time $t$. If we combine these two equations we have

$$E = Pt = IAt.$$

Substituting the data, being careful with units, we find

$$E = (1.6 \times 10^{-6}\ \text{W/cm}^2)(4.70\ \text{cm}^2)(3600\ \text{s}) = 0.0271\ \text{J} = 27.1\ \text{mJ}.$$

20-17

(a) For two different sounds, the difference in sound level is given by Eq. 20:

$$\Delta SL = SL_1 - SL_2 = 10\log(I_1/I_2),$$

or

$$I_1 = I_2 10^{(0.1)\Delta SL}.$$

Therefore, for $\Delta SL = 30$ dB,

$$I_1 = I_2 10^{(0.1)(30)} = 1000 I_2.$$

(b) By Eq. 18,

$$I_1/I_2 = (\Delta p_{m1}/\Delta p_{m2})^2,$$

$$1000 = (\Delta p_{m1}/\Delta p_{m2})^2,$$

$$\Delta p_{m1}/\Delta p_{m2} = 31.6.$$

## 20-21

From Problem 29 in Chapter 19 we have

$$I = uv,$$

where $u$ is the desired energy density. But (see Sample Problem 2)

$$P = 4\pi r^2.$$

Eliminating the intensity between these two equations gives

$$P = 4\pi r^2 uv,$$

$$5200 \text{ J/s} = 4\pi(4820 \text{ m})^2 u(343 \text{ m/s}),$$

$$u = 5.19 \times 10^{-8} \text{ J/m}^3 = 51.9 \text{ nJ/m}^3.$$

## 20-23

(a) The path difference between the two waves going via SBD in the two positions must be half a wavelength; therefore,

$$\tfrac{1}{2}\lambda = 2(1.65 \text{ cm}) \quad \rightarrow \quad \lambda = 6.6 \text{ cm} = 0.066 \text{ m};$$

the factor of 2 reflects the two arms in the U-shaped tube SBD. With the wavelength found, the frequency follows from

$$v = v/\lambda = (343 \text{ m/s})/(0.066 \text{ m}) = 5200 \text{ Hz}.$$

(b) Let $A$ be the amplitude of the wave when at D that went via the route SAD, and $B$ the amplitude at D of the wave that traversed SBD in either of the two positions. Since the intensity is proportional to the square of the resultant combined amplitude (Eq. 18), we have

$$(A + B)^2 = 2\rho v I_9,$$

$$(A - B)^2 = 2\rho v I_1,$$

where $I_9 = 90 \ \mu\text{W/cm}^2$ and $I_1 = 10 \ \mu\text{W/cm}^2$. Solve for $2\rho v$ in one equation, substitute the result into the other (thereby removing this term) and take the square root of the resulting equation, after substituting the numerical values of the intensities. We get

$$A + B = 3(A - B) \quad \rightarrow \quad B = \tfrac{1}{2}A.$$

(c) Waves traveling different distances lose different amounts of energy by, for example, gas friction with the tube walls.

Let $I_1$ be the intensity produced by a single person. By Eq. 19,

$$SL_1 = 65 = 10\log(I_1/I_0),$$

$$I_1 = I_0 10^{6.5}.$$

We will not need to substitute the numerical value of $I_0$, although one could do so. With $N$ persons talking, we have

$$I = NI_1,$$

not $SL = N(SL_1)$, since it is the intensity $I$ that is proportional to the energy of the sound waves. We want $I = 80$ dB; applying Eq. 19 to the crowd gives, therefore,

$$80 = 10\log(I/I_0) = 10\log[(NI_0 10^{6.5})/I_0] = 10\log[N10^{6.5}].$$

Now use the properties of logs to obtain

$$80 = 10[\log N + \log 10^{6.5}] = 10\log N + 10(6.5) = 10\log N + 65,$$

$$\log N = 1.5,$$

$$N = 10^{1.5} = 31.6.$$

Therefore, to get a sound level of 80 dB, we need 32 persons.

20-32

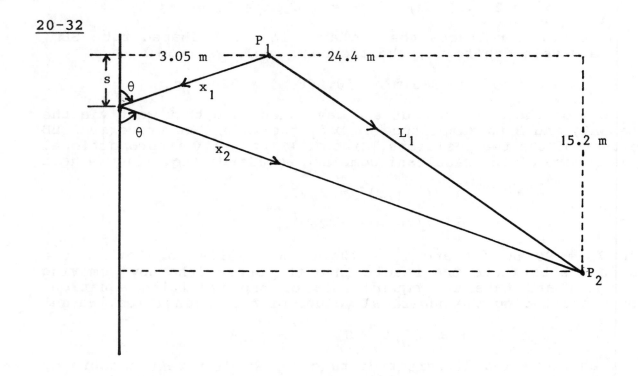

Given that the angle of incidence $\theta$ equals the angle of reflection for the sound waves reflected off the wall we have, by examining two of the right triangles in the sketch on p. 144 and applying the Pythagorean theorem,

$$\tan\theta = \frac{3.05 + 24.4}{15.2 - s} = \frac{3.05}{s},$$

which gives $s = 1.52$ m. The path lengths of the two interfering waves (direct and reflected) are given by

$$L_1 = \sqrt{24.4^2 + 15.2^2} = 28.75 \text{ m},$$

$$L_2 = x_1 + x_2 = \sqrt{3.05^2 + 1.52^2} + \sqrt{27.45^2 + 13.68^2} = 34.08 \text{ m}.$$

Thus the path difference is $L_2 - L_1 = 5.33$ m. For constructive interference this must equal an integral number of wavelengths, there being no phase change on reflection; that is,

$$n\lambda = 5.33 \text{ m}.$$

With $n = 1$ this gives $\lambda = 5.33$ m; the corresponding frequency is $\nu = v/\lambda = (343 \text{ m/s})/(5.33 \text{ m}) = 64.4$ Hz. For $n = 2$ we obtain $\lambda = 2.665$ m and $\nu = 129$ Hz.

## 20-35

Let $d$ be the limiting distance for the whispered conversation to be audible with the reflector. If $P$ is the power output of the conversation, then the intensity at the reflector is $I_s = P/4\pi d^2$. Sound energy enters the reflector at the rate $I_s(\pi R^2)$, where $R$ is the radius of the reflector. The intensity at the tube opening is $I_s \pi R^2/\pi r^2$, where $r$ is the radius of the tube. But only 12% of this actually passes down the tube, so that the actual intensity $I$ available to the ear is $(0.12)I_s R^2/r^2$, or

$$I = (0.12)(P/4\pi d^2)(R^2/r^2) = 1200P/4\pi d^2.$$

The last result follows after substituting $R = 50$ cm and $r = 0.5$ cm. Now consider a whisper at a distance of 1 m. Its intensity at the reflector is $I_w = P_w/4\pi(1)^2 = P_w/4\pi$, where $P_w$ is the power output of the whisper. The sound level of the whisper is

$$SL_w = 10\log(P_w/4\pi I_0) = 20 \text{ dB}.$$

But the conversation is also whispered, so that $P = P_w$. Hence, the sound level of the conversation at the earpiece of the reflector is

$$SL = 10\log(1200P_w/4\pi d^2 I_0) = 10\log(P_w/4\pi I_0) + 10\log(1200/d^2),$$

$$SL = SL_w + 10\log 1200 - 20\log d,$$

$$0 = 20 + 10\log 1200 - 20\log d,$$

$$d = 346 \text{ m}.$$

## 20-36

(a) In the fundamental mode,

$$L = \lambda_0/2 = v/2\nu.$$

After shortening the length of the string is $L - \Delta L$ and the frequency is $r\nu_0$, rather than $\nu_0$. We are still in the fundamental mode, so we have for the shortened string,

$$L - \Delta L = v/2r\nu_0 .$$

Eliminating $\nu_0$ between these two equations gives

$$\Delta L = L(1 - \frac{1}{r}).$$

(b) For $L = 80$ cm, successive substitution of $r = 6/5,\ 5/4,\ 4/3,\ 3/2$ gives $\Delta L = 13.3$ cm, 16 cm, 20 cm, 26.7 cm.

## 20-42

(a) For spherically symmetric pulsations the center of the star remains at rest and is a displacement node.
(b) In the fundamental mode of oscillation, antinodes exist at the surface and nowhere else. Since the distance between adjacent antinodes is $\frac{1}{2}\lambda$ and $\lambda = v/\nu$, with $v$ the speed of sound, we have

$$v/2\nu = 2R,$$

$$1/\nu = T = 4R/v.$$

(c) By Eq. 6, the speed of sound is

$$v = \sqrt{\frac{B}{\rho}} = \sqrt{\frac{1.33 \times 10^{22} \text{ Pa}}{1.0 \times 10^{10} \text{ kg/m}^3}} = 1.15 \times 10^6 \text{ m/s}.$$

The radius of the Sun is found in Appendix C. The radius $R$ of the star is $(0.009)R_{Sun} = (0.009)(6.96 \times 10^8 \text{ m}) = 6.26 \times 10^6$ m. Using these values of $v$ and $R$ in the equation for $T$ found above gives a value $T = 22$ s.

## 20-49

Possible frequencies are given by Eq. 14 in Chapter 19:

$$v = v/2L, \quad v/L, \quad 3v/2L, \text{ etc.,}$$

among which must be 880 Hz and 1320 Hz as successive frequencies. If we notice that $880/1320 = 2/3$, then it is evident from an examination of the possible frequencies that

$$880 = v/L, \quad 1320 = 3v/2L.$$

Either of these relations gives

$$v = 880L = (880 \text{ Hz})(0.30 \text{ m}) = 264 \text{ m/s}.$$

But $\mu = 0.652 \times 10^{-3}$ kg/m , so that by Eq. 18 of Chapter 19,

$$F = \mu v^2 = (6.52 \times 10^{-4} \text{ kg/m })(264 \text{ m/s})^2 = 45.4 \text{ N}.$$

## 20-50

The beat frequency is given by Eq. 33:

$$v_{beat} = |v_1 - v_2|.$$

If $v_{beat} = 3$ and either $v_1$ or $v_2 = 384$ Hz, the frequency of the other fork is $384 \pm 3 = 387$ Hz or 381 Hz. Putting a small piece of wax on the fork will decrease its frequency. Hence, if the fork had a frequency of 387 Hz, the wax brings its frequency closer to 384 Hz. This would decrease the beat frequency as noted. Thus, the fork frequency was 387 Hz.

## 20-55

Apply Eq. 40. We have $v = 15.8$ kHz, $v = 343$ m/s, $v_0 = 246$ m/s, and $v_s = 193$ m/s. The source is moving away from the observer, so use the lower sign on $v_s$; the observer is moving toward the source, so use the upper sign on $v_0$. Thus,

$$v' = v\left(\frac{v + v_0}{v + v_s}\right) = (15.8 \text{ kHz})\left(\frac{343 + 246}{343 + 193}\right) = 17.4 \text{ kHz}.$$

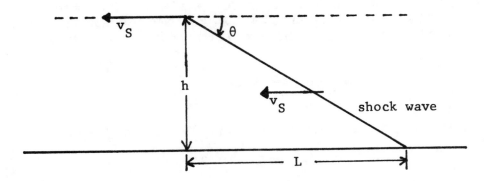

(a) By Eq. 41,

$$\sin\theta = v/v_s = v/1.52v \rightarrow \theta = 41.1°.$$

(b) The shock wave travels with the plane at a speed of Mach 1.52 = (1.52)(331 m/s) = 503.1 m/s. It must cover a horizontal distance $L = h\cot\theta = (5140 \text{ m})\cot 41.1° = 5892 \text{ m}$ to reach the ground observer. This will take a time $t = (5892 \text{ m})/(503.1 \text{ m/s}) = 11.7 \text{ s}$.

20-69

(a) The observer (wall) is at rest so use Eq. 39. The source (trumpet) moves toward the wall so the minus sign applies. We find

$$\nu' = (438 \text{ Hz})\frac{343}{343 - 19.3} = 464 \text{ Hz}.$$

(b) In this part we can consider the source to be the wall emitting sound at a frequency of 464 Hz; i.e., the $\nu'$ in (a) becomes the $\nu$ in (b). The listener (observer) is now the person with the trumpet (presumably he/she has stopped playing momentarily to hear the echo), moving toward the source (wall). We invoke Eq. 36 with the plus sign:

$$\nu' = (464 \text{ Hz})\frac{343 + 19.3}{343} = 490 \text{ Hz}.$$

20-70

(a) All speeds are given in the same units, so no speed unit conversions are needed. With both source and observer moving, we use Eq. 40. Also, we choose the upper signs since the submarines are moving toward each other. In this part, the first submarine (the shaded one) is the source and the white submarine is the observer. We find for the frequency $\nu'$ heard by the white submarine

$$\nu' = (1030 \text{ Hz}) \frac{5470 + 20.2}{5470 - 94.6} = 1050 \text{ Hz}.$$

(*b*) As the white submarine reflects the waves, it acts like a source emitting waves at a frequency of 1050 Hz. Again, use Eq. 40 but with the white submarine as the source; we have

$$\nu' = (1050 \text{ Hz}) \frac{5470 + 94.6}{5470 - 20.2} = 1070 \text{ Hz}.$$

## 21-2

The contracted length is given by Eq. 8. The rest length of the tube is $L_0$ = 2.86 m = 286 cm. For the observer $S'$ the speed $u$ of the laboratory frame $S$ equals the speed $v$ of the electron as seen in the lab, since $S'$ is attached to the electron; that is, $u/c$ = 0.999987$c$/$c$ = 0.999987. We have immediately

$$L = (286 \text{ cm})\sqrt{1 - 0.999987^2} = 1.46 \text{ cm}.$$

## 21-5

The speed of the electron relative to the detector is $v$ = 0.992$c$ = (0.992)(3 X $10^8$ m/s) = 2.976 X $10^8$ m/s. To calculate the proper lifetime we attach a frame $S'$ to the particle, so that we identify the speed $u$ connecting this frame with the detector frame ($S$) with the speed $v$. The decay time relative to the detector is

$$\Delta t = x/v = (1.05 \text{ X } 10^{-3} \text{ m})/(2.976 \text{ X } 10^8 \text{ m/s}) = 3.528 \text{ ps}.$$

By Eq. 3 the proper lifetime $\Delta t_0$ is

$$\Delta t_0 = \Delta t\sqrt{1 - (u/c)^2} = (3.528 \text{ ps})\sqrt{1 - 0.992^2} = 0.445 \text{ ps}.$$

## 21-7

The velocity of the particle in frame $S'$ is $v_0$ = 0.413$c$; the relative velocity of $S'$ with respect to $S$ is $u$ = 0.587$c$; both of these velocities are presumed to be positive. Apply Eq. 12 to get the velocity $v$ of the particle in $S$:

$$v = \frac{0.413c + 0.587c}{1 + (0.413)(0.587)} = 0.805c.$$

## 21-10

We identify the decay lifetime of 26 ns with the proper lifetime $\Delta t_0$, since it is measured in a reference frame in which the pions are at rest. The pions are not at rest relative to the Earth, so the lifetime in the Earth frame is as given by Eq. 3; that is,

$$\Delta t = \frac{\Delta t_0}{\sqrt{1 - u^2/c^2}} = \frac{26 \text{ ns}}{\sqrt{1 - 0.99^2}} = 184 \text{ ns}.$$

If a pion is seen to decay in precisely this time, it will travel a distance relative to the Earth of

$$x = u(\Delta t) = [(0.99)(3 \times 10^8 \text{ m/s})](184 \times 10^{-9} \text{ s}) = 55 \text{ m}.$$

## 21-14

By coordinates is meant $x'$ and $t'$. By Eq. 15 the Lorentz factor is

$$\gamma = \frac{1}{\sqrt{1 - u^2/c^2}} = \frac{1}{\sqrt{1 - 0.95^2}} = 3.203.$$

We will use the Lorentz transformation equations, Eq. 14. Since the data are given in units of km and $\mu$s, let us express the speed of light in these same units. We find that $c = 0.3$ km/$\mu$s. Therefore $u = 0.95c = 0.285$ km/$\mu$s. Turning to Eqs. 14:

$$x' = \gamma[x - ut],$$

$$x' = (3.203)[100 \text{ km} - (0.285 \text{ km/}\mu\text{s})(200 \text{ }\mu\text{s})] = 138 \text{ km};$$

$$t' = \gamma[t - ux/c^2],$$

$$t' = (3.203)[200 \text{ }\mu\text{s} - (0.285 \text{ km/}\mu\text{s})(100 \text{ km})/(0.3 \text{ km/}\mu\text{s})^2],$$

$$t' = -374 \text{ }\mu\text{s}.$$

## 21-17

(a) We use a Lorentz transformation equation, in the "interval" form; see Table 2. First, evaluate $\gamma$; with $u = 0.247c$, Eq. 15 yields $\gamma = 1.032$. We also have

$$\Delta x = x_r - x_b = 30.4 - 0 = 30.4 \text{ km},$$

$$\Delta t = t_r - t_b = 0;$$

the second of these results follows from the flashes being seen simultaneously in $S$ so that $t_r = t_b$. Now we can express the speed of light as $c = 0.3$ km/$\mu$s, so that $u = 0.247c = 0.247(0.3$ km/$\mu$s$) = 0.0741$ km/$\mu$s. We want to find $\Delta t'$. By Table 2, bottom row,

$$\Delta t' = (1.032)[0 - (0.0741 \text{ km/}\mu\text{s})(30.4 \text{ km})/(0.3 \text{ km/}\mu\text{s})^2],$$

giving $\Delta t' = -25.8\ \mu s$.

(b) Since we defined $\Delta x = x_r - x_b$, we also have $\Delta t = t_r - t_b$ and also $\Delta t' = t_r' - t_b'$. Since $\Delta t' < 0$, we have $t_b' > t_r'$. Hence, the red flash is seen first.


## 21-20

We adopt the *Hint*. Attach frame $S$ to the Earth and frame $S'$ to the slower particle (slower as seen from $S$, i.e., the "lower" one). Also, orient the $x, x'$ axes to be directed up (SP to NP). Then,

$$v_x = -0.787c, \quad u = +0.612c.$$

Eq. 18 now gives

$$v_x' = \frac{-0.787c - 0.612c}{1 - (0.612)(-0.787)} = -0.944c.$$

This is the velocity of the "upper" particle as seen by the "lower" particle. The negative sign indicates approach, since we picked $x, x'$ as positive up.


## 21-26

(a) The electron's speed is given by

$$v = \frac{2\pi R}{t} = \frac{2\pi(6370\ \text{km})c}{(1\ \text{s})(3 \times 10^5\ \text{km/s})} = 0.133c.$$

(b) The kinetic energy in relativity is given by Eq. 25 in Chapter 7. For the electron, the quantity $mc^2 = 0.511$ MeV (see p. 166 of RHK). Therefore,

$$K = (0.511\ \text{MeV})\left[\frac{1}{\sqrt{1 - 0.133^2}} - 1\right] = 4.58\ \text{keV}.$$

(Recall that 1 MeV = 1000 keV.)

(c) According to classical physics, the kinetic energy is

$$K = \tfrac{1}{2}mv^2 = \tfrac{1}{2}(mc^2)(v/c)^2 = \tfrac{1}{2}(511\ \text{keV})(0.133)^2 = 4.52\ \text{keV}.$$

Hence the %error is

$$\text{\%error} = [(4.52 - 4.58)/4.58](100\%) = -1.3\%,$$

the classical result being smaller than the correct value.

## 21-30

(a) By straightforward application of Eq. 15 we find $\gamma = 1.28$.
(b) We attach frame $S'$ to the clock so that the relative speed between the frames is $u = (0.622)(3 \times 10^8 \text{ m/s}) = 1.866 \times 10^8 \text{ m/s}$. When the clock passes the origin (event 1) we have $t_1 = t_1' = 0$ and $x_1 = x_1' = 0$. The clock passes the point $x_2 = 183$ m (event 2) at a time in frame $S$ given by

$$t_2 = (183 \text{ m})/(1.866 \times 10^8 \text{ m/s}) = 9.807 \times 10^{-7} \text{ s}.$$

The time $t_2'$ on the clock in $S'$ at event 2 is given by the last of the set of Eqs. 14:

$$t_2' = 1.28[9.807 \times 10^{-7} - (1.866 \times 10^8)(183)/(3 \times 10^8)^2],$$

$$t_2' = 7.70 \times 10^{-7} \text{ s} = 770 \text{ ns}.$$

## 21-36

(a) In the frame attached to the spaceship, the distance to the galactic center is contracted. By Eq. 8, if $u$ is very close to $c$, then $L \approx 0$ even though $L_0 = 23,000$ ly. Hence the answer is yes.
(b) In a frame attached to the Earth the travel time is $\Delta t = L_0/u = L_0/\beta c$. If we use ly and y as units of distance and time, then the speed of light is $c = 1$ ly/y, and $\Delta t = 23,000/\beta$. We want to travel so fast that $\Delta t_0 = 30$ y. By Eq. 3, we have then

$$\Delta t = \frac{\Delta t_0}{\sqrt{1 - \beta^2}},$$

$$\frac{23,000}{\beta} = \frac{30}{\sqrt{1 - \beta^2}},$$

and, by squaring both sides of this last equation, we quickly find that $\beta = 0.99999915$.

## 21-39

(a) By Eq. 26 we can write the magnitude of the momentum as

$$p = \gamma mc\beta,$$

$$pc/mc^2 = \gamma\beta,$$

where $\beta = v/c$ is the speed parameter and $\gamma$ is the Lorentz factor. If we express the Lorentz factor in terms of the speed parameter,

153

then the last equation can be written

$$\frac{pc}{mc^2} = \frac{\text{ß}}{\sqrt{1 - \text{ß}^2}}.$$

Solving this equation for ß gives

$$\text{ß} = \frac{A}{\sqrt{1 + A^2}}, \quad A = pc/mc^2.$$

The quantity $mc^2$ is the rest energy which, for the electron, has the value 0.511 MeV. We also have $pc$ = 12.5 MeV. Hence A = 24.46 and we get ß = 0.999165.
(b) A proton has $mc^2$ = 938 MeV. For $pc$ = 12.5 MeV again, we find A = 0.01333, giving ß = 0.0133.

21-44

(a) We have $K = 2E_0$. By Eq. 27, this gives

$$(\gamma - 1)E_0 = 2E_0,$$

$$\gamma = 3 = \frac{1}{\sqrt{1 - \text{ß}^2}} \rightarrow \text{ß} = 0.943,$$

so that $v = 0.943c$.
(b) In this case $E = 2E_0$ so that $K = E - E_0 = E_0$. From Eq. 27 we conclude that $\gamma = 2$, so that ß = 0.866, or $v = 0.866c$.

21-50

First, calculate the speed of the pion relative to the Earth. By Eq. 29,

$$\gamma = E/mc^2 = (135{,}000 \text{ MeV})/(139.6 \text{ MeV}) = 967.05.$$

Solving for the speed parameter, we find

$$\gamma = 967.05 = 1/\sqrt{1 - \text{ß}^2} \rightarrow \text{ß} = 0.9999994653.$$

The lifetime in the Earth's frame is

$$\Delta t = \gamma(\Delta t_0) = (967.05)(35 \text{ ns}) = 33.85 \ \mu s.$$

During this time, the pion travels a distance as seen from the Earth of

$$x = v(\Delta t) = (0.9999994653)(0.3 \text{ km/}\mu s)(33.85 \ \mu s) = 10.2 \text{ km}.$$

Hence the decay altitude is 120 km - 10.2 km = 110 km.

## 21-52

Apply conservation of total relativistic energy. Let $E$ be the total relativistic energy of each of the colliding particles of mass $m$. By Eq. 29,

$$E = \gamma mc^2.$$

Let $M$ be the mass of the particle resulting from the collision. By momentum conservation, we conclude that this particle is created at rest, so that its total relativistic energy is just the rest energy $Mc^2$. By conservation of total relativistic energy, then,

$$2(\gamma mc^2) = Mc^2,$$

$$M = 2\gamma m = 2(\frac{1}{\sqrt{1 - 0.58^2}})(1.30 \text{ mg}) = 3.19 \text{ mg}.$$

## 21-58

(a) Apply conservation of total relativistic energy. For each particle, $E = E_0 + K$, so we have

$$m_\alpha c^2 + K_\alpha + m_N c^2 = m_p c^2 + K_p + m_o c^2 + K_o,$$

$$3730.4 + 7.70 + 13{,}051 = 939.29 + 4.44 + 15{,}843 + K_o,$$

$$K_o = 2.37 \text{ MeV}.$$

Note that we set the kinetic energy of the $^{14}$N nucleus equal to zero since it was at rest before impact.
(b) Let us suppose that the velocity of the alpha particle before the collision is in the +x direction, and that the velocity of the proton after collision is in the +y direction. Then the oxygen nucleus must move off at an angle $\theta$ below the +x axis to conserve momentum in the collision. Furthermore, the x component must equal, in magnitude, the momentum of the alpha particle and the y component must equal, in magnitude, the momentum of the proton. Now turn to Eq. 32 and solve for momentum $p$; we find

$$p = \frac{1}{c}\sqrt{E^2 - (mc^2)^2}.$$

Hence, the conditions on the momentum of the oxygen nucleus described above gives

$$p_{yO} = p_{\text{proton}} = \frac{1}{c}\sqrt{943.73^2 - 939.29^2} = 91.44 \text{ MeV}/c,$$

and

$$p_{xO} = p_{\alpha} = \frac{1}{c}\sqrt{3738.1^2 - 3730.4^2} = 239.8 \text{ MeV}/c.$$

Therefore,

$$\theta = \tan^{-1}(p_{yO}/p_{xO}) = \tan^{-1}(91.44/239.8) = 20.9°.$$

CHAPTER 22

## 22-3

If the gain $G$ depends linearly on the temperature $T$, we can write

$$G = aT + b,$$

where $a$ and $b$ are constants (that is, independent of $G$ and $T$). To find $a$ and $b$, substitute the given pairs of values of $G$ and $T$ into the equation to obtain

$$30 = a(20°C) + b,$$

$$35.2 = a(55°C) + b.$$

Solving these equations gives $a = 0.1486/C°$ and $b = 27.03$, so that

$$G = 0.1486T + 27.03.$$

Hence, at $T = 28°C$, the gain is

$$G = (0.1486/C°)(28°C) + 27.03 = 31.2.$$

(Note that $C°$ and $°C$ are equivalent units.)

## 22-7

(a) By Eq. 2,

$$T_F = (9/5)T_C + 32,$$

so that setting $T_F = T_C = T$ gives

$$T = (9/5)T + 32 \rightarrow T = -40°.$$

(b) The relation between the Kelvin and Fahrenheit scales is, by Eqs. 2 and 8,

$$T_C = T - 273.15 = (5/9)(T_F - 32).$$

If $T_F = T$, then

$$T - 273.15 = (5/9)(T - 32) \rightarrow T = 575°.$$

(c) The Kelvin and Celsius scales are identical except for the choice of a zero point. Therefore, they can never give the same reading, for their readings will always differ by the difference in their zero points, 273.15°.

## 22-10

Use the result of Problem 9($b$). We can use °C and min as units of temperature and time. The initial temperature difference between the room and the outside, at $t = 0$ the moment the heater breaks down, is $\Delta T_0 = 22 - (-7) = 29$ C°. When $t = 45$ min, the difference is $\Delta T = 18 - (-7) = 25$ C°. Therefore, taking natural logarithms of both sides of the equation displayed in Problem 9($b$) of RHK gives

$$At = \ln(\Delta T_0/\Delta T),$$

$$A(45 \text{ min}) = \ln(29/25) \rightarrow A = 0.00330 \text{ min}^{-1}.$$

We now seek the time $t$ when $\Delta T = 14 - (-7) = 21$ C°. Returning to the first equation above with the value of A inserted gives

$$(0.00330 \text{ min}^{-1})t = \ln(29/21) \rightarrow t = 98 \text{ min},$$

or 98 - 45 = 53 min after the room temperature reached 18°C.


## 22-13

For both thermometers $p_{tr} = 100$ cm Hg. From Fig. 5, we see that the readings on the thermometers will differ by $\Delta T = T_N - T_{He} \simeq 0.2$ K. Now apply Eq. 10 to each thermometer and subtract. We obtain

$$T_N - T_{He} = 0.2 \text{ K} = (273.16 \text{ K})\frac{p_N - p_{He}}{100 \text{ cm Hg}},$$

$$p_N - p_{He} = 0.073 \text{ cm Hg},$$

with the nitrogen thermometer at the greater pressure.


## 22-18

At 10°C the area of the glass is $A = LW = (300 \text{ cm})(200 \text{ cm}) = 6 \times 10^4$ cm². The increases in length and width at 40°C can be found from Eq. 12:

$$\Delta L = \alpha L \Delta T = (9 \times 10^{-6}/\text{C}°)(300 \text{ cm})(30 \text{ C}°) = 0.081 \text{ cm},$$

$$\Delta W = \alpha W \Delta T = (9 \times 10^{-6}/\text{C}°)(200 \text{ cm})(30 \text{ C}°) = 0.054 \text{ cm}.$$

Thus the new length is 300.081 cm and the new width is 200.054 cm. The increase in area therefore is

$$\Delta A = (300.081 \text{ cm})(200.054 \text{ cm}) - 6 \times 10^4 \text{ cm}^2 = 32 \text{ cm}^2.$$

This result can also be obtained from Eq. 14.

(a) A cube has 6 faces so that the total surface area is $A = 6L^2 = 6(33.2$ cm$)^2 = 6613$ cm$^2$ at 20°C. For an increase in temperature $\Delta T = 75 - 20 = 55$ C°, the increase in area is given from Eq. 14 with the value of $\alpha$ taken from Table 3:

$$\Delta A = 2\alpha A\Delta T = 2(19 \times 10^{-6}/C°)(6613 \text{ cm}^2)(55 \text{ C}°) = 13.8 \text{ cm}^2.$$

(b) For the change in volume use Eq. 15. The volume at 20°C is $V = L^3 = (33.2$ cm$)^3 = 36{,}590$ cm$^3$. Hence,

$$\Delta V = 3\alpha V\Delta T = 3(19 \times 10^{-6}/C°)(36{,}590 \text{ cm}^3)(55 \text{ C}°) = 115 \text{ cm}^3.$$

## 22-24

A comparison of Eqs. 12 and 15 suggests that

$$\alpha = \tfrac{1}{3}(3.2 \times 10^{-5}) = 1.07 \times 10^{-5} \text{ /K.}$$

For $L$ in Eq. 12 we can use the current radius $R$ of the Earth. Therefore,

$$\Delta R = \alpha R\Delta T = (1.07 \times 10^{-5} \text{ /K})(6370 \text{ km})(2700 \text{ K}) = 180 \text{ km.}$$

## 22-28

The increase in length of each half of the bar is

$$\Delta L = \alpha(\tfrac{1}{2}L_0)\Delta T = (25 \times 10^{-6}/C°)(1.885 \text{ m})(32 \text{ C}°) = 0.0015 \text{ m.}$$

Thus the new half-length is $1.885 + 0.0015 = 1.8865$ m. Taking note of the right triangle configuration of the buckled half-bars, the buckling distance $x$ is

$$x = \sqrt{1.8865^2 - 1.885^2} = 0.0752 \text{ m} = 7.52 \text{ cm.}$$

## 22-29

Let $\Delta d$ be the change in diameter upon heating. It is desired that

$$d_s + \Delta d_s = d_b + \Delta d_b,$$

when the brass (b) ring just slides onto the steel (s) rod; $d_s$ and $d_b$ are the diameters of the steel rod and brass ring at 25°C. Since for each object

$$\Delta d = \alpha d\Delta T,$$

the first equation becomes

$$d_s(1 + \alpha_s\Delta T) = d_b(1 + \alpha_b\Delta T).$$

Substituting data gives

$$3(1 + 11 \times 10^{-6}\Delta T) = 2.992(1 + 19 \times 10^{-6}\Delta T),$$

$$\Delta T = 335 \text{ C}°.$$

Hence the common temperature sought is 25 + 335 = 360°C.


## 22-40

(a) The period $P$ of a simple pendulum is given on p. 325 of RHK: that is

$$P = 2\pi\sqrt{\frac{L}{g}}.$$

We use $P$ for period rather than the $T$ of Chapter 15 to avoid confusion with temperature. The period at the higher temperature can be written as $P + \Delta P$, where

$$\Delta P \approx \frac{\partial P}{\partial L}(\Delta L) = \frac{\pi}{\sqrt{L/g}}\frac{\Delta L}{g}.$$

But $\sqrt{(L/g)} = P/2\pi$ and $g = 4\pi^2L/P^2$. Also $\Delta L = \alpha L\Delta T$. Putting all this together gives the desired result

$$\Delta P = \tfrac{1}{2}\alpha P\Delta T.$$

(b) Numerically,

$$\Delta P = \tfrac{1}{2}(0.7 \times 10^{-6}/\text{C}°)(0.5 \text{ s})(10 \text{ C}°) = 1.75 \times 10^{-6} \text{ s}.$$

Since there are 86,400 s in one day, the pendulum makes 5,184,000 oscillations in 30 days. This leads to a total correction given by $(5.184 \times 10^6)(1.75 \times 10^{-6} \text{ s}) = 9.07$ s. The clock runs slow since the pendulum is longer at the higher temperature and the period of swing increases.


## 22-42

The capacity $V$ of the cup increases by $\Delta V = 3\alpha V\Delta T$, $\alpha$ the linear expansion coefficient of aluminum; see Eq. 15. The volume of the glycerin, which at 22°C equals $V$, the capacity of the cup,

160

increases by $\Delta V = \beta V \Delta T$. Thus, glycerin will spill out if $\beta > 3\alpha$, as it is. The amount of the spill is the difference in the volume changes listed above; that is,

$$\text{spill} = (\beta - 3\alpha)V\Delta T,$$

$$\text{spill} = (5.1 \times 10^{-4} - 6.9 \times 10^{-5})(110)(6) = 0.29 \text{ cm}^3.$$

22-43

The volume of the liquid at 20°C is $V = Ah$, where $A$ is the cross-sectional area of the liquid column and $h$ is the height of the column ($h = 0.64$ m). At 30°C the corresponding volume is $V' = A'h'$. By Eq. 16,

$$\Delta V = V' - V = \beta V \Delta T.$$

The change in the cross-sectional area of the liquid column is governed by the expansion of the glass. We use Eq. 14 to write

$$A' = A + 2\alpha A \Delta T.$$

Using this result, the first equation can be transformed as follows:

$$A'h' - Ah = \beta Ah\Delta T,$$

$$(A + 2\alpha A\Delta T)h' - Ah = \beta Ah\Delta T,$$

$$h' = h\left[\frac{1 + \beta \Delta T}{1 + 2\alpha \Delta T}\right].$$

Hence the change in height is

$$\Delta h = h' - h = \left[\frac{\beta - 2\alpha}{1 + 2\alpha \Delta T}\right]h(\Delta T).$$

Substituting the numerical data gives $\Delta h = 0.017$ cm $= 0.17$ mm.

22-44

If the rod is allowed to contract freely, then we will have

$$\Delta L = \alpha L \Delta T.$$

The stress that each bolt must support to prevent this contraction is given by Eq. 29 of Chapter 14:

$$\text{stress} = F/A = E(\Delta L/L) = E(\alpha \Delta T),$$

substituting the thermal strain from the first equation; $E$ is Young's modulus for the rod (i.e., of steel). Find Young's modulus in Table 1 of Chapter 14, along with the value of the stress at the yield point (the greatest stress to which Eq. 29 of Chapter 14 is valid). Upon substituting the values the equation above gives

$$250 \text{ X } 10^6 \text{ N/m}^2 = (200 \text{ X } 10^9 \text{ N/m}^2)(11 \text{ X } 10^{-6}/\text{C}°)\Delta T,$$

$$\Delta T = 114°\text{C}.$$

Hence, the temperature at which the rod begins to yield to the thermal stress is 24 - 114 = -90°C.


22-45

For each rod, with ' denoting the expanded length,

$$L' = L(1 + \alpha \Delta T),$$

so that, deleting terms in $\alpha^2$,

$$L'^2 = L^2(1 + 2\alpha \Delta T).$$

Use the obvious notation i = invar, a = aluminum, s = steel. The expanded lengths are related by the law of cosines (Appendix H):

$$L_i'^2 = L_a'^2 + L_s'^2 - 2L_a'L_s'\cos\theta$$

where $\theta$ is the angle between the aluminum and steel rods. Into this expression, substitute the expressions above for $L'$ and $L'^2$, written with subscript for the appropriate rod. Delete terms in $\alpha^2$. Also note that $L_i = L_a = L_s$ (the original triangle was equilateral). Solve the resulting equation for $\Delta T$; you should get

$$\Delta T - ½ \frac{1 - 2\cos\theta}{\alpha_i - (\alpha_a + \alpha_s)(1 - \cos\theta)}.$$

Substituting the numerical data gives $\Delta T = 46.4°\text{C}$, so that the desired temperature is 66.4°C.

23-3

(a) Use the ideal gas law, Eq. 7, with the temperature $T = 273 + 12 = 285$ K; we find

$$pV = nRT,$$

$$(108 \times 10^3 \text{ Pa})(2.47 \text{ m}^3) = n(8.31 \text{ J/mol} \cdot \text{K})(285 \text{ K}),$$

$$n = 113 \text{ mol}.$$

(b) Since there are no leaks, the quantity of gas as expressed by the number of moles does not change, so that $n_i = n_f$. Hence (see Sample Problem 1),

$$\frac{p_i V_i}{p_f V_f} = \frac{T_i}{T_f},$$

$$\frac{(108 \text{ kPa})(2.47 \text{ m}^3)}{(316 \text{ kPa})V_f} = \frac{285 \text{ K}}{304 \text{ K}},$$

$$V_f = 0.900 \text{ m}^3.$$

23-5

See Sample Problem 1. We can express the pressures in cm of Hg since this quantity is directly proportional to the pressure (see Section 17-5). Being careful to put the temperatures in kelvin, we find

$$V_f = (p_i/p_f)(T_f/T_i)V_i,$$

$$V_f = (76 \text{ cm}/36 \text{ cm})(225 \text{ K}/295 \text{ K})(3.47 \text{ m}^3) = 5.59 \text{ m}^3.$$

23-9

Let $p$ be the final gauge pressure. We must use absolute pressure in the ideal gas law. See Section 17-5, where we find that

absolute pressure = gauge pressure + atmospheric pressure.

In British units, atmospheric pressure = 14.7 lb/in². There is no need to convert from in³ to m³ since the conversion is one of multiplication and the conversion factor will cancel in the "ratio form" of the ideal gas law (see Sample Problem 1) that we will use.

However, the temperature conversion from °C to K is additive and would not cancel, so that we must use the kelvin (or absolute) temperatures. With all this in mind we have from the first equation in Sample Problem 1 (you add the units),

$$\frac{(24.2 + 14.7)(988)}{273 - 2.6} = \frac{(p + 14.7)(1020)}{273 + 25.9},$$

$$p = 26.9 \text{ lb/in}^2,$$

for the gauge pressure at the higher temperature.


23-12

Since the temperature does not change we have

$$p_i V_i = p_f V_f,$$

assuming that none of the air initially in the pipe is lost as it is thrust under water. In the initial state, then, $p_i = p_{atm}$ and $V_i = V$, where $V$ is the volume of the pipe. In the final state, as pictured in Fig. 16, the volume occupied by the air is $V_f = \frac{1}{2}V$. Since the system is in equilibrium, by supposition, then the air pressure in the final state is equal to the pressure at the surface of the water inside the pipe. This in turn is equal to the pressure of the water at any other point in the lake at the same depth beneath the surface of the lake. Bearing in mind that we need the absolute pressure, we have therefore

$$p_f = p_{atm} + \rho g(h - \tfrac{1}{2}L),$$

where $\rho$ is the density of water. Our first equation now becomes

$$p_{atm}V = [p_{atm} + \rho g(h - \tfrac{1}{2}L)](\tfrac{1}{2}V),$$

$$h = \tfrac{1}{2}L + p_{atm}/\rho g.$$

Substituting $L = 25$ m, $p_{atm} = 1.01 \times 10^5$ Pa, $\rho = 1000$ kg/m$^3$ and, of course, $g = 9.8$ m/s$^2$, we should find $h = 22.8$ m.


23-15

Use the ideal gas law to find the number of moles of gas per unit volume:

$$pV = nRT,$$

$$[(1.22)(1.01 \times 10^5)]V = n(8.31)(273 + 35),$$

$$n/V = 48.14 \text{ mol/m}^3.$$

If we multiply by Avogadro's constant we find the number of atoms per unit volume:

$$(nN_A/V) = N/V = (48.14)(6.02 \times 10^{23}) = 2.898 \times 10^{25} \text{ atoms/m}^3.$$

Now,

$$N = \frac{\text{total volume of atoms}}{\text{volume of one atom}} = \frac{V_{\text{atoms}}}{V_{\text{atom}}}.$$

But,

$$V_{\text{atom}} = (4/3)\pi r^3 = (4/3)\pi(0.710 \times 10^{-10} \text{ m})^3 = 1.499 \times 10^{-30} \text{ m}^3.$$

Hence,

$$V_{\text{atoms}}/V = (1.499 \times 10^{-30} \text{ m}^3)(2.898 \times 10^{25} \text{ m}^{-3}) = 4.34 \times 10^{-5},$$

and this is the fraction sought.

## 23-18

Combining Eqs. 15 and 20 gives

$$\tfrac{1}{2}mv_{\text{rms}}^2 = (3/2)kT.$$

The mass of one molecule of ammonia $NH_3$ is

$$m = 2.33 \times 10^{-26} + 3(1.67 \times 10^{-27}) = 2.831 \times 10^{-26} \text{ kg.}$$

The numerical value of the Boltzmann constant $k$ is found on p. 511 of RHK; the temperature is $T = 273 + 56 = 329$ K. We have, then,

$$\tfrac{1}{2}(2.831 \times 10^{-26} \text{ kg})v_{\text{rms}}^2 = (3/2)(1.38 \times 10^{-23} \text{ J/K})(329 \text{ K}),$$

$$v_{\text{rms}} = 694 \text{ m/s.}$$

## 23-21

(a) Before using Eq. 15, we must put the data into SI base units. For the pressure we have $p = (0.0123 \text{ atm})(1.01 \times 10^5 \text{ Pa/atm}) = 1242$ Pa; also the density is $\rho = 0.0132$ kg/m$^3$. Direct substitution into Eq. 15 yields $v_{\text{rms}} = 531$ m/s.
(b) The temperature is $T = 273 + 44 = 317$ K. By Eq. 18,

$$M = 3RT/v_{\text{rms}}^2,$$

$$M = 3(8.31 \text{ J/mol} \cdot \text{K})(317 \text{ K})/(531 \text{ m/s})^2 = 0.0280 \text{ kg/mol,}$$

or $M = 28$ g/mol. By examination of Table 1, we conclude that the gas is nitrogen $N_2$.

## 23-26

By Eq. 20, the average translation kinetic energy of a single molecule is $3kT/2$. The value of $k$ given on p. 511 of RHK is in SI base units, so we must express the energy in these units also, i.e., in joules. We obtain

$$K_{av} = (3/2)kT,$$

$$1.6 \times 10^{-19} \text{ J} = (3/2)(1.38 \times 10^{-23} \text{ J/K})T,$$

$$T = 7730 \text{ K}.$$

## 23-28

(a) The number of moles $n$ is given by

$$\text{number of moles} = \frac{\text{bulk mass}}{\text{molar mass}},$$

$$n = \frac{2.56 \text{ g}}{197 \text{ g/mol}} = 0.0130 \text{ mol}.$$

(b) The number $N$ of atoms follows from

$$N = nN_A = (0.013 \text{ mol})(6.02 \times 10^{23} \text{ mol}^{-1}) = 7.23 \times 10^{21}.$$

## 23-32

(a) First, calculate the molar mass of ammonia since this is needed to find the number of moles. Taking atomic molar masses from Appendix D, we obtain

$$M = 14.0067 + 3(1.00797) = 17.03 \text{ g/mol}.$$

Hence the number of moles present initially is

$$n = (315 \text{ g})/(17.03 \text{ g/mol}) = 18.50 \text{ mol}.$$

The ideal gas law, Eq. 7, now gives

$$(1.35 \times 10^6)V = (18.50)(8.31)(273 + 77),$$

$$V = 0.0399 \text{ m}^3.$$

(b) A "steel" tank implies a fixed volume. Again using Eq. 7, we find the number of moles of gas present when the second reading is taken:

$$(8.68 \times 10^5)(0.0399) = n(8.31)(273 + 22),$$

$$n = 14.13 \text{ mol}.$$

Thus, the mass of gas present at the second reading is

$$\text{bulk mass} = (14.13 \text{ mol})(17.03 \text{ g/mol}) = 240.6 \text{ g}.$$

Evidently $315 - 240.6 = 74.4$ g of the gas leaked out.

## 23-35

We find the greatest lifting capacity by setting the buoyant force equal to the total weight:

$$F_B = W_{total}.$$

If $\rho_{surr}$ is the density of the air surrounding the balloon, then by Archimedes' principle,

$$F_B = \rho_{surr}Vg,$$

where $V$ is the volume of the balloon. The total weight of the balloon includes contributions from the hot air, the envelope, and the payload (lifting capacity); that is

$$W_{total} = (\rho_{hot}V + m_{env} + m_{pay})g.$$

Therefore we have by the first equation,

$$\rho_{surr}V = \rho_{hot}V + m_{env} + m_{pay},$$

$$(1.22)(2180) = \rho_{hot}(2180) + 249 + 272,$$

$$\rho_{hot} = 0.9810 \text{ kg/m}^3.$$

Now assume that the pressure of the hot air inside the balloon equals the pressure of the outside surrounding air (reasonable?). But the pressure in an ideal gas can be written in the form (see Sample Problem 3),

$$p = \rho RT/M.$$

Now in our case the gas (air) is the same inside and out, so that the molar mass $M$ is the same for the hot air and the surrounding air. With the pressures equal, the equation above implies that

$$\rho_{hot}T_{hot} = \rho_{surr}T_{surr}.$$

The temperatures must be absolute. Substituting known quantities we get

$$(0.981 \text{ kg/m}^3)T_{\text{hot}} = (1.22 \text{ kg/m}^3)(291 \text{ K}),$$

$$T_{\text{hot}} = 362 \text{ K} = 89.0°C.$$

## 23-40

Convert the gauge pressures to absolute pressures by adding 101 kPa (atmospheric pressure). The volume reached in the isothermal expansion is found from the ideal gas law:

$$p_1V_1 = p_2V_2 = nRT = \text{constant}.$$

$$(204 \text{ kPa})(0.142 \text{ m}^3) = (101 \text{ kPa})V_2,$$

$$V_2 = 0.2868 \text{ m}^3.$$

The work done on the gas in the isothermal expansion is given by Eq. 27:

$$W_{\text{exp}} = -nRT\ln(V_f/V_i) = -p_1V_1\ln(V_2/V_1),$$

$$W_{\text{exp}} = -(204{,}000 \text{ Pa})(0.142 \text{ m}^3)\ln(0.2868/0.142) = -20.36 \text{ kJ}.$$

The work done on the gas in the constant pressure cooling follows from Eq. 26:

$$W_{\text{cool}} = -p_2(V_1 - V_2),$$

$$W_{\text{cool}} = -(101{,}000 \text{ Pa})(0.142 \text{ m}^3 - 0.2868 \text{ m}^3) = 14.62 \text{ kJ}.$$

Hence $W_{\text{total}} = -20.36 + 14.62 = -5.74$ kJ.

## 23-46

(a) Adiabatic processes follow Eq. 28:

$$pV^\gamma - \text{constant}.$$

Our problem concerns pressure and temperature, so combine Eq. 28 with Eq. 7, $pV = nRT$, to obtain

$$p^{1-\gamma}T^\gamma - \text{CONSTANT}.$$

168

Of course, temperatures must be in kelvins in this equation. For air,

$$\gamma = 1.4;$$

this result is buried in Sample Problem 6. We can use atmospheres as the pressure unit since we will not need the numerical value of CONSTANT. Inserting the data, we find

$$p_1^{1-\gamma} T_1^{\gamma} = p_2^{1-\gamma} T_2^{\gamma},$$

$$1^{-0.4} 291^{1.4} = 2.3^{-0.4} T_2^{1.4},$$

$$T_2 = 369 \text{ K} = 96°C.$$

(b) In one second the work done is $W = 230$ J. By Eqs. 30 and 7,

$$W = \frac{1}{\gamma - 1} [p_2 V_2 - p_1 V_1] = \frac{nR}{\gamma - 1} (T_2 - T_1),$$

$$230 = \frac{nR}{0.4} (369 - 291),$$

$$nR = 1.179 \text{ J/K}.$$

Hence the volume of air delivered in one second is

$$V_2 = nRT_2/p_2,$$

$$V_2 = (1.179 \text{ J/K})(369 \text{ K})/[(2.3)(101,000 \text{ Pa})] = 1.87 \text{ L},$$

where 1 L = 1 liter = 0.001 m³.

## 23-49

The internal energy is the sum of the translational and rotational kinetic energies:

$$E_{int} = K_{trans} + K_{rot}.$$

Air is diatomic, so Eq. 36 applies to the internal energy. The translational kinetic energy is given by Eq. 18 regardless of whether the gas is monatomic, diatomic or polyatomic. Hence,

$$K_{rot} = E_{int} - K_{trans} = (5/2)nRT - (3/2)nRT = nRT.$$

For one mole,

$$K_{rot} = (1 \text{ mol})(8.31 \text{ J/mol}\bullet\text{K})(298 \text{ K}) = 2.48 \text{ kJ}.$$

## 23-55

Into the van der Waals equation of state, Eq. 41, substitute the expressions for $a$ and $b$ given in Sample Problem 8. After doing some repositioning of terms, we find the following:

$$1 + (27/64)x^2(1 - \tfrac{1}{8}x) = 9x/8,$$

where

$$x = nRT_{cr}/p_{cr}V_{cr} = 8bn/V_{cr}.$$

If we define

$$y = bn/V_{cr} = \tfrac{1}{8}x,$$

the equation becomes

$$27y^3 - 27y^2 + 9y - 1 = 0,$$

which can be factored as

$$27(y - \tfrac{1}{3})^3 = 0.$$

Hence the (only) solution is

$$y = \tfrac{1}{3} = bn/V_{cr},$$

$$V_{cr} = 3nb.$$

24-6

(a) We associate 76 cm Hg with 1 atm = $1.01 \times 10^5$ Pa of pressure, so that $1.10 \times 10^{-6}$ mm Hg = $(1.10 \times 10^{-7}$ cm/76 cm$)(1.01 \times 10^5$ Pa$)$ = $1.462 \times 10^{-4}$ Pa. Hence, by Eq. 4 of Chapter 23 with $V = 1$ m$^3$,

$$pV = NkT,$$

$$(1.462 \times 10^{-4} \text{ Pa})(1 \text{ m}^3) = N(1.38 \times 10^{-23} \text{ J/K})(295 \text{ K}),$$

$$N = 3.59 \times 10^{16},$$

so that $\rho_n = 3.59 \times 10^{16}$ m$^{-3}$.
(b) Now use Eq. 16 to find the mean free path:

$$\lambda = [\sqrt{2}\pi d^2 \rho_n]^{-1},$$

$$\lambda = [\sqrt{2}\pi(2.2 \times 10^{-10} \text{ m})^2(3.59 \times 10^{16} \text{ m}^{-3})]^{-1} = 130 \text{ m}.$$

24-9

Find the number of molecules per unit volume from the ideal gas law (see Sample Problem 3),

$$p = \rho_n kT,$$

$$(1.02 \text{ atm})(1.01 \times 10^5 \text{ Pa/atm}) = \rho_n(1.38 \times 10^{-23} \text{ J/K})(291 \text{ K}),$$

$$\rho_n = 2.565 \times 10^{25} \text{ m}^{-3}.$$

Now use Eq. 16 to find the mean free path (which we will set equal to the wavelength of sound):

$$\lambda = [\sqrt{2}\pi(315 \times 10^{-12} \text{ m})^2(2.565 \times 10^{25} \text{ m}^{-3})]^{-1},$$

$$\lambda = 8.844 \times 10^{-8} \text{ m}.$$

The speed of sound can be found from the formula implied in Problem 42 of Chapter 23; we find,

$$v = 343 \text{ m/s} - (0.59 \text{ m/s} \cdot \text{C}°)(2 \text{ C}°) = 341.8 \text{ m/s}.$$

Therefore, the frequency of the desired sound follows from Eq. 13 of Chapter 19:

$$\nu = v/\lambda = (341.8 \text{ m/s})/(8.844 \times 10^{-8} \text{ m}),$$

$$\nu = 3.86 \times 10^9 \text{ Hz} = 3.86 \text{ GHz}.$$

(a) Use Eq. 9; also, see Fig. 5. Set the distance $r = \lambda$ the mean free path. The constant $c = 1/\lambda$, as shown in Eq. 13. The desired probability is $I/I_0$, since if a particle does not suffer a collision in the distance $r$ it contributes to the emergent intensity $I$. Hence we have

$$\text{Prob.} = \frac{I}{I_0} = e^{-cr} = e^{-(1/\lambda)\lambda} = e^{-1} = 0.368.$$

(b) In this case set Prob. $= I/I_0 = \frac{1}{2}$; thus,

$$\tfrac{1}{2} = e^{-r/\lambda},$$

$$r = \lambda \ln 2.$$

## 24-14

(a) The total number of particles is $2 + 4 + 6 + 8 + 2 = 22$. Therefore,

$$\bar{v} = \left(\frac{1}{22}\right)[2(1) + 4(2) + 6(3) + 8(4) + 2(5)] = 3.18 \text{ km/s}.$$

(b) The root-mean-square speed is found from

$$v_{rms}^2 = \left(\frac{1}{22}\right)[2(1)^2 + 4(2)^2 + 6(3)^2 + 8(4)^2 + 2(5)^2],$$

$$v_{rms} = 3.37 \text{ km/s}.$$

(c) More particles are traveling at 4 km/s than at any other speed, so this is the most probable speed.

## 24-19

The speed distribution appropriate to particles emerging from an oven in a parallel beam can be found on p.538 of RHK:

$$n(v) = Av^3 e^{-av^2},$$

where $A$ is an unspecified constant and

$$a = m/2kT.$$

The most probable speed is found by setting $dn/dv = 0$; see Sample Problem 6. Using the expression above, we find

$$\frac{dn}{dv} = 0 = Av^2 e^{-av^2}[3 - 2av^2].$$

This is satisfied for $v = 0$ and $v = \infty$, but these yield $n(v) = 0$. We seek the solution that gives a maximum of $n(v)$; this is the third solution, found from

$$3 - 2av_p^2 = 0 = 3 - 2(m/2kT)v_p^2,$$

$$v_p = \sqrt{\frac{3kT}{m}}.$$

## 24-21

(a) Each particle has a speed, so if we integrate (sum) the speed distribution over all possible speeds we must find $N$, the total number of particles:

$$N = \int_0^\infty n(v)\, dv = \int_0^{v_0} Cv^2 dv = C(\tfrac{1}{3}v_0^3),$$

$$C = 3N/v_0^3.$$

We do not have to integrate over speeds greater than $v_0$, since $n(v) = 0$ in this region.
(b) Now follow Sample Problem 6, but with our distribution function instead of the Maxwellian; see Eq. 19; the average speed is

$$\bar{v} = \frac{1}{N}\int_0^{v_0} vCv^2 dv = (\frac{C}{N})(\tfrac{1}{4}v_0^4) = (\frac{3}{v_0^3})(\tfrac{1}{4}v_0^4) = 0.750v_0.$$

(c) We use Eq. 21, but delete the term $3kT/m$, since that particular result requires that the speed distribution be Maxwellian, which ours most certainly is not. The left-hand side of Eq. 21 holds, however, since this is basically the definition of (root-mean-square speed)$^2$; we have, then,

$$v_{rms}^2 = \frac{1}{N}\int_0^{v_0} v^2 Cv^2\,dv = (\frac{C}{N})(v_0^5/5) = (3/v_0^3)(v_0^5/5) = \frac{3}{5}v_0^2,$$

$$v_{rms} = 0.775v_0.$$

## 24-24

If $x$ is small enough, we can set

$$e^x = 1 + x;$$

see Appendix H. We will apply this to Eq. 27 (with $x = -E/kT$). The desired fraction $f$ is given by

$$f = \frac{1}{N}\int_{0.01kT}^{0.03kT} N(E)\,dE = \frac{2}{\sqrt{\pi}\,(kT)^{3/2}}\int_{0.01kT}^{0.03kT}\sqrt{E}\,(1 - \frac{E}{kT})\,dE,$$

$$f = \frac{4}{\sqrt{\pi}}[(0.03)^{3/2}(\tfrac{1}{3} - \frac{0.03}{5}) - (0.01)^{3/2}(\tfrac{1}{3} - \frac{0.01}{5})],$$

$$f = 0.00309.$$

## 24-28

By Eq. 22,

$$\tfrac{1}{2}mv_{rms}^2 = (3/2)kT,$$

$$\tfrac{1}{2}(6.2 \times 10^{-17}\text{ kg})(0.014\text{ m/s})^2 = (3/2)k(299\text{ K}),$$

$$k = 1.35 \times 10^{-23}\text{ J/K}.$$

Now invoke Eq. 8 in Chapter 23:

$$N_A = R/k = (8.31\text{ J/mol}\bullet\text{K})/(1.35 \times 10^{-23}\text{ J/K}) = 6.16 \times 10^{23}.$$

## 24-33

(a) Make a list of the possible ways that three distinguishable particles can be distributed among two distinguishable states; see the table on the next page. We see that there are 8 ways. If the particles are, in fact, distinguishable as we are to assume in (a), then there are 8 "ways".
(b) For indistinguishable particles, ways 3,5,7 are equivalent (that is, are indistinguishable and count as a single way); also ways 4,6,8 are equivalent. We have, then, 4 distinct ways.

| Way | State 1 | State 2 | |
|-----|---------|---------|---|
| 1 | 123 | | |
| 2 | | 123 | |
| 3 | 12 | 3 | ** |
| 4 | 3 | 12 | * |
| 5 | 1 3 | 2 | ** |
| 6 | 2 | 1 3 | * |
| 7 | 23 | 1 | ** |
| 8 | 1 | 23 | * |

\* and ** form two sets of indistinguishable arrangements ("ways").

25-3

The mass of water that freezes as the heat is extracted is given from Eq. 7, using the value of $L_f$, the heat of fusion, found in Table 2. Being attentive to units, we find

$$Q = L_f m,$$

$$50.4 \text{ kJ} = (333 \text{ kJ/kg})m,$$

$$m = 0.151 \text{ kg} = 151 \text{ g}.$$

Hence, the mass of water that does not freeze (i.e., that remains as a liquid) is 258 g - 151 g = 107 g.


25-6

The heat needed is given by Eq. 5 with $T_f = 100°C$, the temperature at which boiling begins (if we do not know this, we can find the boiling temperature in Table 2). From Table 1, the specific heat capacity of liquid water is $c = 4190$ J/kg•K = 4190 J/kg•C° (since differences in temperatures expressed in °C and K are equal). Noting the mass unit in the specific heat, we have (you add the SI base units),

$$Q = mc(T_f - T_i) = (0.136)(4190)(100 - 23.5) = 43,590 \text{ J}.$$

The needed time follows from the definition of power; we find

$$t = Q/P = (43,590 \text{ J})/(220 \text{ W}) = 198 \text{ s},$$

recalling that 1 W = 1 J/s.


25-9

In Table 2 we find that the melting temperature of silver is 1235 K. The silver initially is at a temperature of 16°C = 289 K. Before we can melt the silver we must heat it to the melting point; the heat needed to do this is given by Eq. 5. Once the silver is at the melting temperature, the heat needed to melt it is given by Eq. 7 with $L = L_f$, the heat of fusion. Therefore, the total heat required is

$$Q = mc(T_f - T_i) + mL_f = m[c(T_f - T_i) + L_f].$$

Substituting the data, specific heat and heat of fusion from Tables 1 and 2, using SI base units (kg, J, K), we find that the total heat required is

$$Q = (0.130)[(236)(1235 - 289) + 105,000] = 42,700 \text{ J} = 42.7 \text{ kJ}.$$

## 25-13

(a) The mass of 0.640 L of water is

$$m = \rho V = (1000 \text{ kg/m}^3)(0.640 \times 10^{-3} \text{ m}^3) = 0.640 \text{ kg}.$$

We assume that the heating element heats both the water and the aluminum kettle to 100°C simultaneously. The needed heat for each is given by Eq. 5 with the appropriate mass and specific heat, so that the total heat required is

$$Q_{total} = Q_{aluminum} + Q_{water},$$

$$Q_{total} = (0.560)(900)(100 - 12) + (0.640)(4190)(100 - 12),$$

$$Q_{total} = 2.803 \times 10^5 \text{ J} = 280.3 \text{ kJ}.$$

The time needed is therefore

$$t = Q_{total}/P = (280.3 \text{ kJ})/(2.40 \text{ kW}) = 117 \text{ s}.$$

(b) Once the water and kettle have reached 100°C, we are told to assume that all additional heat enters the water alone. Under this assumption, the additional heat needed to vaporize all the water is $Q = mL_v$, for a total of

$$Q_{TOTAL} = Q_{total} + Q,$$

$$Q_{TOTAL} = 280.3 \text{ kJ} + (0.640 \text{ kg})(2256 \text{ kJ/kg}) = 1724 \text{ kJ}.$$

The total time elapsed from the beginning until all the water is gone is $t = (1724 \text{ kJ})/(2.40 \text{ kW}) = 718 \text{ s}.$

## 25-19

Let $d$ be the initial diameter of the copper (Cu) ring and $D$ the initial diameter of the aluminum (Al) sphere. We work in units of cm and °C. By Eq. 12 of Chapter 22, these diameters when the objects reach thermal equilibrium at temperature $T$ can be written in the form

$$d + \Delta d = d[1 + \alpha_{Cu}(T - 0)] = 2.54[1 + 17 \times 10^{-6}T],$$

$$D + \Delta D = D[1 + \alpha_{Al}(T - 100)] = 2.54533[1 - 23 \times 10^{-6}(100 - T)].$$

Since the sphere just passes through the ring at this temperature, we must have

$$d + \Delta d = D + \Delta D.$$

We have, then, from the expressions above,

$$2.54[1 + 17 \times 10^{-6}T] = 2.54533[1 - 23 \times 10^{-6}(100 - T)],$$

$$T = 34.13°C.$$

Now, the heat lost by the sphere is equal, in absolute value, to the heat gained by the ring as they reach thermal equilibrium at the temperature $T$. In terms of the mass $m$ of the ring and mass $M$ of the sphere, this means that

$$mc_{Cu}(T - 0) = Mc_{Al}(100 - T).$$

Solve this for the mass $M$ of the sphere to find

$$M = \frac{c_{Cu}T}{c_{Al}(100 - T)} \, m.$$

Look up the values of the specific heats in Table 1; we found the temperature $T$ above; the mass $m$ of the ring is given in the problem statement. Substitute all the numbers into the equation above to find that $M = 4.81$ g. Evidently the aluminum sphere is a thin spherical shell since, if it is solid, its mass would have to be about 23 g (see the density of aluminum in Appendix D).

25-20

(a) Use Eq. 5:

$$Q = mc(T_f - T_i),$$

$$320 \text{ J} = (0.0371 \text{ kg})c(42°C - 26.1°C),$$

$$c = 542 \text{ J/kg} \cdot C° = 542 \text{ J/kg} \cdot K.$$

(b) The number of moles present is

$$n = \frac{\text{bulk mass}}{\text{molar mass}} = \frac{37.1 \text{ g}}{51.4 \text{ g/mol}} = 0.722 \text{ mol.}$$

(c) We can now use Eq. 8 to find the molar heat capacity:

$$C = Q/n\Delta T,$$

$$C = (320 \text{ J})/[(0.722 \text{ mol})(15.9 \text{ K})],$$

$$C = 27.9 \text{ J/mol} \cdot K.$$

The heat needed is given by an expression analogous to Eq. 4; that is, by

$$Q = n \int_{T_i}^{T_f} C \, dT.$$

Geometrically, the integral is the area under the curve of $C$ vs. $T$ down to the $C = 0$ axis. Such a curve is given in Fig. 3. To make the evaluation of the integral by area calculation simple, we are instructed to approximate the actual curve with a straight line segment in the region of interest; that is, by,

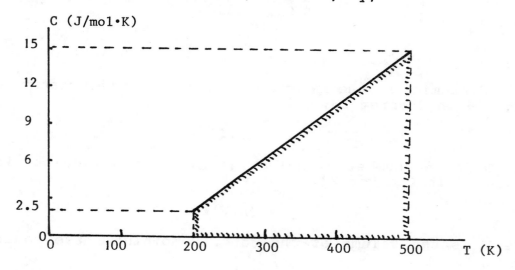

The area can be visualized as a rectangle plus a right triangle, so we find

$$\int_{200 \text{ K}}^{500 \text{ K}} C \, dT = (300)(2.5) + \tfrac{1}{2}(300)(12.5) = 2625 \text{ J/mol}.$$

Hence,

$$Q = (0.45 \text{ mol})(2625 \text{ J/mol}) = 1200 \text{ J}.$$

25-27

(a) The heat required is found from Eq. 17 with the molar heat capacity given in Eq. 21:

$$Q = nC_p\Delta T = (4.34 \text{ mol})(29.1 \text{ J/mol}\cdot\text{K})(62.4 \text{ K}) = 7880 \text{ J}.$$

(*b*) As suggested we turn to Eq. 36 of Chapter 23:

$$\Delta E_{int} = (5/2)nR\Delta T = (5/2)(4.34 \text{ mol})(8.31 \text{ J/mol}\bullet\text{K})(62.4 \text{ K}),$$

$$\Delta E_{int} = 5630 \text{ J}.$$

(*c*) Eq. 18 of Chapter 23 gives the internal translational kinetic energy for one mole; for *n* moles,

$$K_{trans} = (3/2)nRT,$$

so that

$$\Delta K_{trans} = (3/2)nR\Delta T = (3/2)(4.34 \text{ mol})(8.31 \text{ J/mol}\bullet\text{K})(62.4 \text{ K}),$$

$$\Delta K_{trans} = 3380 \text{ J}.$$

## 25-30

Air is diatomic, so the internal energy at any temperature is given by Eq. 36 in Chapter 23,

$$E_{int} = (5/2)NkT,$$

where *N* is the number of molecules in the room. By the ideal gas law, Eq. 4 in Chapter 23,

$$p = (N/V)kT,$$

where *V* is the volume of the room. Combining these equations we find

$$E_{int} = (5/2)pV = (5/2)p_0V_{room},$$

the temperature dropping out. The volume of the room does not change presumably, so that if the pressure remains constant, then the internal energy does not change as the furnace heats the air. Some air escapes, but the remaining air is at a higher temperature and the room feels warmer.

## 25-32

Over a complete cycle $\Delta E_{int} = 0$. Now apply the first law, Eq. 30:

$$Q_{BC} + Q_{CA} + Q_{AB} + W_{BCA} + W_{AB} = 0,$$

$$0 + Q_{CA} + 20 + (-15) + 0 = 0,$$

$$Q_{CA} = -5 \text{ J},$$

that is, 5 J of heat was removed over this part of the cycle.

**25-35**

(a) Use Eq. 38 with $V_2 = 0.10 V_1$; we find

$$T_1 V_1^{\gamma-1} = T_2 V_2^{\gamma-1},$$

$$T_2 = T_1 \left(\frac{V_1}{V_2}\right)^{\gamma-1} = (292 \text{ K})(10)^{\gamma-1}.$$

For a monatomic gas, the ratio of specific heats = 1.67; the equation above yields $T_2$ = 1366 K = 1090°C.
(b) For a diatomic gas the ratio of specific heats = 1.40. The equation in (a) gives $T_2$ = 733 K = 460°C in this case.

**25-40**

Over a complete cycle $\Delta E_{int}$ = 0 so that, by the first law, $Q = -W$. The work done on the gas is the area inside the path, and this is positive since the cycle is traversed counterclockwise. Noting the units, we have then

$$W = \tfrac{1}{2}\pi(1.5 \text{ L})(15 \text{ MPa}) = 35.3 \text{ L}\bullet\text{MPa} = 35.3 \text{ kJ}.$$

Therefore $Q$ = -35.3 kJ, meaning that 35.3 kJ of heat is removed during the cycle.

**25-46**

The rate is given by Eq. 45. The area is $A$ = (6.2 m)(3.8 m) = 23.56 m². The temperature difference is $\Delta T$ = 26°C - (-18°C) = 44 C° = 44 K. Also the thickness (i.e., the distance the heat travels) is $\Delta x$ = 32 cm = 0.32 m. The heat flow rate is

$$H = kA\frac{\Delta T}{\Delta x},$$

$$H = (0.74 \text{ W/m}\bullet\text{K})(23.56 \text{ m}^2)(44 \text{ K})/(0.32 \text{ m}) = 2.4 \text{ kW}.$$

**25-49**

(a) The temperature gradient is

$$\Delta T/\Delta x = (136 \text{ C}°)/(0.249 \text{ m}) = 546 \text{ C}°/\text{m} = 546 \text{ K/m}.$$

(*b*) The rate of heat transfer is given by Eq. 45 with the value of the thermal conductivity of copper as found in Table 5. We obtain

$$H = kA(\Delta T/\Delta x) = (401 \text{ W/m} \bullet \text{K})(1.80 \text{ m}^2)(546 \text{ K/m}),$$

$$H = 3.94 \text{ X } 10^5 \text{ W} = 394 \text{ kW}.$$

(*c*) We put $x = 0$ at the high-temperature end, with $T + \Delta T = -12 + 136 = 124°C$ as the temperature at this end. The temperature at $x = 11$ cm $= 0.11$ m is

$$T_x = (T + \Delta T) - (\Delta T/\Delta x)x = 124 - (546)(0.11) = 63.9°C.$$

We took cognizance of the fact that the temperature decreases as we move into the block from the high-temperature end.


## 25-59

In a steady state the rate of heat transfer through the water must equal that through an equal area of ice:

$$H_{\text{water}} = H_{\text{ice}}.$$

Each term $H$ is given by Eq. 45 with the appropriate thermal conductivity and temperatures. If we let $x$ be the thickness of ice, we have

$$(1.67)A\frac{5.20 - 0}{x} = (0.502)A\frac{3.98 - 0}{1.42 - x},$$

$$x = 1.15 \text{ m}.$$

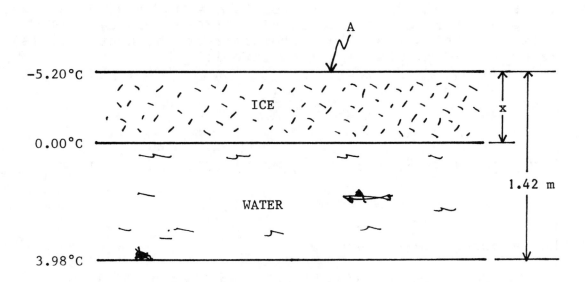

Use the differential form, Eq. 48, of the heat flow equation and substitute the expression for the temperature-dependent thermal conductivity. We get

$$H = -kA\frac{dT}{dx} = -(aT)A\frac{dT}{dx},$$

$$TdT = -\left(\frac{H}{aA}\right)dx.$$

Integrating gives

$$\tfrac{1}{2}T^2 = -\left(\frac{H}{aA}\right)x + C,$$

where $C$ is the constant of integration. We evaluate this by setting $T = T_1$ at $x = 0$; we find $C = \tfrac{1}{2}T_1^2$. Finally, put $T = T_2$ at $x = L$. Making these two changes, and solving for $H$ gives

$$H = \frac{aA}{2L}(T_1^2 - T_2^2).$$

# CHAPTER 26

## 26-1

(a) Use Eq. 4; we have $|Q_{in}| = Q_{in} = 52.4$ kJ and $|Q_{out}| = |-36.2| = 36.2$ kJ, so we find

$$e = 1 - \frac{36.2 \text{ kJ}}{52.4 \text{ kJ}} = 0.309.$$

(b) Eq. 3 applies here:

$$e = |W|/(52.4 \text{ kJ}) = 0.309,$$

$$|W| = 16.2 \text{ kJ}.$$

To determine the sign, note that the first law of thermodynamics applied over one complete cycle requires that

$$\Delta E_{int} = Q + W = 0,$$

so that

$$W = -Q = -(52.4 - 36.2) = -16.2 \text{ kJ}.$$

Now see p. 557 of RHK; $W > 0$ when work is done on the system. Since in this case $W < 0$, this means that work is done by the system (engine) in the amount of +16.2 kJ.

## 26-3

Eq. 3 applies:

$$e = |W|/|Q_{in}|.$$

We are given the <u>rates</u> of energy and work transacted. We will use the energies corresponding to one second of time. Hence the work $|W| = 755$ MJ. Now, 1 metric ton = 1000 kg so that over one hour we have $|Q_{in}| = (382,000 \text{ kg})(28 \text{ MJ/kg}) = 10.70$ TJ. Dividing by 3600, the number of seconds in one hour, yields $|Q_{in}| = 2972$ MJ during one second of operation. The efficiency can now be calculated:

$$e = (755 \text{ MJ})/(2972 \text{ MJ}) = 0.254 = 25.4\%.$$

## 26-4

(a) Process $ab$ occurs at constant volume. By Eqs. 20 and 19 of Chapter 25, the appropriate specific heat is

$$C_V = C_p - R = (5/2)R - R = (3/2)R.$$

so that

$$Q_{ab} = nC_V\Delta T = n(3R/2)\Delta T.$$

For a constant volume process the ideal gas law gives

$$pV = nRT,$$

$$(\Delta p)V = nR(\Delta T),$$

and therefore

$$Q_{ab} = (3/2)(\Delta p)V_0 = (3/2)(p_b - p_a)V_0.$$

Since $bc$ is an adiabatic process we have

$$p_b V_b^\gamma = p_c V_c^\gamma,$$

$$(10.4\ \text{atm})(1.22\ \text{m}^3)^{5/3} = p_c(9.13\ \text{m}^3)^{5/3},$$

$$p_c = p_a = 0.3632\ \text{atm}.$$

Introducing a factor of $1.01 \times 10^5$ Pa/atm, we have finally

$$Q_{ab} = (3/2)(10.4 - 0.3632)(1.01 \times 10^5)(1.22) = 1.86\ \text{MJ}.$$

(b) Process $ca$ is at constant pressure; Eq. 20 of Chapter 25 implies that

$$Q_{ca} = nC_p\Delta T = n(5R/2)(T_a - T_c) = (5/2)p_c(V_a - V_c),$$

$$Q_{ca} = (5/2)(0.3632)(1.01 \times 10^5)(1.22 - 9.13) = -0.725\ \text{MJ}.$$

(c) Over one cycle, $\Delta E_{int} = 0$ so that $W = -Q$. Therefore the work done by the gas $= Q = 1.86 - 0.725 = 1.13$ MJ.
(d) Find the efficiency from Eq. 4. We have $|Q_{in}| = Q_{ab} = 1.86$ MJ and $|Q_{out}| = |Q_{ca}| = 0.725$ MJ; hence

$$e = 1 - (0.725)/(1.86) = 0.610 = 61.0\%.$$


## 26-8

(a) Use Eq. 6; we have immediately

$$K = |Q_L|/|W| = (568\ \text{J})/(153\ \text{J}) = 3.71.$$

(b) Now use the second version of Eq. 6 for the coefficient of performance to find

$$K = \frac{|Q_L|}{|Q_H| - |Q_L|},$$

$$3.71 = \frac{568 \text{ J}}{|Q_H| - 568 \text{ J}},$$

$$|Q_H| = 721 \text{ J}.$$

## 26-11

(a) During an isothermal process the change in internal energy is zero: $\Delta E_{int} = 0$. Now apply the first law and conclude that $W = -Q = -2090$ J. But $W$ is the work done on the gas; the work $W_{gas}$ done by the gas is $W_{gas} = -W = 2090$ J.

(b) The 2090 J of heat transferred to the gas during the isothermal expansion corresponds to $Q_H$ in the terminology of Section 3. We can use Eq. 9; the temperatures are given in kelvins, so we have directly

$$|Q_L| = |Q_H|T_L/T_H = (2090 \text{ J})(297 \text{ K})/(412 \text{ K}) = 1510 \text{ J}.$$

(c) The work done on the gas is $W$ and this follows from the first law applied to the isothermal compression; we find

$$\Delta E_{int} = 0 = Q + W,$$

$$W = -Q_L = -(-1510 \text{ J}) = 1510 \text{ J}.$$

## 26-18

A heat pump operates as a refrigerator (see Sample Problem 4). By combining Eqs. 5 and 6 to eliminate $|Q_L|$ we get

$$|W| = \frac{|Q_H|}{K + 1} = \frac{7.6 \text{ MJ}}{3.8 + 1} = 1.58 \text{ MJ}.$$

The 7.6 MJ used above is delivered in one hour, and therefore the 1.58 MJ found above is performed in one hour. Hence the rate at which work is done (power) is

$$P = W/t = (1.58 \times 10^6 \text{ J})/(3600 \text{ s}) = 440 \text{ W},$$

where W (watt) = J/s.

The "ideal" efficiency is the Carnot efficiency given by Eq. 10. The actual efficiency of the refrigerator is

$$e_{ref} = (0.85)e = (0.85)[1 - 270/299] = 0.0824.$$

In 15 min the work done by the motor is

$$W = Pt = (210 \text{ J/s})[(15 \text{ min})(60 \text{ s/min})] = 189 \text{ kJ}.$$

By Eq. 3,

$$|Q_{in}| = |W|/e_{ref} = (189 \text{ kJ})/(0.0824) = 2.29 \text{ MJ}.$$

Finally, by Eq. 2,

$$|Q_{out}| = |Q_{in}| - |W| = 2.29 \text{ MJ} - 0.189 \text{ MJ} = 2.10 \text{ MJ}.$$

## 26-25

(a) Steam is water vapor, that is $H_2O$, for which the ratio of specific heats is 1.33 (see Eq. 26 in Chapter 25). For the adiabatic expansion, Eq. 36 in Chapter 25 yields

$$p_f(5.6V_i)^{1.33} = (16 \text{ atm})(V_i)^{1.33},$$

$$p_f = 1.62 \text{ atm}.$$

(b) The greatest possible efficiency is the Carnot cycle efficiency, which is given in terms of the temperatures by Eq. 10. However, we can use Eq. 39 of Chapter 25 to write the efficiency in terms of the volumes instead. Doing this, we find

$$e = 1 - \frac{T_f}{T_i} = 1 - (\frac{V_i}{V_f})^{\gamma-1},$$

$$e = 1 - (\frac{V_i}{5.6V_i})^{1.33-1} = 0.437.$$

## 26-30

(a) For an isothermal process the work done on the gas is given by Eq. 27 of Chapter 23; see also Table 4 in Chapter 25. The given temperature is already in kelvins, so we have immediately

$$W = -nRT\ln\frac{V_f}{V_i},$$

$$W = -(4 \text{ mol})(8.31 \text{ J/mol} \cdot \text{K})(410 \text{ K})\ln(3.45) = -16.9 \text{ kJ}.$$

We identified $V_i$ with $V_1$ and $V_f$ with $V_2$, so the volume ratio is 3.45. (b) For an isothermal process $\Delta E_{int} = 0$ so that the first law implies

$$Q = -W = -(-16.9 \text{ kJ}) = 16.9 \text{ kJ}.$$

We therefore have for the entropy change for this isothermal process,

$$\Delta S = Q/T = (16,900 \text{ J})/(410 \text{ K}) = 41.2 \text{ J/K}.$$

(c) For an adiabatic process $Q = 0$ so that $\Delta S = 0$ also.

26-32

(a) Use the specific heat at constant pressure given in Table 1 of Chapter 25 to find

$$Q = mc(T_f - T_i),$$

$$Q = (1.22 \text{ kg})(387 \text{ J/kg} \cdot \text{K})(80 \text{ K}) = 37.8 \text{ kJ}.$$

(b) We cannot write $\Delta S = Q/T$ since the temperature $T$ changes during the heating. Instead, follow the argument of Section 7 leading to Eq. 29. We associate $T_1$ with the initial temperature of 25°C = 298 K and $T_e$ with the final temperature of 105°C = 378 K. We then have

$$\Delta S = mc\ln(T_f/T_i) = (1.22 \text{ kg})(387 \text{ J/kg} \cdot \text{K})\ln(378/298) = 112 \text{ J/K}.$$

26-39

(a) By the definition of a heat reservoir, its temperature does not change even though heat is added or removed. The temperatures of the two reservoirs are $T_H = 403$ K and $T_c = 297$ K. The heat leaves the hot reservoir and enters the cold reservoir; this determines the signs of $Q$ for each. Thus, the entropy changes of the reservoirs are

$$\Delta S_H = Q/T_H = (-1200 \text{ J})/(403 \text{ K}) = -2.978 \text{ J/K},$$

$$\Delta S_c = Q/T_c = (+1200 \text{ J})/(297 \text{ K}) = +4.040 \text{ J/K}.$$

The entropy change of the system is 4.040 - 2.978 = +1.06 J/K.

(b) Since the rod simply transmits the heat, absorbing none itself, its entropy change is zero, as assumed above in calculating the entropy change of the system.

## 26-44

(a) The heat lost by the aluminum must equal, in absolute value, the heat gained by the water. Each of these heats is given by an expression of the form of Eq. 5 in Chapter 25. Extracting specific heats from Table 1 in Chapter 25, and paying attention to units, we have

$$|Q_{Al}| = |Q_{water}|,$$

$$(0.196)(900)(107 - T_e) = (0.0523)(4190)(T_e - 18.6),$$

$$T_e = 58.0°C.$$

(b) Using Eq. 29, we have, after converting the temperatures to kelvins and masses to kg,

$$\Delta S_{Al} = mc\ln(T_e/T_1) = (0.196)(900)\ln(331/380) = -24.4 \text{ J/K}.$$

(c) Similarly to (b),

$$\Delta S_{water} = (0.0523)(4190)\ln(331/291.6) = +27.8 \text{ J/K}.$$

(d) We cannot use Eq. 31 for the total entropy change since that equation was derived for identical objects. However, since we have the separate entropy changes we need only add them to find that the entropy change of the system is +27.8 - 24.4 = +3.4 J/K.

27-2

Rearrange Eq. 1 to solve for the distance $r$. Then insert data in SI base units; for a summary of these as applied to Coulomb's law, see the text below Eq. 4 on p. 597 of HRK. Insert only the absolute value of the charges, since Eq. 1 deals only with the magnitude of the force (see the footnote † on p. 596 of HRK). Recalling that $1 \mu C = 1 \times 10^{-6}$ C, we obtain

$$r = \sqrt{\frac{kq_1 q_2}{F}},$$

$$r = \sqrt{\frac{(8.99 \times 10^9)(26.3 \times 10^{-6})(47.1 \times 10^{-6})}{5.66}} = 1.40 \text{ m.}$$

27-4

(a) By Newton's third law the forces on each particle due to the other particle are equal in magnitude. Since $F = ma$, this requires that

$$F = m_1 a_1 = m_2 a_2,$$

$$(6.31 \times 10^{-7} \text{ kg})(7.22 \text{ m/s}^2) = m_2(9.16 \text{ m/s}^2),$$

$$m_2 = 4.97 \times 10^{-7} \text{ kg.}$$

(b) Now apply Newton's second law to either particle; we choose the particle labeled 1 in (a) above. Use Eq. 1 for the force, with $q_1 = q_2 = q$ and $r = 3.2$ mm $= 0.0032$ m; we find

$$k\frac{q^2}{r^2} = m_1 a_1,$$

$$\frac{(8.99 \times 10^9)q^2}{(0.0032)^2} = (6.31 \times 10^{-7})(7.22),$$

$$q = 7.20 \times 10^{-11} \text{ C} = 72.0 \text{ pC}.$$

The charges may be of the same sign or of opposite signs.

## 27-9

The net force on the negative
charge $F_{net}$ is the vector sum
of the separate forces of
attraction $F$ between the negative
charge and each positive charge.
An equilateral triangle has equal
angles of 60°. By symmetry, the
net force $F_{net}$ will lie along the
angle bisector and has magnitude

$$F_{net} = 2F\cos 30°.$$

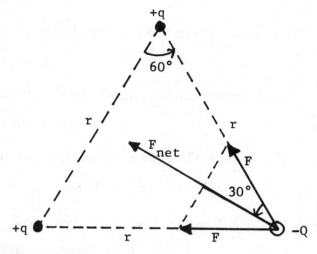

For the forces $F$ we use Eq. 1:

$$F = \frac{kqQ}{r^2} = \frac{(8.99 \times 10^9)(4.18 \times 10^{-6})(6.36 \times 10^{-6})}{0.13^2},$$

$$F = 14.14 \text{ N}.$$

Hence,

$$F_{net} = 2(14.14 \text{ N})\cos 30° = 24.5 \text{ N}.$$

## 27-11

Assume that the spheres are small enough so that the charges are
distributed uniformly over their surfaces, allowing them to be
treated as point charges (radii < 0.5 m). Let the initial charges
be $q_1$, $-q_2$, with $q_1$ and $q_2$ both positive numbers. Before being
connected by the wire,

$$F = kq_1q_2/r^2,$$

where $r = 0.5$ m. Thus, in SI base units,

$$q_1q_2 = (0.108)(0.5)^2/(8.99 \times 10^9) = 3 \times 10^{-12} \text{ C}^2.$$

When the spheres are connected with the wire, they and the wire
form a single conductor carrying charge $(q_1 - q_2)$. When equilibrium
is reached, the wire will carry virtually no charge since its
surface area is very small compared with the surface area of the
spheres. These spheres, being identical, will each carry one half

of the net charge, which is $\frac{1}{2}(q_1 - q_2)$. Coulomb's law, Eq. 4, then gives, using SI base units,

$$F = (8.99 \times 10^9)[\tfrac{1}{2}(q_1 - q_2)]^2/(0.5)^2 = 0.036,$$

$$q_1 - q_2 = \pm 2 \times 10^{-6} \text{ C}.$$

But, we found earlier that

$$q_1 = 3 \times 10^{-12}/q_2.$$

Combining these last two equations yields

$$q_2{}^2 \pm (2 \times 10^{-6})q_2 - 3 \times 10^{-12} = 0.$$

The two solutions we find, with the corresponding values of $q_1$ are

$$q_2 = 1 \ \mu C, \ q_1 = 3 \ \mu C, \quad \text{(upper sign)};$$

$$q_2 = 3 \ \mu C, \ q_1 = 1 \ \mu C, \quad \text{(lower sign)}.$$

($1 \ \mu C = 1 \times 10^{-6}$ C.) Hence, the initial charges on the spheres are $3 \ \mu C$ and $1 \ \mu C$ in magnitude, but they must be of opposite sign since the spheres initially attracted each other.


## 27-12

Let us write $q_1 = 1.07 \ \mu C$ and $-q_2 = -3.28 \ \mu C$. Note that $q_2 > q_1$. Call the third charge $q$. In order that the net force on $q$ be zero, the forces $F_1$ and $F_2$ exerted on $q$ by $q_1$ and $q_2$ must be equal in magnitude and opposite in direction. To be opposite in direction, $q$ must be on the line joining $q_1$ and $q_2$, but not between them. To force the magnitudes to be equal, $q$ must be more distant from $q_2$ than from $q_1$, to compensate for the fact that $q_2 > q_1$. Therefore, we obtain the arrangement sketched below; we have assumed $q > 0$ in drawing the directions of the forces $F_1$ and $F_2$. If we call $x$ the distance between $q$ and $q_1$, then the distance between $q$ and $-q_2$ is $x + d$, where $d = 61.8$ cm is the distance between the given fixed charges. Setting $F_1 = F_2$ (magnitudes) we have

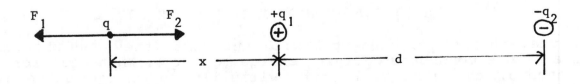

$$k\frac{qq_1}{x^2} = k\frac{qq_2}{(x + d)^2}.$$

Solving for $x$ yields

$$x = \frac{d}{\sqrt{q_2/q_1} - 1}.$$

Substituting the numerical data gives $x = 82.3$ cm.

## 27-18

As far as dependence on the charge $q$ is concerned, the result quoted in Problem 16($a$) can be written

$$x = aq^{\frac{2}{3}},$$

where $a$ is a constant independent of $q$. As $q$ changes at the rate $dq/dt$, the relative speed of approach is $v = dx/dt$. Take the time derivative of the equation above to obtain $v$, and then eliminate the constant $a$ to obtain

$$v = \frac{2}{3} \frac{x}{q} \frac{dq}{dt}.$$

Now evaluate this numerically, using the answer to Problem 16($b$); we find

$$v = \frac{2}{3} \frac{47 \text{ mm}}{22.8 \text{ nC}} (1.20 \text{ nC/s}) = 1.65 \text{ mm/s}.$$

## 27-23

If the charge $q$ is displaced a distance $x$ from the midpoint of the line joining the two positive charges $Q$, which are a distance $d$ apart, then $q$ is a distance $\frac{1}{2}d - x$ from one of the charges and a distance $\frac{1}{2}d + x$ from the other. The forces exerted by the fixed charges $Q$ on $q$ are in opposite directions, so the magnitude of the net force is equal to the difference in the magnitudes of the two individual forces. Factoring out the common term $kqQ$, we find for the magnitude $F$ of the net force,

$$F = kqQ[\frac{1}{(\frac{1}{2}d - x)^2} - \frac{1}{(\frac{1}{2}d + x)^2}],$$

$$F = \frac{kqQ}{(\frac{1}{2}d)^2}[\frac{1}{(1 - 2x/d)^2} - \frac{1}{(1 + 2x/d)^2}].$$

Now apply the second binomial theorem (see Appendix H) with $n = 2$ to the two terms inside the [] above. Include only the first two terms in each expansion; that is, write

$$(1 - 2x/d)^{-2} = 1 + 4x/d,$$

$$(1 + 2x/d)^{-2} = 1 - 4x/d.$$

With these approximations we find

$$F = -\frac{kqQ}{(\frac{1}{2}d)^2}(\frac{8x}{d}) = -(\frac{32qQ}{4\pi\epsilon_0 d^3})x = -kx.$$

This net force is directed toward the midpoint of the line joining the fixed charges; i.e., back to $x = 0$. To take account of this, we have added a minus sign in the equation. This equation now has the same form as Eq. 2 in Chapter 15, for which simple harmonic motion results. By Eq. 8 of Chapter 15, the period is given by

$$T = 2\pi\sqrt{m/k} = \sqrt{\pi^3 m\epsilon_0 d^3/2qQ}.$$

27-26

(a) By Eq. 1, with $q_1 = q_2 = q$,

$$q = r\sqrt{\frac{F}{k}}.$$

Now substitute $r = 5 \times 10^{-10}$ m, $F = 3.7 \times 10^{-9}$ N, and $k = 8.99 \times 10^9$ N•m$^2$/C$^2$ to find $q = 3.2 \times 10^{-19}$ C.
(b) The charge on an electron, in magnitude, is $e = 1.6 \times 10^{-19}$ C. Therefore, by Eq. 9, the number of electrons missing is

$$n = q/e = (3.2 \times 10^{-19} \text{ C})/(1.6 \times 10^{-19} \text{ C}) = 2.$$

<u>27-30</u>

The area of the Earth is

$$A = 4\pi R^2 = 4\pi(6.37 \text{ X } 10^6 \text{ m})^2 = 5.099 \text{ X } 10^{14} \text{ m}^2.$$

Each proton carries one elementary charge $e = 1.6$ X $10^{-19}$ C. Therefore, the current due to the protons is, by Eq. 2,

$$i = (1.60 \text{ X } 10^{-19} \text{ C})(1500 \text{ /m}^2\bullet s)(5.099 \text{ X } 10^{14} \text{ m}^2) = 122 \text{ mA},$$

where 1 mA = 1 milliamp = 0.001 A = 0.001 C/s.

<u>27-36</u>

(*a*) The charge on a nucleus is equal to $Ze$, where $Z$ is the atomic number and $e$ is the elementary charge. Find the atomic number in either Appendix D or Appendix E. We find that for Thorium $Z = 90$ and for Helium $Z = 2$; this means that the respective charges are $q_{Th}$ = 90$e$ and $q_{He}$ = 2$e$; both charges are positive. By Eq. 1, the repulsive electrostatic force on the helium particle due to the thorium nucleus is

$$F = k(90e)(2e)/r^2.$$

Now substitute $e = 1.60$ X $10^{-19}$ C, $r = 12$ X $10^{-15}$ m and $k = 8.99$ X $10^9$ N$\bullet$m$^2$/C$^2$ to find $F = 290$ N.
(*b*) The acceleration is given by $a = F/m$, so we need only find the mass $m$ of the helium particle. From Appendix D we find that the molar mass of helium is $M = 4.0026$ g/mol. Hence, the mass of one helium particle is

$$m = M/N_A = (4.0026 \text{ g/mol})/(6.02 \text{ X } 10^{23} \text{ mol}^{-1}),$$

$$m = 6.65 \text{ X } 10^{-24} \text{ g} = 6.65 \text{ X } 10^{-27} \text{ kg}.$$

Therefore,

$$a = F/m = (290 \text{ N})/(6.65 \text{ X } 10^{-27} \text{ kg}) = 4.36 \text{ X } 10^{28} \text{ m/s}^2.$$

28-1

Consider magnitude first. Newton's second law $F = ma$ can be combined with Eq. 2, which relates electric field with electric force. Writing $e$ for the magnitude of the charge on an electron, we have

$$F = ma = eE,$$

$$E = ma/e = (9.11 \times 10^{-31})(1.84 \times 10^{9})/(1.6 \times 10^{-19}),$$

$$E = 1.05 \times 10^{-2} \text{ N/C.}$$

The net force always acts in the same direction as the acceleration, in this case to the east. Since the electron carries negative charge, the electric force is directed opposite to the electric field. Therefore, the electric field points to the west.

28-5

The magnitude of the electric field set up by a point charge $q$ is given by Eq. 4. We must substitute $r$ in meters, i.e., $r = 0.75$ m. We obtain for the charge $q$,

$$q = (4\pi\epsilon_0)Er^2,$$

$$q = (8.99 \times 10^{9} \text{ N} \cdot \text{m}^2/\text{C}^2)^{-1}(2.3 \text{ N/C})(0.75 \text{ m})^2 = 144 \text{ pC,}$$

where $1 \text{ pC} = 1 \times 10^{-12}$ C.

28-7

We are given the dipole moment $p = qd = 3.56 \times 10^{-29}$ C•m and $x = 25.4$ nm. Although Eq. 8 gives the magnitude of the electric field due to a dipole along a perpendicular bisector, we cannot use it because we do not know the value of the distance $d$ between the charges forming the dipole. But $d = p/q$ and the smallest possible value of the charge is $q = e$, where $e$ is the elementary charge. Hence, the maximum possible value of $d$ is

$$d_{max} = p/e = (3.56 \times 10^{-29} \text{ C} \cdot \text{m})/(1.6 \times 10^{-19} \text{ C}),$$

$$d_{max} = 2.23 \times 10^{-10} \text{ m} = 0.223 \text{ nm.}$$

Since $x = 25.4$ nm, we see that $x \gg d_{max}$. But $d_{max} \geq d$, so that $x \gg d$. But this last condition is precisely the one for which Eq. 10 is valid. So we use Eq. 10 to find the electric field due to the dipole, finding that

$$E = \frac{1}{4\pi\epsilon_0} \frac{p}{x^3} = \frac{(8.99 \text{ X } 10^9)(3.56 \text{ X } 10^{-29})}{(25.4 \text{ X } 10^{-9})^3} = 19.5 \text{ kN/C.}$$

## 28-10

If in Fig. 4 the lower charge is made positive, then the electric field labeled $E_-$ in that figure will be reversed in direction, and we obtain the situation shown in the sketch below. The $z$ components of the two electric field vectors cancel since they are equal and oppositely directed. We are left with

$$E = E_x = 2 \left( \frac{kq}{r^2} \right) \cos\theta = 2 \left( \frac{kq}{r^2} \right) \frac{x}{r},$$

$$E = \frac{2kqx}{[x^2 + (d/2)^2]^{3/2}},$$

where $k = 1/4\pi\epsilon_0$; we also used Eq. 4 for the magnitude of the field due to each charge and the Pythagorean theorem to write $r$ in terms of $x$ and $d$. Assuming now that $x \gg d$, we can delete the $(d/2)^2$ term in the denominator to obtain

$$E = 2 \frac{1}{4\pi\epsilon_0} \frac{qx}{x^3} = \frac{1}{4\pi\epsilon_0} \frac{2q}{x^2}.$$

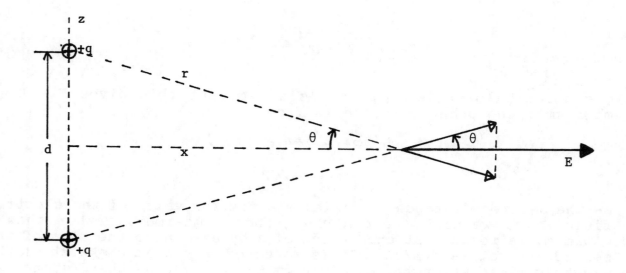

(a) The force due to the electric field on an electron will be directed to the right since the electron is negatively charged so that the electric force and electric field are oppositely directed. In magnitude,

$$F = eE = (1.6 \times 10^{-19} \text{ C})(40 \text{ N/C}) = 6.4 \times 10^{-18} \text{ N}.$$

(b) See the rule (3) on p. 610 of HRK. Imagine two equal areas, perhaps squares, perpendicular to the page (and therefore to the lines of **E**), one through B and the other through A. The spacing of the lines along the edge of the square lying in the page through B is about twice that through A. We are told that the spacing of the lines in the direction perpendicular to the page (along the other edge of the square) is the same at B as at A. Hence, the number of lines through the square at B is about one-half the number through the square at A, so that the field $E_B \approx \frac{1}{2}E_A = \frac{1}{2}(40) = 20$ N/C.

The electric field E, as a function of distance z from the ring, is given by Eq. 23:

$$E = \left(\frac{q}{4\pi\epsilon_0}\right)\frac{z}{(z^2 + R^2)^{3/2}}.$$

To find the value of z that makes E a maximum, set $dE/dz = 0$. The term in ( ) in the equation above can be ignored, since it does not contain z and is multiplicative with the terms that do contain z. After some algebra, we obtain

$$\frac{R^2 - 2z^2}{(R^2 + z^2)^{5/2}} = 0.$$

One solution of this equation is $z = \infty$, but this gives $E = 0$ (a minimum); the other solution is from

$$R^2 - 2z^2 = 0,$$

$$z = \pm R/\sqrt{2},$$

as the desired distance that yields a maximum value of the electric field E. If we wish, we can check that a maximum, rather than a minimum, is reached at this value of z by examining the sign of the second derivative $d^2E/dz^2$. It is worthwhile to make a sketch of the electric field strength, as given by Eq. 23, versus the distance z.

(a) The center of the disk is at $z = 0$. At this location, Eq. 27 reduces to $E = \sigma/2\epsilon_0$. Extracting the value of $E$ from the table, we have therefore

$$\sigma/2\epsilon_0 = 2.043 \times 10^7 \text{ N/C.}$$

Now pick any other $z$ listed in the table; we choose $z = 3$ cm. Again use Eq. 27, substituting the corresponding value of $E$ from the table and the value of $\sigma/2\epsilon_0$ found above. Note that it is not necessary to convert $z$ to meters; we find after cancelling the common factor of $10^7$,

$$1.187 - 2.043\left[1 - \frac{3}{\sqrt{3^2 + R^2}}\right].$$

Solving this for the radius gives $R = 6.50$ cm.
(b) The area of the disk is $\pi R^2$, so that $\sigma = q/\pi R^2$. Hence, the field at the surface of the disk, $z = 0$, is given by

$$\frac{q/\pi R^2}{2\epsilon_0} - E(z-0).$$

Substituting the radius of the disk from (a) and the value of the field at $z = 0$ from the table (all in SI base units), we find

$$\frac{q/\pi(0.0650)^2}{2(8.85 \times 10^{-12})} - 2.043 \times 10^7,$$

$$q = 4.80 \times 10^{-6} \text{ C} = 4.80 \text{ } \mu\text{C.}$$

28-34

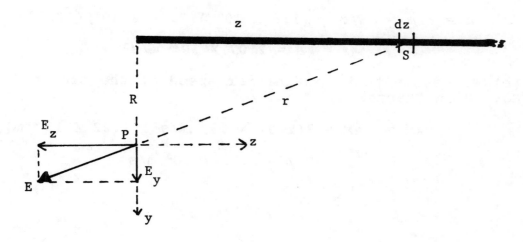

199

Due to the element of charge $dq = \lambda dz$ located at a point $S$ a distance $z$ from the end of the rod as shown, the contribution to the electric field at $P$ is

$$dE = \frac{1}{4\pi\epsilon_0} \frac{\lambda dz}{r^2},$$

directed along $SP$. The components of this field are

$$dE_y = (dE)\cos\theta, \quad dE_z = -(dE)\sin\theta.$$

Also note that

$$\cos\theta = R/r, \quad \sin\theta = z/r, \quad r^2 = R^2 + z^2.$$

Therefore, summing the contributions over the whole rod, we have

$$E_y = \frac{\lambda}{4\pi\epsilon_0} \int_0^\infty \frac{1}{R^2 + z^2} \frac{R}{\sqrt{R^2 + z^2}} dz = \frac{\lambda}{4\pi\epsilon_0 R};$$

$$E_z = -\frac{\lambda}{4\pi\epsilon_0} \int_0^\infty \frac{1}{R^2 + z^2} \frac{z}{\sqrt{R^2 + z^2}} dz = -\frac{\lambda}{4\pi\epsilon_0 R}.$$

In magnitude, then, $E_y = E_z$, and therefore the electric field $\mathbf{E}$ must make an angle of $45°$ with the rod.

## 28-36

(a) The charge on the proton is positive and equals the elementary charge $e$. We have then

$$a = F/m = eE/m = (1.6 \times 10^{-19})(2.16 \times 10^4)/(1.67 \times 10^{-27}),$$

$$a = 2.07 \times 10^{12} \text{ m/s}^2.$$

(b) We assume that the initial speed of the proton is $v_0 = 0$. By Eq. 20 in Chapter 2,

$$v^2 = 2ax = 2(2.07 \times 10^{12} \text{ m/s}^2)(1.22 \times 10^{-2} \text{ m}),$$

$$v = 2.25 \times 10^5 \text{ m/s}.$$

The weight $W$ of the drop is

$$W = mg = (\rho V)g = \rho[4\pi R^3/3]g,$$

$$W = (851 \text{ kg/m}^3)[4\pi(1.64 \times 10^{-6} \text{ m})^3/3](9.8 \text{ m/s}^2),$$

$$W = 1.541 \times 10^{-13} \text{ N}.$$

When the drop is balanced by the electric field against gravity, we have

$$qE = (ne)E = W,$$

$$n(1.6 \times 10^{-19} \text{ C})(1.92 \times 10^5 \text{ N/C}) = 1.541 \times 10^{-13} \text{ N},$$

$$n = 5.02,$$

so that the charge on the drop is $q = 5e$.

We need the value of the vertical component of the velocity $v_y$ of the drop as it exits the plates. The time $t$ required for the drop to pass through the plates is

$$t = L/v_x = (0.016 \text{ m})/(18 \text{ m/s}) = 8.89 \times 10^{-4} \text{ s},$$

so that

$$v_y = at = (qE/m)t = 1.44 \text{ m/s},$$

the numerical value following after substitution of the data. Now, the time of flight from the moment of exiting the plates to the paper is

$$t = (0.0068 \text{ m})/(18 \text{ m/s}) = 3.78 \times 10^{-4} \text{ s},$$

the horizontal component $v_x$ of the velocity being affected neither by the electric field between the plates nor by gravity. The vertical deflection during this part of the trip is, by Eq. 24 of Chapter 2,

$$y = v_y t - \tfrac{1}{2}gt^2,$$

$$y = (1.44)(3.78 \times 10^{-4}) - \tfrac{1}{2}(9.8)(3.78 \times 10^{-4})^2,$$

$$y = 5.44 \times 10^{-4} \text{ m} = 0.544 \text{ mm}.$$

(Note that the second term, representing the effect of gravity, is negligible.) Adding the plate deflection from Sample Problem 6, we find the total deflection is 0.64 + 0.54 = 1.2 mm.

<u>28-46</u>

By Eq. 23, the electric field a distance $z$ from the center of the ring is

$$E = \frac{qz}{4\pi\epsilon_0(z^2 + R^2)^{3/2}}.$$

Since the ring is charged positively, an electron on the axis will be attracted back toward the center ($z = 0$) of the ring (restoring force); we shall insert a minus sign to take account of this. Also, for small oscillations we have $z \ll R$, so ignore the $z^2$ term in the denominator. Taking these factors into account, Newton's second law requires that

$$F = -eE = -\frac{eq}{4\pi\epsilon_0 R^3}z = ma = -m\omega^2 z,$$

so that

$$\omega = \sqrt{\frac{eq}{4\pi\epsilon_0 mR^3}}.$$

We used the relation $a = -\omega^2 z$ valid for simple harmonic motion; see Section 3 in Chapter 15.

<u>28-48</u>

(a) By Eq. 9, being careful to use SI base units, we have

$p = qd = (1.48 \times 10^{-9}$ C$)(6.23 \times 10^{-6}$ m$) = 9.22 \times 10^{-15}$ C•m.

(b) Use Eq. 42:

$$U = -\mathbf{p}\cdot\mathbf{E},$$

$$\Delta U = -pE(\cos 0° - \cos 180°) = -2pE,$$

$$\Delta U = -2(9.22 \times 10^{-15}\text{ C•m})(1100\text{ N/C}) = -1.01 \times 10^{-11}\text{ J}.$$

<u>28-51</u>

See Eq. 39. Now $W$ is the work done by the electric field, so the work $W_{ext}$ needed to be done by an external agent against the field is just

$$W_{ext} = -W = -pE(\cos\theta - \cos\theta_0).$$

To turn the dipole "end for end", set $\theta = \theta_0 + 180°$. But in this event,

$$\cos\theta = \cos(\theta_0 + 180°) = -\cos\theta_0.$$

Hence, the equation for the needed work reduces to

$$W_{ext} = 2pE\cos\theta_0.$$

We see that the needed work depends on the initial orientation of the dipole, as specified by the angle $\theta_0$.

29-1

Since the electric field has the same magnitude and direction at every point on the square, the flux over the square is

$$\Phi_E = \int \mathbf{E} \cdot d\mathbf{A} = EA\cos\theta.$$

The area of the square is $A = (0.0032 \text{ m})^2 = 1.024 \times 10^{-5} \text{ m}^2$. The angle between the normal shown on Fig. 21 with the electric field is $\theta = 180° - 65° = 115°$. Therefore, the flux is

$$\Phi_E = (1800 \text{ N/C})(1.024 \times 10^{-5} \text{ m}^2)\cos 115° = -0.0078 \text{ N} \cdot \text{m}^2/\text{C}.$$

29-5

By Gauss' Law the flux is given by Eq. 8, where $q$ is the net charge enclosed by the Gaussian surface. We have

$$\Phi_E = q/\epsilon_0 = (1.84 \times 10^{-6} \text{ C})/(8.85 \times 10^{-12} \text{ C}^2/\text{N} \cdot \text{m}^2),$$

$$\Phi_E = 2.08 \times 10^5 \text{ N} \cdot \text{m}^2/\text{C}.$$

The size of the Gaussian surface does not enter, as long as the 1.84 $\mu$C charge is inside.

29-9

Along the four vertical sides of the cube **E** is at 90° to $d\mathbf{A}$ so that $\mathbf{E} \cdot d\mathbf{A} = EdA\cos 90° = 0$. On the upper face of the cube **E** and $d\mathbf{A}$ point in opposite directions. Since the electric field has the same magnitude at all points on the upper surface, the flux on the upper surface is

$$\Phi_{Eu} = E_u A\cos 180° = -E_u A = -E_u a^2.$$

Along the bottom surface **E** and $d\mathbf{A}$ point in the same direction, so the flux through this surface is just

$$\Phi_{Eb} = E_b A\cos 0° = E_b a^2.$$

Hence, by Gauss' law, and using SI base units,

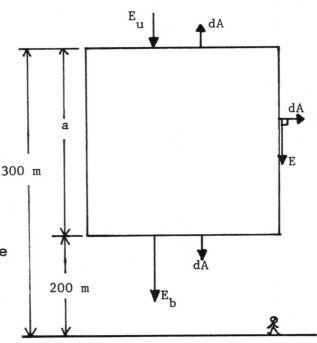

$$\Phi_E = E_b a^2 - E_u a^2 = (E_b - E_u)a^2 = q/\epsilon_0,$$

$$q = \epsilon_0 (E_b - E_u)a^2,$$

$$q = (8.85 \times 10^{-12})(110 - 58)(100)^2 = 4.6 \times 10^{-6} \text{ C} = 4.6 \ \mu\text{C}.$$

## 29-13

(a) The electric field points in the $+x$ direction (i.e., it is parallel to the $x$ axis). Therefore, there is a non-zero flux only through the two faces of the cube that are perpendicular to the $x$ axis. The vector $d\mathbf{A}$ is directed out of the cube, so the angle between $\mathbf{E}$ and $d\mathbf{A}$ is 180° for the surface at $x = a$, and is equal to 0° for the surface at $x = 2a$. The area of each surface is $a^2$. Substituting the appropriate value of $x$ into the formula for the electric field strength $E$, we find for the flux through the cube,

$$\Phi_E = (ba^{\frac{1}{2}})(a^2)\cos 180° + [b(2a)^{\frac{1}{2}}](a^2)\cos 0° = ba^{5/2}(\sqrt{2} - 1),$$

$$\Phi_E = (8830)(0.13)^{5/2}(\sqrt{2} - 1) = 22.3 \text{ N} \bullet \text{m}^2/\text{C}.$$

(b) By Gauss' law,

$$q = \epsilon_0 \Phi_E = (8.85 \times 10^{-12})(22.3) = 197 \text{ pC}.$$

We used SI base units throughout; you may wish to check the units yourself.

## 29-14

(a) The charge is

$$q = \sigma A = \sigma(4\pi r^2) = (8.13 \ \mu\text{C/m}^2)4\pi(1.22 \text{ m})^2 = 152 \ \mu\text{C}.$$

(b) By Gauss' law,

$$\Phi_E = q/\epsilon_0 = (152 \times 10^{-6} \text{ C})/(8.85 \times 10^{-12} \text{ C}^2/\text{N}\bullet\text{m}^2),$$

$$\Phi_E = 1.72 \times 10^7 \text{ N}\bullet\text{m}^2/\text{C}.$$

(c) The electric field at the surface of the sphere is given by Eq. 11; we have

$$E = \sigma/\epsilon_0 = (8.13 \times 10^{-6})/(8.85 \times 10^{-12}) = 919 \text{ kN/C}.$$

You may want to run through the units in (c).

## 29-18

(a) The sketch on the next page shows a conductor with a cavity. Also shown is a Gaussian surface S which lies         inside the

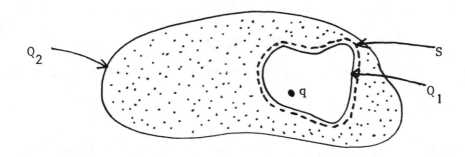

conductor and "parallel" to the cavity surface. $Q_1$ and $Q_2$ are the charges on the inner and outer surfaces and $q$ is the enclosed charge. (We know that there is no free charge in the body of the conductor under electrostatic conditions.) Apply Gauss' law to the surface S; be aware that $E = 0$ everywhere on S since S lies inside the conductor. Thus,

$$\Phi_E = \oint \mathbf{E} \cdot d\mathbf{A} = 0 = (q + Q_1)/\epsilon_0,$$

$$Q_1 = -q = -3.0 \ \mu C.$$

(b) We have been told that the net charge on the conductor is 10 $\mu$C. Therefore,

$$Q_2 = 10 - Q_1 = 10 - (-3) = 13 \ \mu C.$$

### 29-23

The restriction on points to be considered is tantamount to regarding the plates as infinite in extent in two dimensions (the thickness is unaffected). The electric field lines, by symmetry, must pass perpendicular to the plates from the positive to the negative charges. Under conditions of electrostatic equilibrium, no field lines can penetrate into the conducting plates, and since there is no fringing of the lines around the plate edges, there being no edges in effect, we conclude that the field must be zero at all points not between the plates; that is, (a) $E = 0$ and (c) $E = 0$.
(b) To calculate $E$ at points between the plates, construct a right-rectangular cylindrical "pillbox" Gaussian surface (any shape cross-section of area $A$), with end caps

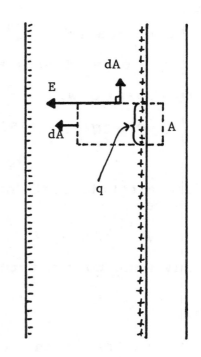

parallel to the plates, as shown. The flux over that part of the Gaussian surface lying inside the plate is zero since $E = 0$ there. The flux is also zero over the curved part of the Gaussian surface found outside the plates; this is because $\mathbf{E}$ and $d\mathbf{A}$ are at 90° to each other. Over the left end cap, however, $\mathbf{E}$ and $d\mathbf{A}$ are parallel, and since all points on the left end cap are at the same distance from the plate, Gauss' law yields

$$\Phi_E = \epsilon_0 \int E dA = \epsilon_0 EA = q.$$

The charge enclosed by the Gaussian surface resides on that part of the plate lying inside the surface; i.e., $q = \sigma A$. Therefore,

$$\epsilon_0 EA = \sigma A,$$

$$E = \sigma/\epsilon_0.$$

This is directed to the left (toward the negative plate). It may appear that the presence of the negative plate has been ignored but this is not so for, if that plate was absent, the positive charge on the remaining plate (now isolated) would then be found distributed over both sides of the plate, making $E = \sigma_{isol}/\epsilon_0 = \frac{1}{2}\sigma/\epsilon_0 = \sigma/2\epsilon_0$ at all points not inside the plate.

## 29-27

By Gauss' law, for the electric field to be zero outside the cylinder, the net charge on the cylinder must be zero. For a length $L$ of the cylinder, we must have

$$q = \lambda L + \rho(\pi r^2 L) = 0,$$

where $\lambda$ is the charge per unit length on the wire ; $r$ is the radius of the cylinder; the volume of the wire has been ignored since the wire is "thin". Solving for the volume charge density, we get

$$\rho = -\lambda/\pi r^2 = -(-3.6 \times 10^{-9} \text{ C/m})/[\pi(0.015 \text{ m})^2],$$

$$\rho = 5.09 \times 10^{-6} \text{ C/m}^3 = 5.09 \ \mu\text{C/m}^3.$$

## 29-31

For uniform circular motion the acceleration is $a = v^2/r$. Outside a charged sphere, the electric field due to the sphere is given by Eq. 14. Let $q$ be the magnitude of the charge on the sphere. An attractive force is required, so the actual charge on the sphere is $-q$. The proton carries one elementary unit $e$ of positive charge. By Newton's second law

$$F = ma = eE.$$

Substituting the expressions for $a$ and $E$ gives

$$\frac{1}{4\pi\epsilon_0} \frac{qe}{r^2} = \frac{mv^2}{r}.$$

Solving for the magnitude of the charge gives

$$q = \frac{(4\pi\epsilon_0)mv^2r}{e}.$$

Now substitute the numerical values: $e = 1.6 \times 10^{-19}$ C, $v = 294,000$ m/s, $r = 0.0113$ m, $m = 1.67 \times 10^{-27}$ kg and $(4\pi\epsilon_0) = (8.99 \times 10^9)^{-1}$ $C^2/N\cdot m^2$; we find $q = 1.13$ nC, so that the charge on the sphere is $-1.13$ nC.

## 29-36

If the length of the wire is $L$ (= 16 cm = 0.16 m) and it carries charge $q$, then the charge per unit length is

$$\lambda = q/L,$$

so that, by Problem 34, the electric field due to the wire is

$$E = \frac{q}{2\pi\epsilon_0 Lr},$$

a distance $r$ from the wire. At the cylinder wall we have $r = 1.4$ cm = 0.014 m; hence, the charge is

$$q = 2\pi\epsilon_0 LrE.$$

Substituting the data gives

$$q = 2\pi(8.85 \times 10^{-12}\ C^2/N\cdot m^2)(0.16\ m)(0.014\ cm)(2.9 \times 10^4\ N/C),$$

$$q = 3.6 \times 10^{-9}\ C = 3.6\ nC.$$

## 29-39

Use the work-energy theorem from Chapter 7:

$$W = \Delta K = K_f - K_i = 0 - K_i.$$

We set $K_f = 0$ since at the distance of closest approach to the sheet the electron is instantaneously at rest and $K = \frac{1}{2}mv^2$. The electric force exerted by the sheet on the electron is directed away from the sheet. Therefore, if the electron travels a distance we shall call $z$ in approaching the sheet, the work done by the electric force is

$$W = Fz\cos\theta = (eE)z\cos 180° = -eEz = -e(\sigma/2\varepsilon_0)z,$$

where we used Eq. 13 for the strength $E$ of the electric field. Therefore, the work-energy theorem becomes

$$ez\sigma/2\varepsilon_0 = K_1.$$

Now, the conversion factor from eV to J (i.e., to the SI base unit of energy) is numerically equal to the elementary charge $e$ in C. Therefore these factors cancel since they are on opposite sides of the equation, and we are left with

$$(2.08 \times 10^{-6})z/2(8.85 \times 10^{-12}) = 115 \times 10^3,$$

$$z = 0.979 \text{ m} = 97.9 \text{ cm}.$$

## 29-42

The electric field due to the sphere, for $a < r < b$, is found from Gauss' law by constructing a spherical Gaussian surface centered at $r = 0$, the center of the sphere. The charge $q_{enc}$ enclosed by the surface is given by

$$q_{enc} = \int \rho \, dV = \int_a^r \left(\frac{A}{r}\right) 4\pi r^2 dr = 2\pi A(r^2 - a^2).$$

Hence, Gauss' law becomes

$$\oint \mathbf{E} \cdot d\mathbf{A} = q_{enc}/\varepsilon_0,$$

$$E(4\pi r^2) = 2\pi A(r^2 - a^2)/\varepsilon_0,$$

$$E = \frac{A}{2\varepsilon_0}\left(1 - \frac{a^2}{r^2}\right),$$

radially in or out. The field $E_q$ due to the point charge $q$ is

$$E_q = \frac{1}{4\pi\varepsilon_0}\frac{q}{r^2}.$$

In order that the net field $\mathbf{E} + \mathbf{E}_q$ be independent of $r$, the sum of the $r$-dependent terms must be zero; that is,

$$-Aa^2/2\epsilon_0 + q/4\pi\epsilon_0 = 0,$$

$$A = q/2\pi a^2.$$

29-46

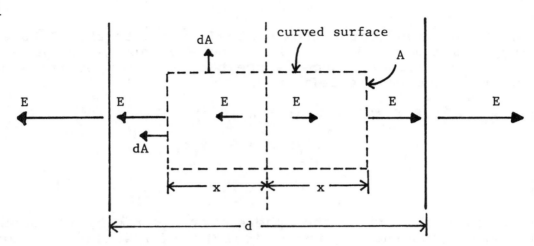

(a) Consider the slab to be of very large area and the charge positive. By symmetry, the field inside the slab will be directed at 90° to the slab surfaces but will vary in strength with the distance $x$. On the median plane ($x = 0$) we expect $E = 0$. Construct a cylindrical Gaussian surface oriented as shown. The flux over the curved part of the surface is zero since $\mathbf{E}$ and $d\mathbf{A}$ are at 90° to each other. Since each end cap is at the same distance from the median plane $E$ is the same on each, so that the flux through the Gaussian surface is $\Phi_E = 2EA$, where $A$ is the area of an end cap. The charge enclosed by the Gaussian surface is $q = \rho V = \rho(2xA)$, since the entire volume of the Gaussian surface contains charge. Therefore, Gauss' law $\Phi_E = q/\epsilon_0$ becomes

$$2EA = \rho(2xA)/\epsilon_0,$$

$$E = \rho x/\epsilon_0.$$

(b) The electric field outside the slab is the same as the field due to a single nonconducting sheet. This is given by Eq. 13 in terms of the surface charge density as $E = \sigma/2\epsilon_0$. But the charge enclosed by the Gaussian surface shown in Fig. 9 of HRK can be written as $q = \sigma A$ or as $q = \rho V = \rho(Ad)$. Thus, $\sigma = \rho d$ and the electric field outside the slab is

$$E = \rho d/2\epsilon_0.$$

## 30-4

The potential energy of a system of 3 point charges is given by Eq. 8. If we choose $q_1 = 25.5$ nC, $q_2 = 17.2$ nC and $q_3 = -19.2$ nC, then we have $r_{12} = 14.6$ cm, $r_{23} = x$ and $r_{13} = x + 14.6$ where $x$ is in cm. Now we require that $V = 0$. The common factor of $1/4\pi\epsilon_0$ cancels out, along with the conversion factors from nC to C and cm to m; that is, we can use nC and cm as units of charge and distance. With these considerations in mind, we have

$$\frac{(25.5)(17.2)}{14.6} + \frac{(25.5)(-19.2)}{14.6 + x} + \frac{(17.2)(-19.2)}{x} = 0,$$

$$30.041x^2 - 381.24x - 4821.5 = 0,$$

$$x = 20.5 \text{ cm}.$$

## 30-6

Use the work-energy theorem,

$$W = \Delta U = q\Delta V = K_f - K_i,$$

$$q\Delta V = 0 - \tfrac{1}{2}mv^2,$$

$$(-1.6 \times 10^{-19} \text{ C})(10.3 \times 10^3 \text{ V}) = -\tfrac{1}{2}(9.11 \times 10^{-31} \text{ kg})v^2,$$

$$v = 6.01 \times 10^7 \text{ m/s}.$$

You may want to try this problem using the relativistic kinetic energy expression, Eq. 27 in Chapter 21.

## 30-7

(a) The energy is $U = q\Delta V = (30 \text{ C})(1 \times 10^9 \text{ V}) = 3 \times 10^{10} \text{ J} = 30$ GJ.
(b) The energy released by the lightning flash is presumed to be converted to translational kinetic energy of the car, so we will have

$$3 \times 10^{10} \text{ J} = \tfrac{1}{2}(1200 \text{ kg})v^2,$$

$$v = 7.1 \text{ km/s}.$$

(c) Use the heat of fusion of water found in Table 2 of Chapter 25. By Eq. 7 of that same chapter, the mass of ice $m$ melted is given from

$$m = Q/L_f = U/L_f = (3 \text{ X } 10^{10} \text{ J})/(333,000 \text{ J/kg}) = 9.0 \text{ X } 10^4 \text{ kg.}$$

30-11

Apply conservation of energy

$$U_1 + K_1 = U_2 + K_2.$$

The potential energy of a set of point charges is given by Eq. 7. In our case $q_1 = q_2 = q$. Also $v_1 = 0$ since the first particle is released from rest. Therefore, the conservation of energy becomes

$$\frac{1}{4\pi\epsilon_0} \frac{q^2}{r_1} + 0 = \frac{1}{4\pi\epsilon_0} \frac{q^2}{r_2} + \tfrac{1}{2}mv^2,$$

$$v^2 = \frac{2}{m} \frac{q^2}{4\pi\epsilon_0} \left[ \frac{1}{r_1} - \frac{1}{r_2} \right].$$

Now substitute $q = 3.1 \text{ X } 10^{-6}$ C, $m = 18 \text{ X } 10^{-6}$ kg, $r_1 = 0.9 \text{ X } 10^{-3}$ m and $r_2 = 2.5 \text{ X } 10^{-3}$ m, to find that $v = 2600$ m/s.

30-15

The work that must be done is equal to the difference in the potential energies of the assembly in the initial and final configurations. We call each charge $q$ and the length of each side of the equilateral triangle $a$. The potential energies are based on Eq. 8. In the initial configuration, all pairs of charges are separated by the same distance $a$, so that

$$U_i = 3\frac{1}{4\pi\epsilon_0} \frac{q^2}{a}.$$

In the final configuration, one pair of charges are still separated by the distance $a$ (since they did not move), but the third charge, the one that did move, is now at a distance of $a/2$ from each of the others. Hence,

$$U_f = \frac{1}{4\pi\epsilon_0} \left[ \frac{q^2}{a} + 2\frac{q^2}{a/2} \right] = 5\frac{1}{4\pi\epsilon_0} \frac{q^2}{a}.$$

Therefore,

$$\Delta U - 2\frac{1}{4\pi\epsilon_0}\frac{q^2}{a},$$

$$\Delta U = 2(8.99 \times 10^9)(0.122)^2/(1.72) = 1.556 \times 10^8 \text{ J}.$$

Since one day = 86,400 s, the required time in days is

$$t = \Delta U/P = (1.556 \times 10^8 \text{ J})/[(831 \text{ J/s})(86,400 \text{ s/d})],$$

$$t = 2.17 \text{ d}.$$

## 30-16

The electric field due to an infinite sheet of charge is uniform, and given by Eq. 13 of Chapter 29 as follows:

$$E = \sigma/2\epsilon_0,$$

$$E = (0.12 \times 10^{-6} \text{ C/m}^2)/2(8.85 \times 10^{-12} \text{ C}^2/\text{N}\bullet\text{m}^2) = 6780 \text{ V/m}.$$

In a uniform electric field, the difference in potential between two points a distance $L$ apart along a field line is given by Eq. 13 (of Chapter 30); that is,

$$V_b - V_a = EL.$$

For $V_b - V_a = 48$ V, we find for the required distance

$$L = (48 \text{ V})/(6780 \text{ V/m}) = 0.0071 \text{ m} = 7.1 \text{ mm}.$$

## 30-18

Use Eq. 13, since the field is presumed to be uniform between the plates. Noting that 1 N/C = 1 V/m, we find

$$\Delta V = EL,$$

$$\Delta V = (1.92 \times 10^5 \text{ V/m})(0.015 \text{ m}) = 2880 \text{ V} = 2.88 \text{ kV}.$$

## 30-22

Follow the reasoning of Sample Problem 6 in Chapter 16. We need only replace the gravitational potential energy with the electric potential energy. Writing $Q$ and $R$ for the charge on and the radius of the sphere, we have

$$K + U = 0,$$

$$\tfrac{1}{2}mv^2 + \frac{1}{4\pi\epsilon_0}\frac{(-e)Q}{R} = 0,$$

$$v = \sqrt{\frac{2eQ}{4\pi\epsilon_0 mR}} .$$

Substitute numerical values ($R = 0.0122$ m, $e = 1.6 \times 10^{-19}$ C, etc.) and verify that $v = 21.3$ km/s.

### 30-27

(a) Find the radius $R$ of the drop from Eq. 18 with $r = R$:

$$R = q/4\pi\epsilon_0 V = (8.99 \times 10^9)(32 \times 10^{-12})/(512),$$

$$R = 5.62 \times 10^{-4} \text{ m}.$$

(b) Apply Eq. 18 to the combined (new) drop. Its charge is $2(32) = 64$ pC. Its radius is found by setting its volume equal to twice the volume of an original drop; i.e.,

$$4\pi R_{new}{}^3/3 = 2[4\pi(5.62 \times 10^{-4} \text{ m})^3/3)],$$

$$R_{new} = 7.081 \times 10^{-4} \text{ m}.$$

Hence,

$$V_{new} = (1/4\pi\epsilon_0)q_{new}/R_{new},$$

$$V_{new} = (8.99 \times 10^9)(64 \times 10^{-12})/(7.081 \times 10^{-4}) = 813 \text{ V}.$$

It goes without saying that we substituted the data in SI base units.

### 30-33

(a) Consider a point a distance $d_1$ from charge $q_1$ and $d_2$ from charge $q_2$. The electric potential at this point due to the two charges is

$$V = k\left(\frac{q_1}{d_1} + \frac{q_2}{d_2}\right).$$

In this expression $k = 1/4\pi\epsilon_0$. If $V = 0$ at this point, then

$$\frac{q_1^2}{d_1^2} = \frac{q_2^2}{d_2^2}.$$

Let $x, y$ be the coordinates of this point. Expressing the distances $d_1$ and $d_2$ in terms of $x$, $y$ and $a$ (= 9.60 nm, the distance between the charges), we have

$$\frac{36}{x^2 + y^2} = \frac{100}{(a - x)^2 + y^2}.$$

(We substituted the values $q_1 = 6e$ and $q_2 = -10e$, and cancelled the factor $e^2$.) This last equation can be rearranged, using our algebra skills, into the form

$$(x + 9a/16)^2 + y^2 = (15a/16)^2.$$

Now, the equation of a circle of radius $R$ with its center at $x_c$, $y_c$ is

$$(x - x_c)^2 + (y - y_c)^2 = R^2.$$

Therefore, our circle has its center at $x_c = -9a/16 = -5.40$ nm and $y_c = 0$.
(b) The radius of our $V = 0$ circle is $R = 15a/16 = 9.00$ nm.
(c) If $V \neq 0$, then our original equation cannot be put into the form of the equation of a circle; you should verify this.


30-37

By Eqs. 3 and 11, the work that must be done by the electric force is given by

$$W_{elect} = -\Delta U = -q\Delta V = -q(V_f - V_i),$$

where $q$ is the charge on the particle. The work done by the external agent is the negative of the work done by the electric field. The potential due to a ring of charge, the $V$ in the equation above, is given by Eq. 25; we will write $Q$ for the charge on the ring (to avoid confusion with the charge on the particle); we also write $x$ instead of $z$ since the axis of our ring lies along the $x$, rather than $z$, axis. In the initial position of the particle we have $x = 3.07$ m; in the final position at the center of the ring we have $x = 0$. In this latter position, the potential of the ring of charge reduces to $V = (1/4\pi\epsilon_0)Q/R$. The work done by the external agent becomes, then,

$$W_{ext} = \frac{qQ}{4\pi\varepsilon_0}\left[\frac{1}{R} - \frac{1}{\sqrt{R^2 + x^2}}\right].$$

Substituting the numerical data (recall that 1 nC = 1 X $10^{-9}$ C) gives $W_{ext}$ = 186 pJ.

30-41

(a) Use Eq. 18 to find the radius of the 30 V equipotential (i.e., the distance from the point charge to any point where the potential due to the charge is 30 V):

$$r = q/4\pi\varepsilon_0 V,$$

$$r = (8.99 \text{ X } 10^9 \text{ V}\bullet\text{m/C})(1.5 \text{ X } 10^{-8} \text{ C})/(30 \text{ V}) = 4.495 \text{ m}.$$

(b) For V = 31 V we find that r = 4.350 m; also, for V = 29 V we get r = 4.650 m. Comparing this with the result for 30 V, we see that the spacing for a potential difference of 1 V changes from 0.155 m to 0.145 m over this range from 29 V to 31 V. Thus the surfaces are not evenly spaced. (We do not expect that they would be since $V \propto 1/r$; only if $V \propto r$ would they be evenly spaced.)

30-49

Use Eq. 29 to obtain

$$E_x = -\partial V/\partial x = -3060x = -3060(0.0128) = -39.2 \text{ V/m},$$

$$E_y = E_z = 0.$$

Therefore the electric field is **E** = -39.2**i**, V/m. Evidently, it points in the negative x direction.

30-50

(a) With $dq = \lambda dx$ = the charge on an element of length $dx$, the potential at point P is found from Eq. 24:

$$4\pi\varepsilon_0 V = \int \frac{\lambda\,dx}{r} = \lambda\int_0^L \frac{dx}{(L + y) - x}.$$

The integral on the right is equal to $\ln(L + y - x)$; after evaluating between the limits and solving for the potential V one finds that

$$V - \frac{\lambda}{4\pi\epsilon_0} \ln(1 + \frac{L}{y}).$$

(b) Now call the axis the $y$ axis
and move the origin from the
lower end to the upper end of
the charged segment, as shown on
the sketch. Then calculate the
$y$ component of the electric field
from Eq. 29:

$$E_y = -dE/dy,$$

$$E_y = \frac{\lambda}{4\pi\epsilon_0} \frac{L}{y(L + y)}.$$

(c) The direction of the electric
field at $P$ due to any element of
the segment of charge is directed
along the $y$ axis; hence $E_z = 0$.
This result cannot be obtained from
the potential derived in (a) above, for that expression is valid
only at $z = 0$ and therefore $\partial V/\partial z$, which requires knowledge of $V$ in
a neighborhood in $z$ about $P$, cannot be computed from it.

30-53

(a) With the conducting spheres connected by a conducting wire, we
have in effect a single isolated conductor. The potential has a
constant value at all points on, or in, the conductor so that, in
particular, $V_1 = V_2$.
(b) We can assume that the charge on the wire is very small. By Eq.
32 we have

$$q_1 = (R_1/R_2)q_2 = (R_1/2R_1)q_2 = \tfrac{1}{2}q_2.$$

If we assume that no charge has leaked off the conductors, then

$$q_1 + q_2 = q + 0 = q.$$

Substituting the result for $q_1$ found above we find that the charges
are $q_2 = \tfrac{2}{3}q$ and $q_1 = \tfrac{1}{3}q$.

30-56

(a) By Eq. 18,

217

$$q = (4\pi\epsilon_0)Vr = (8.99 \times 10^9)^{-1}(215)(0.152) = 3.64 \text{ nC.}$$

(b) The corresponding surface charge density is

$$\sigma = q/A = q/4\pi R^2,$$

$$\sigma = (3.64 \text{ nC})/[4\pi(0.152 \text{ m})^2] = 12.5 \text{ nC/m}^2.$$

## 30-66

(a) The potential of the shell is $V = q/4\pi\epsilon_0 r$ and the electric field near the shell's surface is $E = q/4\pi\epsilon_0 r^2$, so that $E = V/r$. For $E < 100$ MV/m, it is therefore necessary that

$$r > V/E = (9.15 \text{ MV})/(100 \text{ MV/m}) = 0.0915 \text{ m} = 9.15 \text{ cm.}$$

(b) The work done in bringing up to the machine a charge $Q$ is $W = QV$. Therefore, the power $P$ supplied must be

$$P = dW/dt = V(dQ/dt),$$

$$P = (9.15 \text{ MV})(320 \text{ } \mu C/s) = 2930 \text{ W} = 2.93 \text{ kW.}$$

(c) If the surface charge density is $\sigma$ and $x$ denotes a length of the belt, then since

$$Q = \sigma A = \sigma(wx),$$

we have

$$dQ/dt = \sigma d(wx)/dt = \sigma w(dx/dt) = \sigma wv,$$

$$320 \text{ } \mu C/s = \sigma(0.485 \text{ m})(33 \text{ m/s}),$$

$$\sigma = 20.0 \text{ } \mu C/m^2.$$

31-3

Use Eq. 1:

$$q = CV = (26 \times 10^{-6} \text{ F})(125 \text{ V}) = 3.25 \times 10^{-3} \text{ C} = 3.25 \text{ mC}.$$

31-4

(a) The capacitance $C$ is given by Eq. 7. Before we can use this equation, however, the plate area A must be calculated. Since the plates are circular, we have

$$A = \pi r^2 = \pi(0.0822 \text{ m})^2 = 2.123 \times 10^{-2} \text{ m}^2.$$

Therefore,

$$C = \epsilon_0 A/d = (8.85 \times 10^{-12} \text{ F/m})(2.123 \times 10^{-2} \text{ m}^2)/(0.00131 \text{ m}),$$

$$C = 1.43 \times 10^{-10} \text{ F} = 143 \text{ pF}.$$

(b) By Eq. 1,

$$q = CV = (1.43 \times 10^{-10} \text{ F})(116 \text{ V}) = 1.66 \times 10^{-8} \text{ C} = 16.6 \text{ nC}.$$

31-7

(a) See Eq. 13 for the capacitance of a spherical capacitor:

$$C = 4\pi\epsilon_0 \frac{ab}{b-a} = 4\pi(8.85 \text{ pF/m}) \frac{(0.038 \text{ m})(0.040 \text{ m})}{0.040 \text{ m} - 0.038 \text{ m}} = 84.5 \text{ pF}.$$

(b) Use Eq. 7:

$$C = \epsilon_0 A/d,$$

$$84.5 \text{ pF} = (8.85 \text{ pF/m})A/(0.002 \text{ m}),$$

$$A = 0.0191 \text{ m}^2 = 191 \text{ cm}^2.$$

31-12

The capacitors labeled $C_1$ and $C_2$ are in parallel, so that their equivalent capacitance $C_{12}$, as given by Eq. 18, is

$$C_{12} = C_1 + C_2 = 10.3 \text{ } \mu\text{F} + 4.80 \text{ } \mu\text{F} = 15.1 \text{ } \mu\text{F}.$$

The capacitor $C_{12}$, replacing $C_1$ and $C_2$, and $C_3$ are in series. Their equivalent capacitance $C_{eq}$, which is the equivalent capacitance of the original combination, is give from Eq. 23. We have

$$\frac{1}{C_{eq}} = \frac{1}{C_{12}} + \frac{1}{C_3} = \frac{1}{15.1} + \frac{1}{3.90},$$

$$C_{eq} = 3.10 \ \mu F.$$

31-15

(a) For capacitors in series,

$$\frac{1}{C_{eq}} = \frac{1}{6} + \frac{1}{4} \rightarrow C_{eq} = 2.4 \ \mu F.$$

(b) By Eq. 22,

$$q = C_{eq}V = (2.4 \ \mu F)(200 \ V) = 480 \ \mu C.$$

The same charge appears on both capacitors, since they are connected in series.
(c) Apply Eq. 1 to the 6 $\mu F$ capacitor; we find

$$V_6 = q/C_6 = (480 \ \mu C)/(6 \ \mu F) = 80 \ V.$$

By Eq. 21, the potential across the 4 $\mu F$ capacitor is

$$V_4 = V - V_6 = 200 \ V - 80 \ V = 120 \ V.$$

31-21

The capacitance of a parallel-plate capacitor is $C = \epsilon_0 A/d$, by Eq. 7, where $d$ is the plate separation. If the plates of the upper capacitor are a distance $D$ apart, then the plates of the lower capacitor must be a distance $[a - (b + D)]$ apart. Therefore, the equivalent capacitance as given by Eq. 23 leads to

$$\frac{1}{C_{eq}} = \frac{1}{\epsilon_0 A/D} + \frac{1}{\epsilon_0 A/(a - b - D)} = \frac{D + a - b - D}{\epsilon_0 A} = \frac{a - b}{\epsilon_0 A},$$

and therefore

$$C_{eq} = \epsilon_0 A/(a - b).$$

(a) Let primed quantities be the
final values (switches closed).
The final polarities of the two
capacitors must be the same since
the capacitors are then connected
in parallel. By the conservation
of charge, we must have

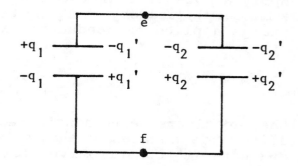

$$-q_1 + q_2 = q_1' + q_2'.$$

But we are also given that

$$q_1/C_1 = q_2/C_2 = 96.6 \text{ V}.$$

Since $C_1 = 1.16 \ \mu F$ and $C_2 = 3.22 \ \mu F$, the equation above gives $q_1 = 112.1 \ \mu C$ and $q_2 = 311.1 \ \mu C$. Substituting these results into the first equation yields

$$q_1' + q_2' = 199.0 \ \mu C.$$

However, in the final arrangement we also have

$$V' = q_1'/C_1 = q_2'/C_2,$$

so that

$$q_2' = (C_2/C_1)q_1' = 2.776q_1',$$

just as $q_2 = 2.776q_1$. Solving the two equations for $q_1'$ and $q_2'$ we find that $q_1' = 52.7 \ \mu C$ and $q_2' = 146.3 \ \mu C$. Therefore,

$$V_{ef} = V' = q_2'/C_2 = (146.3 \ \mu C)/(3.22 \ \mu F) = 45.4 \text{ V}.$$

(b) From (a), $q_1' = 52.7 \ \mu C$.
(c) Also from (a), $q_2' = 146 \ \mu C$.

31-28

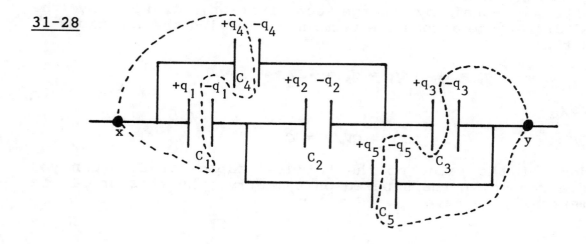

Apply a potential difference $V$ between $x$ and $y$ (say, by connecting a battery), thereby charging the capacitors. By the conservation of charge applied to conductors $x$ and $y$,

$$-q_4 - q_2 + q_3 = 0,$$

$$-q_1 + q_2 + q_5 = 0,$$

the capacitors being considered originally uncharged. The potential difference between $x$ and $y$ is independent of the path taken to evaluate it (conservative field). Hence, with $C_1 = C_3 = C_4 = C_5 = C$, we have

$$\frac{q_4}{C} + \frac{q_3}{C} = V,$$

$$\frac{q_1}{C} + \frac{q_2}{C_2} + \frac{q_3}{C} = V,$$

$$\frac{q_1}{C} + \frac{q_5}{C} = V.$$

From these five equations $q_1$, $q_2$, $q_3$, $q_4$, $q_5$ can be found in terms of $C$, $C_2$, $V$. The equivalent capacitance $C_{eq}$ is

$$C_{eq} = \frac{q_1 + q_4}{V} = \frac{q_3 + q_5}{V};$$

that is, the magnitude of the charge on either terminal $x$ or $y$ (they will be equal by charge conservation) divided by the potential difference across the terminals. Solving the equations it is found that

$$q_1 = q_3 = q_4 = q_5 = \tfrac{1}{2}CV, \quad q_2 = 0.$$

Therefore,

$$C_{eq} = CV/V = C,$$

independent of the value of the "middle" capacitor $C_2$. (Can you think of a reason, perhaps based on symmetry, why this should be so?) Numerically we have $C_{eq} = C = 4 \ \mu F$.

By Eq. 26, the total stored energy is

$$U = U_1 + U_2 = q^2/2C_1 + q^2/2C_2 = \tfrac{1}{2}q^2(1/C_1 + 1/C_2) = q^2/2C_{eq}.$$

The equivalent capacitance follows from Eq. 23:

$$C_{eq} = (1/C_1 + 1/C_2)^{-1} = (1/2.12 + 1/3.88)^{-1} = 1.371 \ \mu F.$$

By Eq. 22, $q = C_{eq}V$, so that

$$U = \tfrac{1}{2}C_{eq}V^2 = \tfrac{1}{2}(1.371 \times 10^{-6} \ F)(328 \ V)^2 = 0.0737 \ J = 73.7 \ mJ.$$

## 31-37

(a) With the plates disconnected, no charge can be transferred from one plate to the other. If the plate separation remains small compared with the dimensions of the plates, the electric field will remain uniform and independent of the distance between the plates (see Problem 23 in Chapter 29). Hence,

$$V_f = E_f d_f = E(2d) = 2(Ed) = 2V.$$

(b) Initially,

$$U_i = \tfrac{1}{2}C_i V_i^2 = \tfrac{1}{2}(\epsilon_0 A/d)V^2.$$

Both the capacitance and the potential difference change as the plates are pulled apart. Thus,

$$U_f = \tfrac{1}{2}C_f V_f^2 = \tfrac{1}{2}(\epsilon_0 A/d_f)(V_f)^2 = \tfrac{1}{2}(\epsilon_0 A/2d)(2V)^2 = \epsilon_0(A/d)V^2.$$

Note that $U_f = 2U_i$.
(c) The work $W$ required to pull the plates apart is

$$W = U_f - U_i = U_i = \tfrac{1}{2}(\epsilon_0 A/d)V^2.$$

## 31-40

Suppose that one plate of the capacitor is removed. Then, the charge on the other plate will move so as to distribute itself uniformly on both sides (rather than on one side) of the plate, so that there is now $q/2$ of charge on each side of the plate. The electric field set up by this isolated plate, by Eq. 11 of Chapter 29, is

$$E_{isol} = \sigma_{isol}/\epsilon_0 = q/2A\epsilon_0,$$

where $q$ is the total charge on the plate. The electric field between the plates of the assembled capacitor is, from Eq. 3,

$$E_{cap} = q/A\epsilon_0.$$

Hence, the electric field acting on one plate due to the presence of the other plate is

$$E = E_{cap} - E_{isol} = q/A\epsilon_0 - q/2A\epsilon_0 = q/2A\epsilon_0.$$

Thus, the force exerted by one plate on the other is

$$F = qE = q^2/2A\epsilon_0.$$

## 31-42

Consider the forces acting on a small area $A$ of the bubble's surface. These forces are due to

(*i*)   gas pressure $p_g$ acting outward,
(*ii*)  atmospheric pressure $p$ acting inward,
(*iii*) electrostatic stress acting outward (see Problem 41).

Thus, for equilibrium,

$$p_gA + \tfrac{1}{2}\epsilon_0E^2A = pA.$$

The area $A$ will cancel. We are told to assume that

$$p_g = p(V_0/V).$$

The electric field due to the spherically symmetric distribution of charge on the bubble is given by Eq. 14 of Chapter 29:

$$E = q/4\pi\epsilon_0R^2.$$

Substituting the expression for $p_g$ and multiplying by $V$ gives

$$pV_0 + \tfrac{1}{2}\epsilon_0E^2V = pV.$$

Now put in the expression for $E$; also, express the volumes in terms of the corresponding radii to get

$$p\left(\frac{4}{3}\pi R_0^3\right) + \tfrac{1}{2}\epsilon_0\left(\frac{q}{4\pi\epsilon_0R^2}\right)^2\left(\frac{4}{3}\pi R^3\right) - p\left(\frac{4}{3}\pi R^3\right).$$

Cancel the common factor of $4\pi/3$ and rearrange to solve for the charge; we obtain

$$q^2 = 32\pi^2\epsilon_0pR(R^3 - R_0^3).$$

31-48

The capacitance of the capacitor is

$$C = 68.4 \times 10^{-9} \text{ F} = \kappa_e \epsilon_0 A/d = 2.8\epsilon_0 A/d.$$

The dielectric strength is the maximum potential gradient possible without breakdown, so that

$$(V/d)_{max} = (4130 \text{ V})/d = 18.2 \times 10^6 \text{ V/m},$$

$$d = 2.269 \times 10^{-4} \text{ m}.$$

Therefore, from the first equation,

$$A = (68.4 \times 10^{-9} \text{ F})(2.269 \times 10^{-4} \text{ m})/[(2.8)(8.85 \times 10^{-12} \text{ F/m})],$$

$$A = 0.626 \text{ m}^2.$$


31-55

Let $q$, $-q$ be the charges on the plates separated by a distance $d$. If $V$ is the potential difference across the plates, then

$$C = q/V.$$

The uniform electric fields in the dielectrics each are $\sigma/\kappa_e \epsilon_0$, so that, in terms of the charge,

$$E_1 = q/\kappa_{e1}\epsilon_0 A, \quad E_2 = q/\kappa_{e2}\epsilon_0 A.$$

Now $V = V_1 + V_2 = E_1 d_1 + E_2 d_2$, and $d_1 = d_2 = d/2$, so that

$$C = \frac{q}{E_1(d/2) + E_2(d/2)} = \frac{2q/d}{q/\kappa_{e1}\epsilon_0 A + q/\kappa_{e2}\epsilon_0 A} = \frac{2A\epsilon_0}{d} \frac{\kappa_{e1}\kappa_{e2}}{\kappa_{e1} + \kappa_{e2}}.$$


31-57

(a) The free charge is $q = CV = (112 \times 10^{-12} \text{ F})(55 \text{ V}) = 6.16 \text{ nC}$. By Eq. 32, in the dielectric,

$$E = \sigma/\kappa_e \epsilon_0 = q/\kappa_e \epsilon_0 A,$$

$$E = (6.16 \times 10^{-9} \text{ C})/[(5.4)(8.85 \times 10^{-12} \text{ F/m})(96.5 \times 10^{-4} \text{ m}^2)],$$

$$E = 1.34 \times 10^4 \text{ V/m} = 13.4 \text{ kV/m}.$$

(b) See (a) above; $q = 6.16$ nC.

(c) By Eq. 36,

$$q' = q(1 - \frac{1}{\kappa_e}) = (6.16 \text{ nC})(1 - \frac{1}{5.4}) = 5.02 \text{ nC}.$$

## 31-62

Let the electric field in the gaps be $E$ and in the dielectric be $E_d$. Now $E_d = E/\kappa_e$ and therefore the potential difference across the plates is

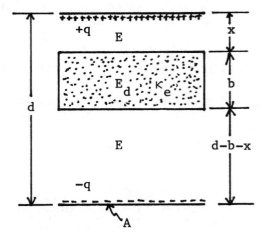

$$V = Ex + E_d b + E(d - b - x),$$

$$V = E(d - b) + E_d b,$$

$$V = E(d - b + b/\kappa_e).$$

But, for a parallel-plate capacitor,

$$E = \sigma/\epsilon_0 = q/A\epsilon_0,$$

$A$ the area of each plate. Thus,

$$V = \frac{q}{A\epsilon_0}[d + b(\frac{1}{\kappa_e} - 1)] = \frac{q}{A\epsilon_0\kappa_e}[\kappa_e d - b(\kappa_e - 1)],$$

and therefore the capacitance is

$$C = \frac{q}{V} = \frac{\kappa_e \epsilon_0 A}{\kappa_e d - b(\kappa_e - 1)}.$$

The cases $b = 0$ and $\kappa_e = 1$ each correspond to an air capacitor for which $C = \epsilon_0 A/d$ by the formula above, as expected. If $b = d$, the dielectric fills entirely the space between the plates, giving $C = \kappa_e \epsilon_0 A/d$, again as anticipated.

# CHAPTER 32

32-5

Use Eq. 1 to find the cross-sectional area A of the wire; we obtain

$$A = i/j = (0.552 \text{ A})/(440 \text{ A/cm}^2) = 1.255 \text{ X } 10^{-3} \text{ cm}^2,$$

$$A = 0.1255 \text{ mm}^2.$$

From the area we can find the wire diameter $d$, as follows,

$$A = \tfrac{1}{4}\pi d^2 = 0.1255 \text{ mm}^2,$$

$$d = 0.400 \text{ mm}.$$

32-13

By Eq. 1, the current density is

$$j = i/A = (115 \text{ A})/(31.2 \text{ X } 10^{-6} \text{ m}^2) = 3.686 \text{ X } 10^6 \text{ A/m}^2.$$

From Sample Problem 2, the number density of conduction electrons in copper is

$$n = 8.49 \text{ X } 10^{28} \text{ m}^{-3}.$$

The drift speed $v_d$ can now be calculated from Eq. 6:

$$v_d = j/ne,$$

$$v_d = (3.686 \text{ X } 10^6 \text{ A/m}^2)/[(8.49 \text{ X } 10^{28} \text{ m}^{-3})(1.6 \text{ X } 10^{-19} \text{ C})],$$

$$v_d = 2.713 \text{ X } 10^{-4} \text{ m/s}.$$

It is now straightforward to find the required time:

$$t = L/v_d = (0.855 \text{ m})/(2.713 \text{ X } 10^{-4} \text{ m/s}) = 3150 \text{ s} = 52.5 \text{ min}.$$

32-14

(a) The current carried is $i = 250 \text{ nA} = 250 \text{ X } 10^{-9}$ C/s. Since each particle carries a charge of $2e$, the number $n$ that strike in 2.9 s is given by

$$n = q/2e = it/2e,$$

$$n = [(250 \text{ X } 10^{-9} \text{ C/s})(2.9 \text{ s})]/[2(1.6 \text{ X } 10^{-19} \text{ C})],$$

$$n = 2.27 \text{ X } 10^{12}.$$

(b) Let the number of particles in a length $L$ of the beam be $N$. The beam current is

$$i = \frac{q}{t} = \frac{N(2e)}{t} = \frac{N(2e)}{L/v} = \frac{2evN}{L},$$

and therefore

$$N = \frac{iL}{2ev}.$$

We need to calculate the speed $v$ from the kinetic energy $K = \frac{1}{2}mv^2$. It is given that

$$K = (22.4 \text{ MeV})(1.6 \times 10^{-13} \text{ J/MeV}) = 3.584 \times 10^{-12} \text{ J}.$$

An alpha particle is a helium nucleus. From Appendix D, the molar mass of helium is $M = 4.0026$ g/mol. Disregarding the mass of the two electrons that a helium atom normally carries, we have for the mass of an alpha particle,

$$m = M/N_A = (4.0026 \times 10^{-3} \text{ kg/mol})/(6.02 \times 10^{23} \text{ /mol}),$$

$$m = 6.649 \times 10^{-27} \text{ kg}.$$

Therefore,

$$K = 3.584 \times 10^{-12} \text{ J} = \frac{1}{2}(6.649 \times 10^{-27} \text{ kg})v^2,$$

$$v = 3.283 \times 10^7 \text{ m/s}.$$

(Note that this is a nonrelativistic calculation.) Substituting this and the other numerical data into the equation above for $N$ yields

$$N = \frac{(250 \times 10^{-9} \text{ C/s})(0.18 \text{ m})}{2(1.6 \times 10^{-19} \text{ C})(3.283 \times 10^7 \text{ m/s})} = 4280.$$

(c) The required potential $V$ is given from

$$K = qV.$$

Expressing the charge as $q = 2e$, where $e$ is the elementary charge, we obtain the potential directly in electron volts,

$$22.4 \text{ MeV} = (2e)V \quad \rightarrow \quad V = 11.2 \text{ MV}.$$

It appears as though the "e" has cancelled! Explain.

## 32-17

The length unit in the resistivity $\rho$ is the meter m, so we must convert the length unit in the other data to m also. We have $A = 56$ cm$^2$ = 56 X $10^{-4}$ m$^2$ and $L = 11$ km = 11 X $10^3$ m. Eq. 13 yields for the resistance $R$,

$$R = \rho \frac{L}{A} = \frac{(3.0 \text{ X } 10^{-7} \text{ } \Omega\text{m}) (11,000 \text{ m})}{56 \text{ X } 10^{-4} \text{ m}^2} = 0.59 \text{ } \Omega.$$

## 32-20

The capacitance of a parallel-plate capacitor, by Eq. 7 of Chapter 31, is

$$C = \epsilon_0 A/d.$$

In calculating the resistance use Eq. 13, associating the length $L$ with the plate separation $d$ so that

$$R = \rho d/A.$$

Multiplying the two equations yields

$$RC = \rho \epsilon_0,$$

and therefore

$$R = \rho \epsilon_0 / C = (9.40 \text{ } \Omega \bullet \text{m})(8.85 \text{ pF/m})/(110 \text{ pF}) = 756 \text{ m}\Omega.$$

## 32-26

From Problem 6, the diameter of the wire is

$$d = (129 \text{ X } 10^{-3} \text{ in})(0.0254 \text{ m/in}) = 3.277 \text{ X } 10^{-3} \text{ m}.$$

Therefore, the cross-sectional area of the wire is

$$A = \tfrac{1}{4}\pi d^2 = \tfrac{1}{4}\pi(3.277 \text{ X } 10^{-3} \text{ m})^2 = 8.434 \text{ X } 10^{-6} \text{ m}^2.$$

The length $L$ of the wire must be

$$L = (250)(2\pi r) = (250)[2\pi(0.122 \text{ m})] = 191.6 \text{ m}.$$

From Table 1, we extract the value of the resistivity of copper:

$$\rho = 1.69 \text{ X } 10^{-8} \text{ } \Omega \bullet \text{m}.$$

Finally, the resistance of the coil (wire) follows directly from Eq. 13; we obtain

$$R = \rho\frac{L}{A} = (1.69 \times 10^{-8}\ \Omega\mathrm{m})\ \frac{191.6\ \mathrm{m}}{8.434 \times 10^{-6}\ \mathrm{m}^2} = 0.384\ \Omega.$$

## 32-27

Let i denote initial quantities. In order that the density of the material not be altered by the stretching, the volume of the wire must likewise remain unchanged (since the mass is unaffected by the process); i.e., $V = V_i$. If the length increases by a factor of 3 (meaning that $L = 3L_i$), then since $V = AL$ and $V_i = A_iL_i$, the cross-sectional area must decrease by a factor of 3; that is, $A = \frac{1}{3}A_i$. Hence, the new resistance $R$ is

$$R = \rho\frac{L}{A} = \rho\frac{3L_i}{\frac{1}{3}A_i} = 9\left(\rho\frac{L_i}{A_i}\right) = 9R_i = 9(6\ \Omega) = 54\ \Omega.$$

## 32-32

When operating, the resistance is

$$R = V/i = (2.9\ \mathrm{V})/(0.31\ \mathrm{A}) = 9.355\ \Omega.$$

Now use the result of Problem 21, for the temperature variation of resistance; i.e.,

$$R - R_0 = \alpha R_0(T - T_0).$$

Extracting the value of $\alpha$ for tungsten from Table 1, this equation gives

$$9.355 - 1.12 = (4.5 \times 10^{-3})(1.12)(T - T_0),$$

$$T - T_0 = 1630\ \mathrm{C}°,$$

and therefore $T = 20 + 1630 = 1650°\mathrm{C}$.

## 32-36

(a) Upward moving negative charges constitute a downward directed current. Therefore, the total current, in magnitude, is

$$j = (n_- + n_+)ev_d,$$

where, as instructed, we assumed that the drift speeds of the positive and negative ions are equal. By Eq. 12, in magnitude,

230

$$j = \sigma E.$$

Combining these equations gives

$$v_d = \frac{\sigma E}{(n_- + n_+)e}.$$

Recalling that $1 \text{ m}^3 = 1 \times 10^6 \text{ cm}^3$, we find, upon substitution of the numerical data, that $v_d = 1.73$ cm/s.
(b) The current density follows immediately from Eq. 12:

$$j = \sigma E = (2.7 \times 10^{-14} \text{ } /\Omega \bullet m)(120 \text{ V/m}) = 3.24 \times 10^{-12} \text{ A/m}^2.$$

(The units follow from $1 \text{ A} = 1 \text{ V}/\Omega$.) We could write the current density more compactly as $j = 3.24 \text{ pA/m}^2$.

## 32-39

Assume that the resistance of the conductor is the sum of the resistances of the iron (Fe) and carbon (C) sections:

$$R = R_{Fe} + R_c.$$

We presume, in the lack of information on this point, that the cross-section of the conductor is uniform (the same for the iron and carbon sections), so that applying Eq. 13, we have

$$R = (\rho_{Fe}L_{Fe} + \rho_c L_c)/A.$$

We want $\Delta R = 0$ as the temperature changes. Assuming that the changes in the dimensions of the sections are negligible, we have

$$\Delta\rho_{Fe}L_{Fe} + \Delta\rho_c L_c = 0.$$

Now apply Eq. 14 to each section. Since $\Delta T$ is assumed to be the same for the iron and carbon sections, we get

$$\rho_{0Fe}\alpha_{Fe}L_{Fe} + \rho_{0c}\alpha_c L_c = 0,$$

$$L_c/L_{Fe} = -(\rho_{0Fe}/\rho_{0c})(\alpha_{Fe}/\alpha_c) = 0.036.$$

The numerical result above follows after substituting the values of the parameters from Table 1.

## 32-40

(a) Consider a thin slice perpendicular to the axis of the cone, the slice having a thickness $dx$ and being situated a distance $x$ from the narrow end of the cone (see the sketch on p. 232). The cross-sectional area of this slice is $A = \pi r^2$ and its

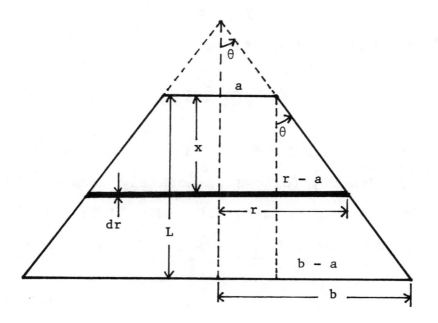

resistance is

$$dR = \rho dL/A = \rho dx/\pi r^2.$$

Let θ be the semiangle of the cone from which the resistor was cut. Then, from the diagram above we see that

$$\tan\theta = \frac{r - a}{x} = \frac{b - a}{L},$$

$$r = \left(\frac{b - a}{L}\right)x + a.$$

Thus, the resistance of the object is

$$R = \int dR = \frac{\rho}{\pi}\int_0^L \frac{dx}{\left[\left(\frac{b - a}{L}\right)x + a\right]^2} = \frac{\rho L}{\pi ab}.$$

(b) For zero taper a = b, and the formula derived above gives R = ρL/πa² = ρL/A, as expected.

32-41

(a) From the given formula, the potential difference is

232

$$V = 3.55(2.4)^2 = 20.45 \text{ V.}$$

The resistance follows from Eq. 8:

$$R = V/i = (20.45 \text{ V})/(0.0024 \text{ A}) = 8.52 \text{ k}\Omega.$$

(b) If now we want $i$ to represent current in amperes (A), then the given equation for the potential difference must be written as

$$V = 3.55(i \text{ X } 10^3)^2 = 3.55 \text{ X } 10^6 \, i^2,$$

where now $i$ is the current in A. By Eq. 8, the potential difference when the resistance is 16 $\Omega$ must be related to the current by

$$V = iR = i(16).$$

Hence,

$$V = 16i = 3.55 \text{ X } 10^6 \, i^2,$$

$$i = 4.51 \ \mu\text{A.}$$

## 32-45

Combine Eqs. 21 and 1 to obtain

$$P = iV = qV/t.$$

We have assumed that the current was constant. Since 1 W = 1 J/s, we need the time in seconds; i.e., $t$ = 6 h = (6 h)(3600 s/h) = 2.16 X $10^4$ s. Therefore, the charge is

$$q = Pt/V = (7.5 \text{ J/s})(2.16 \text{ X } 10^4 \text{ s})/(9 \text{ V}) = 18 \text{ kC.}$$

We used the unit equivalence 1 V = 1 J/C.

## 32-50

For the two temperatures 200°C and 800°C, Eq. 23 gives

$$P_8 = V_8^2/R_8, \quad P_2 = V_2^2/R_2.$$

But $V_2 = V_8$ (= 110 V), so that

$$P_8/P_2 = R_2/R_8.$$

Now $R = \rho L/A$; if we can ignore changes in $L$ and $A$ with changes in temperature, then

$$R_2/R_8 = \rho_2/\rho_8,$$

and therefore

$$P_8/P_2 = \rho_2/\rho_8.$$

By Eq. 14,

$$\rho_2 - \rho_8 = \rho_8\alpha(T_2 - T_8),$$

$$\rho_2/\rho_8 = 1 + \alpha(T_2 - T_8) = 1 + (0.0004)(200 - 800) = 0.76.$$

Therefore, by the third equation,

$$P_2 = P_8(\rho_8/\rho_2) = (500 \text{ W})/0.76 = 660 \text{ W}.$$

## 32-54

The heat $Q$ needed to vaporize a mass $m$ of the liquid follows from Eq. 7 of Chapter 25:

$$Q = mL_v.$$

The rate at which heat must be delivered to obtain a rate of evaporation $dm/dt$ is

$$dQ/dt = P = (dm/dt)L_v.$$

But $P = iV$ by Eq. 21, so that

$$iV = (dm/dt)L_v,$$

$$(5.2 \text{ A})(12 \text{ V}) = (21 \text{ X } 10^{-6} \text{ kg/s})L_v,$$

$$L_v = 3.0 \text{ X } 10^6 \text{ J/kg}.$$

## 32-60

(a) By Eq. 1 of Chapter 31, $q = CV$. Evaluate the potential $V$ at $t = 0.5$ s by substituting 0.5 s for $t$ in the formula for $V(t)$; one finds that $V = 7.5$ V. Hence,

$$q = CV = (32 \text{ X } 10^{-6} \text{ F})(7.5 \text{ V}) = 2.4 \text{ X } 10^{-4} \text{ C} = 240 \text{ } \mu\text{C}.$$

(b) The capacitance does not change as the potential varies with time. Therefore,

$$i = dq/dt = C(dV/dt).$$

Differentiating $V(t)$ we obtain

$$dV/dt = 4 - 4t.$$

Thus at $t = 0.5$ s, $dV/dt = 2$ V/s. Therefore, at this instant, the current is

$$i = C(dV/dt) = (32 \ \mu F)(2 \ V/s) = 64 \ \mu A.$$

(c) Use Eq. 21 and substitute the values of the current and potential difference at $t = 0.5$ s found above; we obtain

$$P = iV = (64 \times 10^{-6} \ A)(7.5 \ V) = 4.8 \times 10^{-4} \ W = 480 \ \mu W.$$

33-3

The work $W$ done in time $t$ is $W = Pt$. By Eq. 1, then,

$$W = Pt = q\mathscr{E},$$

$$(110 \text{ W})t = (125 \text{ A}\bullet\text{h})(12 \text{ V}),$$

$$(110 \text{ J/s})t = (125 \text{ C}\bullet\text{h/s})(12 \text{ V}).$$

Since 1 J = 1 C$\bullet$V, this becomes

$$110t = (125 \text{ h})(12),$$

$$t = 13.64 \text{ h} = 13 \text{ h } 38 \text{ min}.$$

33-7

(*a*) The resistors are in series. The batteries are arranged with their polarities in the opposite sense with $\mathscr{E}_2 > \mathscr{E}_1$. Note that 1 mA = 0.001 A. By the loop rule,

$$\mathscr{E}_2 - \mathscr{E}_1 - ir_1 - iR - ir_2 = 0,$$

$$i(r_1 + r_2 + R) = \mathscr{E}_2 - \mathscr{E}_1,$$

$$(0.050)(3 + 3 + R) = 3 - 2,$$

$$R = 14 \text{ }\Omega.$$

(*b*) The rate at which internal energy is produced is

$$P = i^2R = (0.050 \text{ A})^2(14 \text{ }\Omega) = 0.035 \text{ W} = 35 \text{ mW}.$$

33-10

(*a*) Internal energy is generated at the rate of

$$P = i^2R,$$

so we have

$$9.88 \text{ W} = i^2(0.108 \text{ }\Omega),$$

$$i = 9.565 \text{ A}.$$

By Eq. 3,

$$i(R + r) = \mathcal{E},$$

$$(9.565)(0.108 + r) = 1.50,$$

$$r = 0.0488 \ \Omega = 48.8 \ \text{m}\Omega.$$

(b) By Eq. 23 of Chapter 32,

$$P = V^2/R,$$

$$9.88 \ \text{W} = V^2/(0.108 \ \Omega) \quad \rightarrow \quad V = 1.03 \ \text{V}.$$

## 33-14

(a) The current in the circuit is $i = \mathcal{E}/(r + R)$, so that the rate of internal energy production is

$$P = i^2R = \frac{\mathcal{E}^2 R}{(r + R)^2}.$$

To find the value of $R$ that maximizes $P$, set $dP/dR = 0$ and solve the resulting equation for $R$, as follows:

$$\frac{dP}{dR} = 0 = \mathcal{E}^2 \frac{r - R}{(r + R)^3} \quad \rightarrow \quad R = r.$$

(b) The maximum rate at which internal energy is developed is

$$P(R = r) = \mathcal{E}^2 r/(r + r)^2 = \mathcal{E}^2/4r.$$

## 33-16

The equivalent resistance is given from Eq. 9:

$$\frac{1}{R_{eq}} = \frac{1}{R} + \frac{1}{R} + \frac{1}{R} + \frac{1}{R} = \frac{4}{R},$$

$$R_{eq} = R/4 = (18 \ \Omega)/4 = 4.5 \ \Omega.$$

Thus, the current through the battery is

$$i = \mathcal{E}/R_{eq} = (27 \ \text{V})/(4.5 \ \Omega) = 6 \ \text{A}.$$

33-20

For $n$ equal resistances $R$ connected in parallel, Eq. 9 for the equivalent resistance reduces to

$$R_{eq} = R/n.$$

Therefore, the current in the fuse is

$$i_{fuse} = \mathscr{E}/R_{eq} = n\mathscr{E}/R.$$

Now let $P$ be the power of each lamp. Since the lamps are wired in parallel, the potential drop across each is equal to the potential of the power line, so that for each lamp

$$R = V^2/P = \mathscr{E}^2/P,$$

and therefore

$$i_{fuse} = nP/\mathscr{E}.$$

Inserting the numbers, we find

$$15 \text{ A} = n(500 \text{ W})/(120 \text{ V}) \rightarrow n = 3.6,$$

or a maximum of 3 operating lamps.


33-24

(a) The resistors labeled 2, 3, and 4 are in parallel. By Eq. 9, the equivalent resistance of this set is found to be 18.7 Ω. The resistor $R_1$ = 112 Ω and the 18.7 Ω resistor that represents the parallel set are in series; hence, by Eq. 15, $R_{eq}$ = 112 Ω + 18.7 Ω = 131 Ω is the equivalent resistance of the four resistors.
(b) Label each current by the resistor through which that current passes. We have immediately

$$i_1 = \mathscr{E}/R_{eq} = (6.22 \text{ V})/(131 \text{ Ω}) = 47.48 \text{ mA}.$$

The potential drops across $R_2$, $R_3$, $R_4$ are equal (resistors in parallel); therefore,

$$i_2 R_2 = i_3 R_3,$$

$$i_2 R_2 = i_4 R_4.$$

Substituting the values of the resistances, we find that

$$i_3 = 0.6818 i_2,$$

$$i_4 = 0.5600 i_2.$$

By the junction rule,

$$i_1 = i_2 + i_3 + i_4.$$

We found the numerical value of $i_1$ above; also we have the currents $i_3$ and $i_4$ in terms of $i_2$. Putting all these into the junction equation above yields

$$47.48 = i_2 + 0.6818i_2 + 0.5600i_2,$$

$$i_2 = 21.18 \text{ mA}.$$

Using the equations derived from setting the potential drops across the parallel set equal, we find $i_3 = 14.44$ mA and $i_4 = 11.86$ mA.

## 33-31

(a) The brighter bulb indicates a higher value of the power $P = V^2/R$. In a parallel connection, $V_1 = V_2$. Since $R_2 < R_1$ though, we will have $P_2 > P_1$, so that bulb #2 is brighter.
(b) When connected in series the currents are equal: $i_1 = i_2$. Hence, to compare the powers, it is easiest to use the relation $P = i^2R$. With $R_1 > R_2$, we see that bulb #1 will be the brighter.

## 33-33

Let the currents be directed as shown, with the currents already arranged and labeled to satisfy the junction rule. Applying the loop rule to sector I, we find

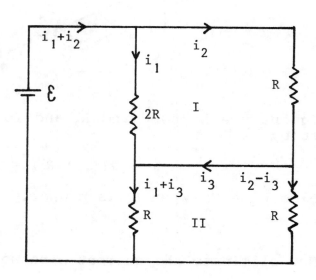

$$i_1(2R) - i_2(R) = 0,$$

$$i_2 = 2i_1.$$

Applied to sector II, the loop rule gives

$$(i_1 + i_3)R - (i_2 - i_3)R = 0,$$

$$i_2 - i_1 = 2i_3.$$

Eliminating $i_1$ between the resulting two current equations above gives $i_2 = 4i_3$. But, the loop rule applied to the outside (boundary) loop yields

$$\mathscr{E} - i_2R - (i_2 - i_3)R = 0.$$

Finally, putting the relation $i_2 = 4i_3$ into this equation yields the ammeter reading $i_3 = \mathscr{E}/7R$.

With the voltmeter removed, we are left with a single-loop circuit with all the resistors in series. The current is

$$i = \mathscr{E}/R_{eq},$$

where, by Eq. 15, $R_{eq}$ is the sum of all the resistances present. Without the ammeter, $R_{eq} = 2 + 5 + 4 = 11\ \Omega$. Hence $i = (5\ \text{V})/(11\ \Omega) = 0.45455\ \text{A}$. With the ammeter, $R_{eq} = 11 + 0.1 = 11.1\ \Omega$ and the current will be $i = (5\ \text{V})/(11.1\ \Omega) = 0.45045\ \text{A}$. Therefore, the percent error committed in ignoring the ammeter resistance is

$$\text{percenterror} = \frac{0.45455 - 0.45045}{0.45455}\ (100\%) = 0.90\%.$$

33-46

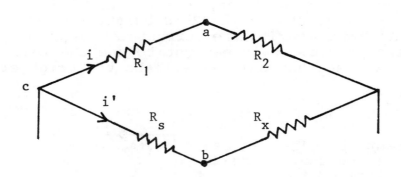

Ignore the battery and $R_0$ and focus on the rhombus. By the loop rule, we have

$$i(R_1 + R_2) = i'(R_x + R_s).$$

Also $V_{ac} = V_{bc}$ when $R_s$ is properly adjusted, so that

$$iR_1 = i'R_s.$$

Now eliminate the current $i$ between these two equations to find

$$i'R_s(R_1 + R_2) = i'R_1(R_x + R_s),$$

$$R_x = R_s(R_2/R_1).$$

33-49

The equilibrium charge is $C\mathscr{E}$. We want to find the time $t$ when the charge $q = (1 - 0.01)C\mathscr{E} = 0.99C\mathscr{E}$. By Eq. 31,

$$0.99 C\mathscr{E} = C\mathscr{E}(1 - e^{-t/\tau_c}),$$

$$-t/\tau_c = \ln(0.01),$$

$$t = 4.61\tau_c.$$

## 33-50

(a) The potential across a capacitor, by Eq. 1 of Chapter 31, is given by $V = q/C$. Therefore, by Eq. 31,

$$V = \mathscr{E}(1 - e^{-t/\tau_c}).$$

Therefore, at $t = 1.28$ $\mu$s,

$$5 \text{ V} = (13 \text{ V})(1 - e^{-1.28/\tau_c}),$$

$$\tau_c = 2.64 \ \mu\text{s}.$$

(b) Use Eq. 33:

$$C = \tau_c/R = (2.64 \times 10^{-6} \text{ s})/(15.2 \times 10^3 \ \Omega) = 174 \text{ pF}.$$

## 33-53

The lamp and the capacitor are in parallel so that at breakdown we will have $V_L = V_C = 72$ V. But

$$V_C = \mathscr{E}(1 - e^{-t/\tau_c}),$$

so that at $t = 0.5$ s,

$$72 \text{ V} = (95 \text{ V})(1 - e^{-0.5/RC}),$$

$$RC = 0.3525 \text{ s} = R(0.15 \times 10^{-6} \text{ F}),$$

$$R = 2.35 \times 10^6 \ \Omega = 2.35 \text{ M}\Omega.$$

(a) The time constant is $\tau_c = RC = (3 \text{ M}\Omega)(1 \text{ }\mu F) = 3$ s. The charge and current are given by Eqs. 31 and 32. By Eq. 32, we have at time $t = 1$ s,

$$dq/dt = i = (4 \text{ V}/3 \text{ M}\Omega)e^{-\frac{1}{3}} = 0.95538 \text{ }\mu A.$$

(b) The rate at which energy is being stored in the capacitor is, by Eq. 26 of Chapter 31,

$$dU_c/dt = d(q^2/2C)/dt = iq/C.$$

Now $q(1) = (1 \text{ }\mu F)(4 \text{ V})(1 - e^{-\frac{1}{3}}) = 1.1339 \text{ }\mu C$; the current $i(1)$ is found in (a) and therefore

$$dU_c/dt = (0.95538 \text{ }\mu A)(1.1339 \text{ }\mu C)/(1 \text{ }\mu F) = 1.0833 \text{ }\mu W.$$

(c) For the resistor,

$$dU_R/dt = i^2R = (0.95538 \text{ }\mu A)^2(3 \text{ M}\Omega) = 2.7383 \text{ }\mu W.$$

(d) Finally, the battery delivers energy at the rate

$$dU_B/dt = i\mathscr{E} = (0.95538 \text{ }\mu A)(4 \text{ V}) = 3.8215 \text{ }\mu W.$$

Note that, within round-off error,

$$dU_B/dt = dU_c/dt + dU_R/dt,$$

as required by the loop rule.

# CHAPTER 34

## 34-2

(a) By Eq. 6, the minimum possible force is $F = 0$, which arises if the angle $\phi = 0°$ or $180°$. For the maximum force put $\phi = 90°$ to get

$$F_{max} = evB\sin 90° = evB,$$

$$F_{max} = (1.6 \times 10^{-19} \text{ C})(7.2 \times 10^{6} \text{ m/s})(0.083 \text{ T}) = 95.6 \text{ fN},$$

where $1 \text{ fN} = 1 \times 10^{-15} \text{ N}$.
(b) Use Eq. 6 and Newton's second law to find

$$F = evB\sin\phi = ma,$$

$$a = (e/m)vB\sin\phi.$$

An electron has mass $m = 9.11 \times 10^{-31}$ kg. Substituting this and the other data into the equation above gives $\phi = 28°$.

## 34-8

The mass of the ion is

$$m = (6.01 \text{ u})(1.66 \times 10^{-27} \text{ kg/u}) = 9.977 \times 10^{-27} \text{ kg}.$$

Find the speed from

$$K = qV,$$

$$\tfrac{1}{2}(9.977 \times 10^{-27} \text{ kg})v^2 = (1.6 \times 10^{-19} \text{ C})(10.8 \times 10^{3} \text{ V}),$$

$$v = 5.886 \times 10^{5} \text{ m/s}.$$

Eq. 10 gives the zero-deflection condition:

$$E = vB = (5.886 \times 10^{5} \text{ m/s})(1.22 \text{ T}) = 7.18 \times 10^{5} \text{ V/m}.$$

## 34-12

(a) By the work-energy relation $K = qV$, we have

$$(1.6 \times 10^{-19} \text{ C})(1220 \text{ V}) = \tfrac{1}{2}(9.11 \times 10^{-31} \text{ kg})v^2,$$

$$v = 2.07 \times 10^{7} \text{ m/s}.$$

(b) By Eq.14,

$$B = mv/er,$$

$$B = \frac{(9.11 \times 10^{-31} \text{ kg})(2.07 \times 10^{7} \text{ m/s})}{(1.6 \times 10^{-19} \text{ C})(0.247 \text{ m})} = 477 \text{ } \mu\text{T}.$$

(c) Find the frequency from Eq. 16:

$$\nu = \frac{eB}{2\pi m} = \frac{v}{2\pi r} = \frac{2.07 \times 10^{7} \text{ m/s}}{2\pi(0.247 \text{ m})} = 13.3 \text{ MHz}.$$

(d) The period $T$ is related to the frequency $\nu$ by

$$T = 1/\nu = 1/(1.33 \times 10^{7} \text{ s}^{-1}) = 75.2 \text{ ns}.$$

## 34-17

The radius of the circular path, by Eq. 14, is

$$r = \frac{mv}{qB'},$$

where we write $q$ for the magnitude of the charge; $B'$ is the magnetic field that causes the ions to move in the circular path. The speed $v$ of the ions that pass through the velocity selector is given by $v = E/B$ (see Eq. 10), where $B$ is the magnetic field in the velocity selector. Substitute this $v$ into the equation for the radius of the path to find

$$r = \frac{mE}{qBB'};$$

rearranging this gives

$$\frac{q}{m} = \frac{E}{rBB'},$$

as asserted.

## 34-22

The ion enters the spectrometer with a speed $v$ related to the accelerating potential $V$ by Eq. 34 of Chapter 30, $W = K = qV$, so that

$$\tfrac{1}{2}mv^2 = qV.$$

Inside the instrument the ion undergoes uniform circular motion with the speed $v$ remaining constant. By Newton's second law

$$F = qvB = mv^2/r,$$

the angle $\phi$ in Eq. 6 being 90°. Now, from the first equation above, we have $v^2 = 2qV/m$. Also $r = x/2$ (see Fig. 32). Putting these relations into the equation above gives

$$\frac{m\sqrt{2qV/m}}{x/2} = qB,$$

$$m = \frac{B^2qx^2}{8V}.$$

## 34-24

(a) Our data consists of

$$m = (238\ u)(1.66 \times 10^{-27}\ kg/u) = 3.950 \times 10^{-25}\ kg,$$

$$q = 2e = 2(1.6 \times 10^{-19}\ C) = 3.2 \times 10^{-19}\ C,$$

$$V = 105\ kV = 105 \times 10^3\ V.$$

Also, from Fig. 32,

$$x = 2r = 2(0.973\ m) = 1.946\ m.$$

Now substitute these quantities into the result quoted in Problem 22 and solve for the magnetic field strength; we find $B = 0.523$ T. (b) The number $N$ of uranium ions collected in one hour is

$$N = (90 \times 10^{-6}\ kg)/(3.95 \times 10^{-25}\ kg/ion) = 2.278 \times 10^{20}.$$

The total charge on these ions is

$$Q = Nq = (2.278 \times 10^{20})(3.2 \times 10^{-19}\ C) = 72.90\ C,$$

and therefore the current is

$$i = Q/t = (72.90\ C)/(3600\ s) = 20.2\ mA.$$

(c) The internal energy arises from the kinetic energy of the ions deposited in the cup. The magnetic field does not change the kinetic energy given to the ions by the accelerating potential. Therefore, the internal energy delivered in one hour is

$$E = N(\tfrac{1}{2}mv^2) = N(qV) = QV = (72.9 \text{ C})(0.105 \text{ MV}) = 7.65 \text{ MJ}.$$

## 34-27

(a) Use Eq. 14 in the form $r = p/qB$, where $p$ is the relativistic momentum given in Eq. 23 of Chapter 21; we obtain

$$r = \frac{mv/qB}{\sqrt{1 - v^2/c^2}}.$$

Rearranging this to solve for the speed $v$ yields

$$v/c = \frac{qBr/mc}{\sqrt{1 + (qBr/mc)^2}}.$$

Now insert the numerical data to verify that

$$qBr/mc = 83.408.$$

Using this in the previous equation we find $v/c = 0.9999281$ so that $v = 2.999784 \times 10^8$ m/s.
(b) The arrangement of the vectors is shown on the sketch. Their directions must satisfy the right-hand rule; see Fig. 4. In actuality, it is the magnetic south pole that is near the north geographic pole: see Fig. 11 in Chapter 37. We label the geographic poles in our sketch.

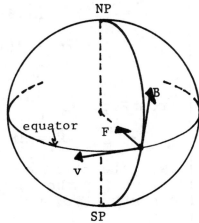

## 34-29

Use Eq. 14 but substitute the relativistic momentum $p$; since $q = 2e$ we obtain

$$r = \frac{mv}{2eB\sqrt{1 - (v/c)^2}}.$$

We have $v = 0.71c = 0.71(3 \times 10^8 \text{ m/s}) = 2.13 \times 10^8$ m/s. Also $B = 1.33$ T and $r = 4.72$ m. The equation above can now be solved for the mass $m$ of the particle, with the result $m = 6.641 \times 10^{-27}$ kg. Expressing this in terms of the atomic mass unit $u = 1.66 \times 10^{-27}$ kg, we find $m = 4.00$ u. We recall that an alpha particle (helium nucleus) has a mass of 4.00 u and charge $2e$.

## 34-33

We need to calculate the final deuteron energy. Combining Eqs. 16 and 18 we find

$$K = 2m(\pi \nu R)^2.$$

The mass of the deuteron is $m = 2$ u $= 2(1.66 \times 10^{-27}$ kg$)$; we also have $\nu = 12 \times 10^6$ Hz and $R = 0.53$ m. Putting these into the equation above yields

$$K = 2.65 \times 10^{-12} \text{ J} = 16.6 \text{ MeV}.$$

There are two "kicks" per revolution; each kick imparts an additional kinetic energy

$$\Delta K = qV = (e)(80 \text{ kV}) = 80 \text{ keV}.$$

Therefore the number of revolutions completed during the acceleration process is

$$\#\text{rev} = (16.6 \text{ MeV})/(0.160 \text{ MeV/rev}) = 104.$$

Now consider a single kick. As the deuteron passes from one dee to the other its speed increases from $v$ to $v'$, say, and the radius of its semicircular path increases from $r$ to $r'$. These quantities are related by

$$\tfrac{1}{2}mv^2 + qV = \tfrac{1}{2}mv'^2,$$

$$\tfrac{1}{2}m(r\omega)^2 + qV = 2\pi^2 m \nu^2 r^2 + qV = 2\pi^2 m \nu^2 r'^2,,$$

$$r'^2 = r^2(1 + qV/2\pi^2 m \nu^2 r^2).$$

If we can write $r' = r + \Delta r = r(1 + \Delta r/r)$, with $\Delta r/r \ll 1$, then by the BINOMIAL THEOREM (see Appendix H)

$$r'^2 \approx r^2(1 + 2\Delta r/r).$$

Comparison of the last two equations suggests that $\Delta r \propto 1/r$. We see that the increase in orbit radius per kick is largest at small radii and decreases toward the edge of the dees. Thus the deuterons perform more of the 104 revolutions at radii greater than the average radius, $\approx(0.53 \text{ m})/2$, than at smaller radii. As an actual average radius, then, we pick $\tfrac{3}{4}(0.53 \text{ m}) = 0.40$ m, for which the circumference of the corresponding circle is $2\pi(0.40 \text{ m}) = 2.5$ m.

Since the deuteron makes 104 revolutions, the path traveled must be approximately (104)(2.5 m) = 260 m. This answer is approximate because we did not make a careful calculation of the actual average radius.

## 34-39

Charge carriers move with speed $v$ through the magnetic field not because of an electric field set up by a battery, but because of the motion of the strip itself. Hence, we associate the speed of the strip $v$ with the drift speed $v_d$ of the electrons in Section 4. We still have, by Eq. 22,

$$v_d = v = E/B.$$

In terms of the Hall potential difference $V$, the Hall electric field is given by $E = V/w$, where $w$ is the width of the strip. Therefore,

$$v = V/wB = (3.9 \times 10^{-6} \text{ V})/[(0.0088 \text{ m})(0.0012 \text{ T})],$$

$$v = 0.37 \text{ m/s} = 37 \text{ cm/s}.$$

## 34-43

(a) With the rail horizontal Eq. 29 applies, so we get

$$F = iLB,$$

$$10 \times 10^3 \text{ N} = i(3 \text{ m})(10 \times 10^{-6} \text{ T}),$$

$$i = 3.3 \times 10^8 \text{ A} = 330 \text{ MA}.$$

(b) For $R = 1 \Omega$ the power loss (rate of appearance of internal energy) is

$$P = i^2R = (3.3 \times 10^8 \text{ A})^2(1 \Omega) = 1.1 \times 10^{17} \text{ W}.$$

(c) When we realize that the internal energy in (b) will appear as heat in the rail, it is difficult to see what the rail can be made of to withstand this rate of heat production. Regretfully, we must judge this project as being totally unrealistic.

## 34-46

Let **B** make an angle $\alpha$ with the vertical. The magnetic force is $F = iLB$, since the magnetic field is at 90° to the rod. For the rod to be on the verge of sliding we must have

$$iLB\cos\alpha - \mu_sN = 0,$$

where $\mu_s$ is the coefficient
of static friction and $N$ is
the normal force exerted by
the rails on the rod. In the
vertical direction the forces
also must sum to zero:

$N + iLB\sin\alpha - mg = 0.$

Eliminating the normal force
between these two equations
gives

$iLB\cos\alpha - \mu_s(mg - iLB\sin\alpha) = 0.$

Solving for the magnetic field
we get

$$B = \frac{\mu_s mg/iL}{\cos\alpha + \mu_s\sin\alpha}.$$

To find the minimum field needed, set $dB/d\alpha = 0$; this gives

$$\alpha = \tan^{-1}\mu_s = \tan^{-1}(0.58) = 30.11°.$$

Numerically,

$\mu_s mg/iL = (0.58)(1.15 \text{ kg})(9.8 \text{ m/s}^2)/(53.2 \text{ A})(0.95 \text{ m}) = 129.3 \text{ mT}.$

Using the equation for $B$ above with the value of $\alpha$ just found
yields $B = 112$ mT.


34-48

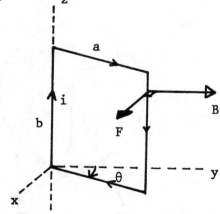

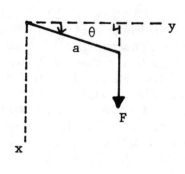

Let the hinge lie along the z axis. It is evident that the only force that can exert a torque along the hinge is the one on the side of the rectangle opposite the hinge. (The forces on the 5 cm sides are parallel to the z axis and, by the cross-product rule involved in the definition of torque, their torques are at 90° to the z axis.) The magnitude of the force F is

$$F = ibB\sin 90° = ibB.$$

From the sketch, the moment arm is seen to be $a\cos\theta$, so that the torque for N loops is

$$\tau = N(ibB)(a\cos\theta),$$

$$\tau = (20)(0.1 \text{ A})(0.12 \text{ m})(0.5 \text{ T})(0.05 \text{ m})\cos 33° = 0.00503 \text{ N•m}.$$

From the right-hand rule, it is seen that this is in the $-z$ direction, that is, down.

## 34-51

If N closed loops are formed from the wire of length L, each will have a circumference of $L/N$, a radius of $L/2\pi N$, and an area

$$A = \pi(L/2\pi N)^2 = L^2/4\pi N^2.$$

For maximum torque, orient the planes of the loops parallel to the magnetic field lines, so that $\theta = 90°$ in Eq. 34. In this case,

$$\tau = NiAB = Ni(L^2/4\pi N^2)B = iL^2B/4\pi N.$$

Since N appears in the denominator, the maximum torque occurs for minimum N, i.e., $N = 1$ so that $\tau_{max} = iL^2B/4\pi$.

## 34-55

If the cylinder rolls, the instantaneous axis of rotation will pass through P, the point on the rim in contact with the plane (see Fig. 24 in Chapter 12). Neither the normal force N nor the friction force f will exert any torque about P since their lines of action pass through P. The torque due to gravity is

$$\tau_g = mgr = mgR\sin\theta,$$

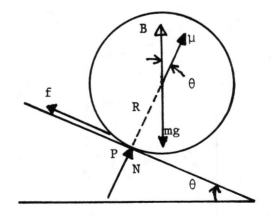

directed into the page. The torque due to the magnetic field on the current loop is

$$\tau_m = \mu B \sin\theta = NiAB\sin\theta = Ni(2RL)B\sin\theta,$$

directed out of the page. For no rotation, the net torque must vanish, so that

$$2NiRLB\sin\theta = mgR\sin\theta,$$

$$i = \frac{mg}{2NLB} = \frac{(0.262 \text{ kg})(9.8 \text{ m/s}^2)}{2(13)(0.477 \text{ T})(0.127 \text{ m})} = 1.63 \text{ A}.$$

## 34-58

(a) The area of a circle is $A = \pi r^2$, so Eq. 36 yields

$$\mu = NiA = Ni[\pi r^2] = (1)(2.58 \text{ A})[\pi(0.16 \text{ m})^2] = 0.207 \text{ A} \bullet \text{m}^2.$$

(b) Taking the magnitude of Eq. 37, we have

$$\tau = \mu B \sin\theta = (0.207 \text{ A} \bullet \text{m}^2)(1.20 \text{ T})\sin 41° = 0.163 \text{ N} \bullet \text{m}.$$

(Although 1 N•m = 1 J, it is not usual to express a torque in joules.)

### 35-3

Eq. 11 applies since the value of the magnetic field is sought at a point outside the confines of the current; specifically, at a distance $R$ = 1.5 mm from the axis of the beam. (This point is outside the beam, which constitutes the current, since the radius of the beam is only 0.11 mm.) The current $i$ must first be computed; we have

$$i = q/t = [ne]/t = [(5.6 \times 10^{14})(1.6 \times 10^{-19} \text{ C})]/(1 \text{ s}),$$

$$i = 8.96 \times 10^{-5} \text{ A}.$$

By Eq. 11, the magnetic field is

$$B = \frac{\mu_0 i}{2\pi R} = \frac{(4\pi \times 10^{-7} \text{ Tm/A})(8.96 \times 10^{-5} \text{ A})}{2\pi(1.5 \times 10^{-3} \text{ m})} = 12 \text{ nT}.$$

### 35-7

The currents must flow in opposite directions. The line midway between the wires is at $x$ = 0 in the equation for $B$ in Sample Problem 1. We have $d$ = 4.05 cm = 0.0405 m. Hence,

$$B = \mu_0 i/\pi d,$$

$$i = \pi B d/\mu_0 = [\pi(296 \times 10^{-6})(0.0405)]/(4\pi \times 10^{-7}) = 30.0 \text{ A}.$$

### 35-11

(a) The area of the coil is $A = \frac{1}{4}\pi d^2 = \frac{1}{4}\pi(0.048 \text{ m})^2 = 1.810 \times 10^{-3}$ m². By Eq. 36 of Chapter 34, the magnetic moment is

$$\mu = NiA = (320)(4.2 \text{ A})(1.81 \times 10^{-3} \text{ m}^2) = 2.43 \text{ A} \cdot \text{m}^2.$$

(b) Eq. 17 applies here. Set $B$ = 5 $\mu$T and $\mu$ = 2.43 A•m² (the $\mu$'s are different) and solve for $z$; we find

$$B = \frac{\mu_0}{2\pi} \frac{\mu}{z^3},$$

$$5 \times 10^{-6} = (2 \times 10^{-7})(2.43)/z^3,$$

$$z = 0.46 \text{ m}.$$

See p. 763 of HRK where the field due to an infinitely long current-carrying wire is determined. For a wire of finite length $L$, it is only necessary to replace the limits on the integral of $\pm\infty$ with $\pm L/2$ (the origin $x = 0$ is at the midpoint of the wire). With this done we find

$$B = \frac{\mu_0 i}{4\pi}\int_{-L/2}^{L/2}\frac{R\,dx}{(x^2 + R^2)^{3/2}} = \frac{\mu_0 i}{2\pi R}\frac{L}{\sqrt{L^2 + 4R^2}}.$$

If $L \to \infty$, we can ignore the $4R^2$ in the denominator, obtaining

$$B = \mu_0 i/2\pi R,$$

which is Eq. 11, the field due to a very long wire.

## 35-20

Use the result of Problem 18, but put $z = 0$ to get the field at the center of the square and set $a = \frac{1}{4}L$ (make one square out of the wire with no wire left over). We obtain

$$B_{sq} = 8\sqrt{2}(\mu_0 i/\pi L) = 3.60(\mu_0 i/L).$$

For the circle, use Eq. 16; since $L = 2\pi R$, this becomes

$$B_{ci} = \pi(\mu_0 i/L) = 3.14(\mu_0 i/L).$$

Thus, as asserted, the square yields the larger magnetic field at its center.

## 35-21

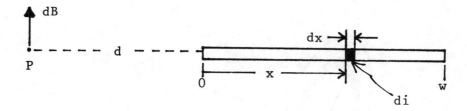

Adopt the *Hint*. Consider the thin strip of width $dx$; it carries a current $di$ given by

$$di = i(dx/w),$$

i.e., the current in the ribbon times the fraction of the area of the ribbon occupied by the strip. We put the origin $x = 0$ at the left edge of the ribbon. The magnetic field $dB$ due to the thin strip, which forms a very long thin wire in effect, is given by Eq. 11, if we replace $i$ with $di$ and $R$ with $d + x$; that is,

$$dB = \frac{\mu_0 (i/w)\, dx}{2\pi (d + x)}.$$

Find the total magnetic field by integrating this result over the ribbon; one finds

$$B = \int dB = \frac{\mu_0 i}{2\pi w}\int_0^w \frac{dx}{d + x} = \frac{\mu_0 i}{2\pi w}\ln\left(\frac{d + w}{d}\right).$$

The field is directed "up", as can be seen by application of the right-hand rule.

## 35-28

(a) The field is given by Eq. 16:

$$B = \tfrac{1}{2}\mu_0 i/R = \tfrac{1}{2}(4\pi \times 10^{-7}\ \text{T}\bullet\text{m/A})(13\ \text{A})/(0.12\ \text{m}) = 68\ \mu\text{T}.$$

(b) The magnetic moment of the small loop is

$$\mu = Ni'A = (50)(1.3\ \text{A})[\pi(0.82 \times 10^{-2}\ \text{m})^2] = 0.0137\ \text{A}\bullet\text{m}^2.$$

In Eq. 34 of Chapter 34, the angle $\theta = 90°$ so that the torque on the small loop is

$$\tau = \mu B\sin 90° = (0.0137\ \text{A}\bullet\text{m}^2)(68 \times 10^{-6}\ \text{T}) = 9.3 \times 10^{-7}\ \text{N}\bullet\text{m}.$$

## 35-32

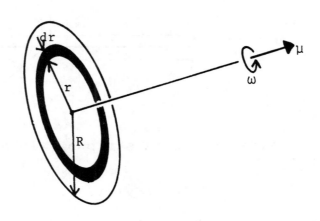

(*a*) Consider a narrow ring of radius *r* and width *dr*; it carries charge

$$dq = q(2\pi r dr/\pi R^2);$$

i.e., the charge on the disk times the fraction of the area of the disk occupied by the ring. In time $T = 2\pi/\omega$ all this charge passes through any cross-section cutting the ring so that the current is

$$di = dq/T = qr\omega dr/\pi R^2.$$

By Eq. 16 (with a difference in notation), this ring sets up a field *dB* at the center of the disk given by

$$dB = \frac{\mu_0 di}{2r} = \tfrac{1}{2}\mu_0 q\omega dr/\pi R^2.$$

Thus, the total field due to the whole disk is

$$B = \int dB = (\tfrac{1}{2}\mu_0 q\omega/\pi R^2)\int_0^R dr = \frac{\mu_0 q\omega}{2\pi R}.$$

(*b*) The dipole moment of the whole disk is the sum (integral) of the dipole moments of the rings from which it can be imagined as assembled. Each ring is like a single loop ($N = 1$) of wire. We have, then,

$$\mu = \int A di = \int_0^R (\pi r^2)\left(\frac{q\omega r dr}{\pi R^2}\right) = \left(\frac{q\omega}{R^2}\right)\int_0^R r^3 dr = \tfrac{1}{4}q\omega R^2.$$

## 35-36

Each side of the rectangle that is perpendicular to the wire experiences a force parallel to the wire, but these are equal and opposite in direction (since these two sides are identically positioned with respect to the wire but carry opposite currents). Hence, the two forces cancel. Each long side feels a force perpendicular to the wire, each force of the form $i_2 LB$. The force on the side nearer the wire is attractive and greater in magnitude than the force on the far side, which is repulsive. Therefore, the net force on the loop is attractive (toward the wire) and has a magnitude

$$F = i_2 L[B_{\text{near}} - B_{\text{far}}].$$

The fields *B* due to the wire, current $i_1$, each have the form of Eq.

11 with the appropriate distance: for the near side $R = a$ and for the far side $R = a + b$. Putting all this together gives

$$F = \frac{\mu_0 i_1 i_2 L}{2\pi} \left[ \frac{1}{a} - \frac{1}{a + b} \right].$$

Substituting numerical values (SI base units please) yields $B$ = 3.27 mN. As discussed above, this net force is attractive, i.e., toward the wire.

### 35-39

(a) The net current enclosed by the dotted path is 2 A, out of the page. A net current out of the page gives rise to a field directed in the counterclockwise sense (the right-hand rule). Since the dotted path is traversed clockwise, the term $\mathbf{B} \cdot d\mathbf{s}$ gives rise to a factor of $\cos 180° = -1$ in the integral part of Ampère's law, so that the integral equals $\mu_0(-2) = -2\mu_0 = (-2 \text{ A})(4\pi \times 10^{-7} \text{ T} \cdot \text{m/A}) = -2.5 \ \mu\text{T} \cdot \text{m}$.
(b) The net current enclosed by the dashed path is zero (4 A into the page, 4 A out). Hence, the value of the line integral is zero, by Ampère's law.

### 35-43

Use Ampère's law by integrating around the dashed rectangle on the sketch. Since there is no current cutting the rectangle, it is anticipated that

$$\oint \mathbf{B} \cdot d\mathbf{s} = 0.$$

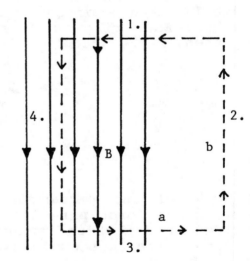

Now examine the sides of the rectangle in turn. On side 2, $\int \mathbf{B} \cdot d\mathbf{s} = 0$ since $B = 0$ on this side. On sides 1 and 3 the integral is zero either because $B = 0$ or $\mathbf{B} \perp d\mathbf{s}$. On side 4,

$$\int \mathbf{B} \cdot d\mathbf{s} = B_4 b,$$

where $B_4$ is the value of the magnetic field along side 4. Thus, adding up the separate values of the integrals along the four sides we find that

$$\oint \mathbf{B} \cdot d\mathbf{s} = B_4 b \neq 0.$$

This contradicts the expectation from Ampère's law. We conclude that the geometry drawn for the magnetic field must be in error.

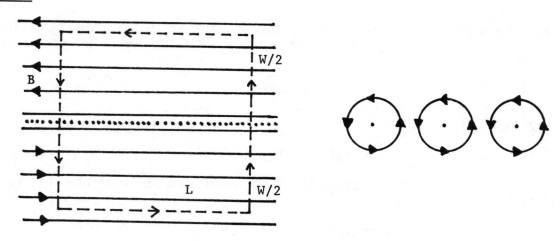

Consider a large number of long wires carrying identical currents and arranged in a plane as in the sketch above. By the right-hand rule, the magnetic fields above the plane due to the wires all are to the left and below the plane all are to the right. Between the wires the fields due to adjacent wires are in the opposing directions. It seems reasonable that as the wires are brought close together the already opposing fields between them will cancel, yielding the pattern shown on the sketch above on the left. Select as a path of integration a rectangle of length $L$ and width $W$ aligned as shown, perpendicular to the sheet of wires which evenly divides the rectangle. We see that $\mathbf{B}$ is at 90° to $ds$ along the sides $W$, so the line integral in Ampère's law gives zero on these sides. Along the top of length $L$, however,

$$\int \mathbf{B} \cdot ds = \int B \, ds \cos 0° = B \int ds = BL.$$

Since both of the sides (length $L$, parallel to the sheet) are at the same distance ($W/2$) from the sheet, $B$ has the same value at each of them, so that Ampère's law gives

$$\oint \mathbf{B} \cdot ds = 2BL = \mu_0 i_{enc} = \mu_0 (n i_0 L),$$

$$B = \tfrac{1}{2} \mu_0 n i_0.$$

To obtain this result from the magnetic field of a strip of width $a$, let $a \to \infty$. In this case (see Sample Problem 3),

$$\tan^{-1}(a/2R) \to \tan^{-1}(\infty) = \tfrac{1}{2}\pi,$$

so that

$$B \to (\mu_0 i / \pi a)(\tfrac{1}{2}\pi) = \mu_0 i / 2a.$$

But, in this formula $i$ is the total current in the strip. Hence $i/a$ is the current per unit length (width). This equals $n i_0$ in Problem 46. Making this identification reconciles the two results.

35-51

Let $x$ be the length of the wire, $D$ the diameter and $L$ the length of the solenoid and $N$ the number of turns of wire on the solenoid. Then

$$x = \pi DN.$$

Also, by Eq. 22,

$$B = \mu_0 i_0 n = \mu_0 i_0 (N/L).$$

Solve for $N$ in the second equation and substitute into the first to find that

$$x = \pi BDL/\mu_0 i_0 = 109 \text{ m},$$

the numerical result following after substitution of the data.


35-53

Combine Eq. 14 of Chapter 34 for the radius of the circular path of the electron with Eq. 22 for the magnetic field $B$ inside a solenoid to obtain

$$r = \frac{mv}{eB} = \frac{mv}{eni_0\mu_0}.$$

Solve for the current to find

$$i_0 = \frac{mv}{enr\mu_0}.$$

We are told that $r = 0.023$ m, $n = 100$ cm$^{-1} = 1 \times 10^4$ m$^{-1}$, and also that $v = 0.046(3 \times 10^8$ m/s$)$. The charge and mass of an electron are, by now, well-known quantities. With these data, the equation above yields $i_0 = 0.272$ A.


35-55

(a) The force due to the magnetic field must be directed to the center of curvature; by the right-hand rule, then, the particle must carry negative charge.
(b) The changing radius of curvature of the path is given by Eq. 14 of Chapter 34 (using $R$ for this radius),

$$R = mv/qB,$$

where $q$ is the magnitude of the charge on the particle. Now, the magnetic field does no work on the particle (see Section 2 in Chapter 34). By the work-energy theorem, the kinetic energy of the particle, and therefore its speed, does not change. Hence, at any two points 1 and 2 on the trajectory, the quantity

$$RB = mv/q,$$

remains constant, so that

$$R_1 B_1 = R_2 B_2.$$

But, for a toroid, by Eq. 23,

$$B = \mu_0 i_0 N / 2\pi r,$$

where $r$ is distance from the axis of the toroid. Writing this expression for the two locations 1 and 2 and substituting into the previous equation gives

$$R_1 / r_1 = R_2 / r_2.$$

Putting in the numbers gives

$$11/125 = R_2/110,$$

$$R_2 = 9.7 \text{ cm}.$$

## 36-2

At any instant the flux through the antenna is

$$\Phi_B = \int \mathbf{B} \cdot d\mathbf{A} = \int B dA \cos 0° = \int B dA = B \int dA = BA = B(\tfrac{1}{4}\pi D^2),$$

$D$ being the antenna diameter. The flux changes as the magnetic field $B$ changes ($D$ remains fixed) so that, disregarding the sign in Lenz' law (the direction, or "sense", of the emf is unimportant), we have by Eq. 4,

$$\mathscr{E} = N[d\Phi_B/dt] = N[d(B\tfrac{1}{4}\pi D^2)/dt] = N(\tfrac{1}{4}\pi D^2)(dB/dt),$$

$$\mathscr{E} = (1)[\tfrac{1}{4}\pi(0.112\text{ m})^2](0.157\text{ T/s}) = 1.55\text{ mV}.$$

## 36-6

By Eq. 23 of Chapter 32, $P = \mathscr{E}^2/R$. The induced emf is, in absolute value,

$$\mathscr{E} = d(BA)/dt = A(dB/dt) = A(B/\Delta t).$$

Hence, the energy dissipated is

$$E = P(\Delta t) = \frac{\mathscr{E}^2}{R}(\Delta t) = \frac{(AB/\Delta t)^2}{R}(\Delta t) = \frac{A^2 B^2}{R \Delta t}.$$

## 36-8

(a) By Faraday's law, Eq. 4, the current $i$ induced in the coil is

$$i = \frac{\mathscr{E}}{R} = \left(\frac{N}{R}\right) d\Phi_B/dt.$$

Now $\Phi_B = \int \mathbf{B} \cdot d\mathbf{A}$, the integral taken over the cross-section of the coil, and $\mathbf{B}$ is the magnetic field due to the solenoid. Since $B = \mu_0 n i_s$ inside the solenoid ($i_s$ is the current in the solenoid) and $B = 0$ outside the solenoid, the integral for the flux between the coil and solenoid is zero, so that

$$\Phi_B = \mu_0 n i_s A_s,$$

where $A_s$ is the cross-sectional area of the solenoid. Thus,

$$i = (\frac{N}{R})d(\mu_0 n i_s A_s)/dt = (\frac{N}{R})\mu_0 n A_s di_s/dt = (\frac{N}{R})\mu_0 n A_s i_0/t.$$

Numerically, $N = 120$, $R = 5.3$ Ω, $\mu_0 = 4\pi \times 10^{-7}$ T•m/A, $n = 22,000$ m$^{-1}$, $A_s = \frac{1}{4}\pi(0.032$ m$)^2$, $i_0 = 1.5$ A and $t = 0.16$ s; these data put into the equation above yield $i = 4.72$ mA.
(b) As the magnetic field changes, electric fields appear and these affect the conduction electrons in the coil.

## 36-12

Since the cross-sectional area of the cylinder does not change, we can write the result of Problem 11(a), adapted for $N$ turns, as

$$q = N\Delta\Phi_B/R = NA(\Delta B)/R.$$

The change in the magnetic field is

$$\Delta B = B_f - B_i = 1.57 \text{ T} - (-1.57 \text{ T}) = 3.14 \text{ T}.$$

Therefore, the charge sought is

$$q = (125)(12.2 \times 10^{-4} \text{ m}^2)(3.14 \text{ T})/(13.3 \text{ Ω}) = 36.0 \text{ mC}.$$

## 36-14

The flux at any time is, (with $L$ = length of each side of the square),

$$\Phi_B = B(\tfrac{1}{2}L^2),$$

only half the square being in the field. Since $dB/dt = -0.87$ T/s we have, by Faraday's law,

$$\mathscr{E} = -d\Phi_B/dt = -(\tfrac{1}{2}L^2)(dB/dt) = -\tfrac{1}{2}(2.3 \text{ m})^2(-0.87 \text{ T/s}) = 2.30 \text{ V}.$$

The field inside the square is directed out of the page and is decreasing in strength. Therefore, the induced current will flow so as to produce an outward flux inside the square. To do this a counterclockwise current is needed. This is in the same direction as the current set up by the battery. Thus the induced emf must be in the same sense as the battery emf, for a total emf of 2.0 + 2.3 = 4.3 V.

## 36-20

(a) The emf induced is $\mathscr{E} = d\Phi_B/dt = d(BA)/dt = d(BLx)/dt = BL(dx/dt) = BLv$, where $x$ is the distance of the rod from the closed end of the rails. Thus,

$$\mathscr{E} = BLv = (1.18 \text{ T})(0.108 \text{ m})(4.86 \text{ m/s}) = 0.619 \text{ V}.$$

(b) The induced current is $i = \mathscr{E}/R = (0.619 \text{ V})/(0.415 \text{ }\Omega) = 1.49 \text{ A}$. As the rod moves to the left, an increasing upward directed flux becomes contained within the rod-rail circuit. The induced current will flow so that the magnetic field that it sets up will point downward in this same region bounded by the circuit. By the right-hand rule, this requires an induced current flowing clockwise.
(c) Internal energy is generated at the rate

$$P = i^2R = (1.49 \text{ A})^2(0.415 \text{ }\Omega) = 0.912 \text{ W}.$$

(d) The force $F$ exerted by the external agent must overcome the $i\mathbf{L} \times \mathbf{B}$ force exerted by the external magnetic field $B$ on the induced current in the rod. The latter force points to the right; for zero acceleration, set $F = iLB$ exactly (zero net force); then we have

$$F = iLB = (1.49 \text{ A})(0.108 \text{ m})(1.18 \text{ T}) = 0.190 \text{ N}.$$

(e) The rate that this force $F$ does work is $Fv = (0.190 \text{ N})(4.86 \text{ m/s}) = 0.923 \text{ W}$; this equals, within round-off error, the rate of generation of internal energy.


36-24

(a) Far away from the larger loop the field due to this loop can be considered as approximately uniform over the area of the smaller loop, and equal to the value on the common axis. From Eq. 17 in Chapter 35, writing $x$ for $z$, this field is

$$B = \mu_0 iR^2/2x^3,$$

so that the flux through the small loop, by the simplifications listed above, is

$$\Phi_B = B(\pi r^2) = \mu_0 iR^2 \pi r^2/2x^3.$$

(b) For a single loop the emf is, by Eq. 3,

$$\mathscr{E} = d\Phi_B/dt = (\tfrac{1}{2}\mu_0 iR^2 \pi r^2)d(x^{-3})/dt.$$

But

$$d(x^{-3})/dt = -3x^{-4}dx/dt = -3v/x^4.$$

Hence, except for sign,

$$\mathscr{E} = 3\mu_0 \pi iR^2 vr^2/2x^4.$$

(c) The field due to the current $i$ in the larger loop, near its axis, is directed away from the larger loop toward the smaller loop. Therefore, as the smaller loop moves away from the larger, the small loop will see a steadily decreasing upward directed flux cutting through it. To oppose this, the induced emf will seek to set up an induced current the magnetic field of which will be directed upward also, within the area bounded by the small loop. By the right-hand rule, this will require a counterclockwise current as seen looking down from above the small loop (i.e., in the same direction as the current in the large loop).

36-27

(a) By Faraday's law $\mathscr{E} = -Nd\Phi_B/dt$ and therefore the flux through the rectangular loop at any time $t$ must first be found. But

$$\Phi_B = \int \mathbf{B} \cdot d\mathbf{A} = \int BdA\cos\theta = B\cos\theta \int dA = abB\cos\theta.$$

This last result follows because $B$ is uniform and $\theta$, the angle between $\mathbf{B}$ and $d\mathbf{A}$, has at any instant the same value at all points on the rectangle. If the loop rotates at the constant rate $v$, then $\theta = \omega t = 2\pi v t$ and, by Faraday's law,

$$\mathscr{E} = -Nd\Phi_B/dt = -Nd(abB\cos2\pi vt)/dt = 2\pi vNabB\sin2\pi vt,$$

$$\mathscr{E} = (2\pi vNabB)\sin2\pi vt = \mathscr{E}_0\sin2\pi vt.$$

(b) If $\mathscr{E}_0 = 150$ V, $v = 60$ Hz and $B = 0.50$ T, then

$$2\pi Nab = \mathscr{E}_0/vB = 150/[(60)(0.5)] = 5 \text{ m}^2,$$

and any loop built according to the specification $Nab = 5/2\pi = 0.796$ m² will produce the desired effect.

36-33

(a) Consider a small element of the rod of length $dx$, situated at a distance $x$ from the wire. In a time $t$, the rod moves a distance $vt$, so that this small element sweeps out a narrow rectangle of length $vt$, thickness $dx$, at a distance $x$ from the wire. Hence, the flux through the area swept out by the rod is

$$\Phi_B = \int_a^{L+a} \frac{\mu_0 i}{2\pi r} x dr = \frac{\mu_0 ix}{2\pi} \ln\frac{L + a}{a}.$$

Therefore,

$$\mathscr{E} = \frac{d\Phi_B}{dt} = \frac{\mu_0 iv}{2\pi} \ln\frac{L + a}{a},$$

which yields $\mathscr{E}$ = 253 $\mu$V upon substitution of the numerical data.
(b) The induced current is

$$i = \mathscr{E}/R = (253 \ \mu V)/(0.415 \ \Omega) = 610 \ \mu A.$$

(c) The required rate is

$$P = \mathscr{E}^2/R = (253 \ \mu V)^2/(0.415 \ \Omega) = 154 \ nW.$$

(d),(e) The rate at which the force does work must equal the rate of generation of internal energy; therefore,

$$F = P/v = (154 \ nW)/(4.86 \ m/s) = 31.7 \ nN.$$

36-37

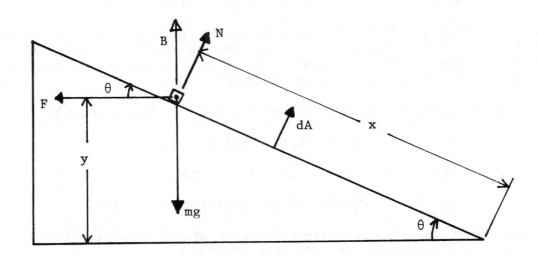

(a) The forces acting on the rod are its weight $mg$, the normal force $N$, and the force $F$ exerted by the magnetic field on the current induced in the rod by virtue of its motion induced by gravity. Newton's second law requires that

$$mg\sin\theta - F\cos\theta = ma.$$

But $F = iLB = (\mathscr{E}/R)LB$, $L$ the length of the rod. Since the angle between $dA$ (the normal to the inclined plane) and the magnetic field $B$ is $\theta$, the angle of the incline, then $\mathscr{E} = BLv\cos\theta$, and therefore Newton's second law becomes

$$mg\sin\theta - (B^2L^2v/R)\cos^2\theta = ma.$$

Initially $v = 0$; as the rod accelerates down the plane the speed $v$ will increase until it reaches a value $v_t$ given by

$$v_t = \frac{mgR\sin\theta}{B^2L^2\cos^2\theta},$$

at which time, by the preceding equation, the acceleration $a = 0$. After this, the rod will slide with constant speed $v_t$.
(b) The kinetic energy is $K = \frac{1}{2}mv^2$ and the gravitational potential energy is $U = mgy$. Let $x$ be the distance of the rod from the bottom of the incline measured along the incline. Then $y = x\sin\theta$. By the conservation of energy it is expected that

$$\frac{dU}{dt} = \frac{dK}{dt} + P,$$

the last term $P$ being the rate of generation of internal energy. Since $v = dx/dt$ and the acceleration is $a = dv/dt$, this equation can be recast as

$$mgv\sin\theta = mv\frac{dv}{dt} + i^2R,$$

$$mg\sin\theta = ma + \frac{B^2L^2v}{R}\cos^2\theta,$$

since $i = \mathscr{E}/R = BLv\cos\theta/R$. But this last equation agrees with (a); hence the result in (a) is consistent with energy conservation. After the rod has reached its terminal speed we have $dK/dt = 0$ so that $P = dU/dt$, as asserted.
(c) If $\mathbf{B}$ is directed down instead of up, the induced current will flow in the direction opposite to that in (a) above, but the force $\mathbf{F} = (-i)\mathbf{L} \times (-\mathbf{B})$ will be in the same direction as in (a), so the motion of the rod will be unaffected.

36-39

As the small area $a^2$ leaves the magnetic field, the emf induced will be

$$\mathscr{E} = d\Phi_B/dt = Bav = Bar\omega,$$

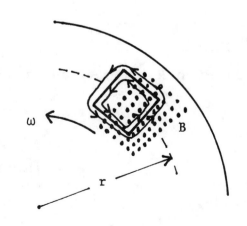

by Eq. 5 (different notation). The eddy current is $i = \mathscr{E}/R$, $R$ the resistance of the current path. If the current "hugs" the area, the resistance will be

$$R = \rho L/A \approx \rho(4a)/(at) = 4/(\sigma t).$$

Thus,

$$i - \frac{\mathscr{E}}{R} - \frac{Bra\omega}{4/\sigma t} - \tfrac{1}{4}Bra\omega\sigma t.$$

The force opposing the motion is $F \approx iaB$ and is felt twice each revolution, as the area enters and leaves the field; hence,

$$\tau = 2rF = \tfrac{1}{2}B^2 a^2 r^2 \omega \sigma t.$$

## 36-45

By Sample Problem 4, the induced electric field is

$$E = \tfrac{1}{2}r(dB/dt),$$

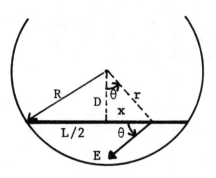

at points a distance $r$ from the center of the cylindrical region. The direction of $\mathbf{E}$ is perpendicular to the radius, as shown, normal to the axis. Thus, the emf induced in a length $dL$ of the rod is

$$d\mathscr{E} - E(dx)\cos\theta - (\tfrac{1}{2}r\frac{dB}{dt})(dx)(\frac{D}{r}) - \tfrac{1}{2}D\frac{dB}{dt}dx.$$

Hence, the emf induced between the ends of the rod is

$$\mathscr{E} - \frac{D}{2}\frac{dB}{dt}\int_{-\frac{1}{2}L}^{\frac{1}{2}L}dx - \frac{D}{2}\frac{dB}{dt}L - \frac{L}{2}\frac{dB}{dt}\sqrt{R^2 - (L/2)^2}.$$

## 36-47

(a) With $R$ the radius of the electron orbit Eq. 13 yields

$$E\bullet 2\pi R = d\Phi_B/dt.$$

But

$$\Phi_B = \int \mathbf{B}\bullet d\mathbf{A} = B_{av}(\pi R^2) = 2B_{orb}\pi R^2.$$

Hence,

$$E = R(dB_{orb}/dt) = (0.32)(0.28)(120\pi)\cos 120\pi t.$$

At time $t = 0$, this gives

$$E = (0.32 \text{ m})(0.28 \text{ T})(120\pi \text{ rad/s}) = 34 \text{ V/m}.$$

(*b*) The acceleration is

$$a = F/m = eE/m = 6.0 \times 10^{12} \text{ m/s}^2,$$

the numerical result following from substituting the values of $E$ from (*a*) and the charge and mass of an electron.

# CHAPTER 37

<u>37-2</u>

Through one end, call it end 1, the flux is -25 $\mu$Wb, where a minus sign is present to account for the flux over this end being inwardly directed. The flux through the other end, end 2, can readily be calculated since the field is uniform over this end. The flux is outward (positive) and given by

$$\Phi_{end2} = BA = B(\pi R^2) = (0.0016 \text{ T})[\pi(0.13 \text{ m})^2] = 85 \ \mu\text{Wb}.$$

Now apply Gauss' law for magnetism, Eq. 2:

$$\Phi_B = \Phi_{end1} + \Phi_{end2} + \Phi_{curved} = 0,$$

$$-25 + 85 + \Phi_{curved} = 0,$$

$$\Phi_{curved} = -60 \ \mu\text{Wb};$$

that is, an inward flux of 60 $\mu$Wb.

<u>37-11</u>

The volume of the rod is

$$V = \tfrac{1}{4}\pi D^2 L = \tfrac{1}{4}\pi(0.011 \text{ m})^2(0.048 \text{ m}) = 4.56 \text{ X } 10^{-6} \text{ m}^3.$$

By Eq. 12, the dipole moment of the rod is

$$\mu = MV = (5300 \text{ A/m})(4.56 \text{ X } 10^{-6} \text{ m}^3) = 0.024 \text{ A} \bullet \text{m}^2 = 24 \text{ mJ/T}.$$

<u>37-12</u>

From Appendix D, the density of iron is $\rho = 7.87$ g/cm$^3$ = 7870 kg/m$^3$ and the molar mass of iron is $M = 55.8$ g/mol = 0.0558 kg/mol (here $M$ = molar mass, not magnetization). Hence, in iron there are

$$n = \rho N_A/M = (7870 \text{ kg/m}^3)(6.02 \text{ X } 10^{23} \text{ mol}^{-1})/(0.0588 \text{ kg/mol}),$$

$$n = 8.491 \text{ X } 10^{28} \text{ atoms/m}^3.$$

The volume of the bar is $V = (1.31 \text{ cm}^2)(4.86 \text{ cm}) = 6.367 \text{ X } 10^{-6} \text{ m}^3$. Therefore the bar contains

$$N = nV = (8.491 \text{ X } 10^{28} \text{ atoms/m}^3)(6.367 \text{ X } 10^{-6} \text{ m}^3),$$

$$N = 5.406 \text{ X } 10^{23} \text{ atoms.}$$

The dipole moment of each atom is

$$\mu_{atom} = 2.22\mu_B = 2.22(9.27 \times 10^{-24} \text{ J/T}) = 2.058 \times 10^{-23} \text{ J/T}.$$

It follows that the dipole moment of the bar, assuming complete alignment of the atoms, is

$$\mu = N\mu_{atom} = (5.406 \times 10^{23})(2.058 \times 10^{-23} \text{ J/T}) = 11.1 \text{ J/T}.$$

(b) By Eq. 37 of Chapter 34,

$$\tau = \mu B\sin 90° = (11.1 \text{ J/T})(1.53 \text{ T}) = 17.0 \text{ N} \bullet \text{m}.$$

## 37-14

(a) By Eq. 5, the magnetic dipole moment of an orbiting charge can be written as

$$\mu = \tfrac{1}{2}evr,$$

assuming that the magnitude of the charge is e. Now, for motion in a circle perpendicular to a uniform magnetic field, Eq. 14 in Chapter 34 gives the radius of the orbit as

$$r = mv/eB,$$

so that

$$\mu = \tfrac{1}{2}ev(mv/eB) = (\tfrac{1}{2}mv^2)/B = K_e/B.$$

An electron circulates clockwise in a magnetic field that is directed into the paper (to choose an example). The resulting angular velocity vector ω is into the paper also. But the electron charge is negative, so that $\mu$ is antiparallel to ω and therefore is directed out of the paper, antiparallel to **B**.
(b) The charge cancels out in the calculation of $\mu$ in (a). Thus, for a positive ion, the same relation holds: $\mu = K_i/B$. A positive ion circulates counterclockwise in a magnetic field pointing into the paper, so ω is directed out of the paper. Since the ion carries positive charge, $\mu$ is parallel to ω, and therefore antiparallel to **B**, just as for the electron.
(c) The directions of the dipole moments due to the electrons and ions are the same, by (b) above, so that

$$\mu = N_e\mu_e + N_i\mu_i = N_eK_e/B + N_iK_i/B = (N_eK_e + N_iK_i)/B,$$

where $N_e$, $N_i$ are the numbers of electrons and ions. Now these numbers are equal, so put $N_e = N_i = N$. In this event, the magnetization becomes

$$M = \mu/V = N(K_e + K_i)/BV.$$

Now put $N = 5.28 \times 10^{21}$, $V = 1 \text{ m}^3$, $B = 1.18 \text{ T}$, $K_e = 6.21 \times 10^{-20}$ J and $K_i = 7.58 \times 10^{-21}$ J to find $M = 312$ A/m.

## 37-15

By Sample Problem 8(*b*) in Chapter 34, the energy needed to turn a dipole end for end is $2\mu B$. For the mean kinetic energy of translation of a gas atom, $3kT/2$, see Eq. 20 in Chapter 23. Setting these equal and solving for temperature $T$ gives

$$2\mu B = 3kT/2,$$

$$T = 4\mu B/3k = 4(1.2 \text{ X } 10^{-23} \text{ J/T})(0.50 \text{ T})/3(1.38 \text{ X } 10^{-23} \text{ J/K}),$$

$$T = 0.580 \text{ K.}$$

## 37-16

From p. 813 of HRK we find the Curie temperature of iron to be 770°C. Since the assumed surface temperature is 20°C, we must descend to a depth of

$$d = \frac{770°C - 20°C}{30 \text{ C}°/\text{km}} = 25 \text{ km,}$$

before iron ceases to be ferromagnetic and becomes paramagnetic.

## 37-26

In Arizona the vertical component of the Earth's magnetic field is directed down, so that the flux through Arizona is

$$\Phi_{Ariz} = \int \mathbf{B} \cdot d\mathbf{A} = B_v A \cos 180°,$$

$$\Phi_{Ariz} = (43 \text{ X } 10^{-6} \text{ T})(295,000 \text{ X } 10^6 \text{ m}^2)(-1) = -12.7 \text{ MWb.}$$

Through the whole Earth, $\Phi_E = 0$; therefore, the flux through the rest of the Earth (Arizona omitted) is

$$\Phi_{rest} = -\Phi_{Ariz} = +12.7 \text{ MWb.}$$

## 37-27

(*a*) The magnetic moment of a current loop is given by Eq. 36 of Chapter 34; we have

$$\mu = NiA = Ni[\pi R^2],$$

$$8.0 \text{ X } 10^{22} \text{ J/T} = (1)i[\pi(6370 \text{ X } 10^3 \text{ m})^2],$$

$$i = 6.3 \text{ X } 10^8 \text{ A} = 630 \text{ MA.}$$

(b) The field due to any magnetic dipole at a great enough distance is given by Eq. 7 in Chapter 35. Therefore, by choosing the correct direction for the current in (a), the magnetic field of the Earth at a great distance could be canceled.

(c) Close to a dipole, the dipole's magnetic field depends sensitively on the geometry of the dipole's currents (for example, square loop vs. circular loop). Since the Earth's magnetic field is due to currents in the core of the Earth, far beneath the surface, it cannot be cancelled near the Earth by a wire around the equator.

## 37-31

See the equation for the magnetic field given in Problem 29(a). We are dealing with two points at the same magnetic latitude $L_m$, so the term containing the latitude will cancel. With $R$ = radius of the Earth and $h$ = altitude of the point sought, it is required that

$$\frac{1}{r^3} = \frac{1}{(R + h)^3} = \frac{1}{2} \frac{1}{R^3},$$

$$R + h = 2^{1/3}R,$$

$$h = R(2^{1/3} - 1) = (6370 \text{ km})(2^{1/3} - 1) = 1660 \text{ km}.$$

## 38-2

(a) The number of flux linkages $N\Phi_B$ induced by the external magnetic field $B$ is $NBA$. Here $A$ is the cross-sectional area of the coil, so that

$$A = \pi(0.103 \text{ m})^2 = 3.333 \text{ X } 10^{-2} \text{ m}^2.$$

Therefore

$$N\Phi_B = (34)(2.62 \text{ mT})(3.333 \text{ X } 10^{-2} \text{ m}^2) = 2.97 \text{ mWb}.$$

(b) If the net flux vanishes with a non-zero current in the coil, it must be that the flux due to this current is opposite in direction and equal in magnitude to the flux produced by the external field. By Eq. 6, then,

$$L = N\Phi_B/i = (0.00297 \text{ Wb})/(3.77 \text{ A}) = 788 \text{ } \mu\text{H}.$$

## 38-5

(a) By Eq. 2,

$$\mathscr{E}_L = L(di/dt),$$

$$3 \text{ X } 10^{-3} \text{ V} = L(5 \text{ A/s}) \quad \rightarrow \quad L = 600 \text{ } \mu\text{H}.$$

(b) Using Eq. 6, we find

$$N = Li/\Phi_B,$$

$$N = (600 \text{ X } 10^{-6} \text{ H})(8 \text{ A})/(40 \text{ X } 10^{-6} \text{ Wb}) = 120.$$

## 38-7

The number of windings per unit length is

$$n = 1870/(1.26 \text{ m}) = 1484 \text{ m}^{-1},$$

and the cross-sectional area of the solenoid is

$$A = \tfrac{1}{4}\pi d^2 = \tfrac{1}{4}\pi(0.0545 \text{ m})^2 = 2.333 \text{ X } 10^{-3} \text{ m}^2.$$

Now find the inductance of the solenoid from Eq. 16:

$$L = \kappa_m\mu_0 n^2 lA,$$

$$L = (968)(4\pi \text{ X } 10^{-7})(1484)^2(1.26)(2.333 \text{ X } 10^{-3}) = 7.87 \text{ H}.$$

You should check the units in the last equation; note that 1 H =
1 V•s/A = 1 T•m²/A.

## 38-16

The steady-state, or equilibrium, value of the current is $\mathscr{E}/R$ (the
value of $i$ for $t \to \infty$). At time $t = 5.22$ s, we have $i = \frac{1}{3}(\mathscr{E}/R)$. By
Eq. 21, this leads to

$$\frac{1}{3} = 1 - e^{-5.22/\tau_L},$$

$$e^{-5.22/\tau_L} = \frac{2}{3},$$

$$-5.22/\tau_L = \ln(2/3),$$

$$\tau_L = 12.9 \text{ s.}$$

## 38-19

(a) The number of flux linkages is $N\Phi_B$. Hence, by Eq. 6,

$$L = N\Phi_B/i = (26.2 \text{ mWb})/(5.48 \text{ A}) = 4.78 \text{ mH.}$$

(b) Calculate the inductive time constant from Eq. 23; we find

$$\tau_L = L/R = (4.78 \text{ mH})/(0.745 \text{ }\Omega) = 6.416 \text{ ms.}$$

The current is given by Eq. 21. Substitute $i = 2.53$ A, $\mathscr{E} = 6$ V, $R$
= 0.745 $\Omega$ and the value of the time constant found above to get

$$i = \frac{\mathscr{E}}{R}(1 - e^{-t/\tau_L}),$$

$$2.53 = \frac{6}{0.745}(1 - e^{-t/6.416}),$$

$$t = 2.42 \text{ ms;}$$

the unit is ms since $\tau_L$ is in ms.

First, use Eq. 23 to calculate the time constant:

$$\tau_L = L/R = (50 \text{ mH})/(180 \text{ }\Omega) = 0.278 \text{ ms.}$$

Now find $di/dt$ from Eq. 22:

$$\frac{di}{dt} = \frac{\mathcal{E}}{L}e^{-t/\tau_L} = \frac{45 \text{ V}}{0.050 \text{ H}}e^{-1.2/0.278} = 12.0 \text{ A/s.}$$

Note that in the exponent, $t$ and $\tau_L$ must be in the same units, in this case ms.

38-25

(a) The inductor "breaks" the right-hand branch, so that $i_1 = i_2 = i$. But

$$i = \mathcal{E}/(R_1 + R_2) = (100 \text{ V})/(10 \text{ }\Omega + 20 \text{ }\Omega) = 3.33 \text{ A.}$$

(b) Now the inductor has no effect and therefore, by the loop rule,

$$\mathcal{E} - i_1R_1 - i_1R_{eq} = 0,$$

where, by Eq. 10 in Chapter 33,

$$R_{eq} = R_2R_3/(R_2 + R_3) = 12 \text{ }\Omega.$$

Thus $i_1 = (100 \text{ V})/(10 \text{ }\Omega + 12 \text{ }\Omega) = 4.545$ A. In addition,

$$\mathcal{E} - i_1R_1 - i_2R_2 = 0,$$

$$100 - (4.545)(10) - i_2(20) = 0,$$

$$i_2 = 2.727 \text{ A.}$$

(c) The left-hand branch is broken now so that $i_1 = 0$. The current through $R_2$ equals the current through $R_3$ since the remaining elements form a series circuit. The initial value of this current equals the current through $R_3$ a long time after switch S was originally closed. From (b) this is $i_2 = 4.545 - 2.727 = 1.82$ A. (d) There are now no sources of emf in the circuit and hence all currents vanish.

38-32

(a) The energy density is given by Eq. 32, $u_B = B^2/2\mu_0$. The magnetic field inside a solenoid is $B = \mu_0 i_0 n$, where $n = N/l$, $N$ the number of turns. Combining these relations yields

$$u_B = (\mu_0 i_0 n)^2/2\mu_0 = \tfrac{1}{2}\mu_0[Ni_0/l]^2,$$

$$u_B = \tfrac{1}{2}(4\pi \times 10^{-7})[(950)(6.57)/(0.853)]^2 = 33.6 \text{ J/m}^3.$$

(b) The volume of the solenoid is

$$V = (17.2 \times 10^{-4} \text{ m}^2)(0.853 \text{ m}) = 1.467 \times 10^{-3} \text{ m}^3.$$

With the energy density being uniform inside the solenoid, the stored energy $U_B$ is given by

$$U_B = (33.6 \text{ J/m}^3)(1.467 \times 10^{-3} \text{ m}^3) = 49.3 \text{ mJ}.$$

## 38-35

(a) By Eq. 31,

$$U_B = \tfrac{1}{2}Li^2 = \tfrac{1}{2}(152 \text{ H})(32 \text{ A})^2 = 7.80 \times 10^4 \text{ J}.$$

(b) From Appendix D we find the molar mass of helium to be 4.00 g/mol. Hence

$$L_v = (85 \text{ J/mol})/(4 \text{ g/mol}) = 21.25 \text{ J/g}.$$

The mass $m$ of helium that boils off follows from Eq. 7 in Chapter 25. We identify the heat $Q$ with the energy stored in the inductor; this energy will ultimately appear as heat when the resistance is no longer zero. Therefore

$$m = Q/L_v = (78{,}000 \text{ J})/(21.25 \text{ J/g}) = 3700 \text{ g} = 3.7 \text{ kg}.$$

## 38-38

(a) The energy delivered by the battery is $U_{\text{batt}} = q\mathscr{E}$, where $q$ is the charge that passes through the battery in the time $t = 2$ s. This charge is found from the current, Eq. 21. We have

$$q = \int i\,dt = \frac{\mathscr{E}}{R}\int_0^t (1 - e^{-t/\tau_L})\,dt = \frac{\mathscr{E}}{R}[t + \tau_L(e^{-t/\tau_L} - 1)].$$

Use Eq. 23 to find the time constant: we find $\tau_L = L/R = 5.48/7.34 = 0.7466$ s. Substituting this, and the other data, into the equation for the charge gives $q = 2.168$ C. With this, the energy delivered by the battery is found to be

$$U_{\text{batt}} = q\mathscr{E} = (2.168 \text{ C})(12.2 \text{ V}) = 26.4 \text{ J}.$$

(b) By Eq. 21, we find that the current at $t = 2$ s is $i = 1.548$ A. Eq. 31 then gives for the energy stored in the inductor at this same instant

$$U_B = \tfrac{1}{2}Li^2 = \tfrac{1}{2}(5.48\ \text{H})(1.548\ \text{A})^2 = 6.57\ \text{J}.$$

(c) By conservation of energy, the energy that appeared in the resistor during the two seconds must be 26.4 - 6.57 = 19.8 J, there being no other element in the circuit than those accounted for above.

## 38-44

There is no energy stored in the inductor when the charge on the capacitor is a maximum. By Eq. 34,

$$U_E = U = q_m^2/2C.$$

Being careful with the numerical data, we find for the value of the capacitance $C$,

$$C = q_m^2/2U = (1.63 \times 10^{-6}\ \text{C})^2/2(142 \times 10^{-6}\ \text{J}) = 9.36\ \text{nF},$$

where 1 nF = 1 X $10^{-9}$ F.

## 32-47

Solve Eq. 36 for the inductance $L$; recalling that kHz refers to frequency $\nu$ (not to angular frequency $\omega$) you should obtain

$$L = [4\pi^2\nu^2C]^{-1}.$$

In substituting the data, watch for the numerical prefixes; one finds for the required inductance

$$L = [4\pi^2(10 \times 10^3\ \text{s}^{-1})^2(6.7 \times 10^{-6}\ \text{F})]^{-1} = 38\ \mu\text{H}.$$

## 38-52

The initial situation is depicted in Fig. 9c; the capacitor is fully charged in Fig. 9e; we see that one quarter of a cycle has elapsed. Thus the required time is $t = \tfrac{1}{4}T$, where $T$ is the period. For any sinusoidal oscillation, $T = 2\pi/\omega$; for $LC$ oscillations, $\omega$ is given by Eq. 44. Therefore

$$T = \frac{2\pi}{\omega} = 2\pi\sqrt{LC} = 2\pi\sqrt{(0.0522\ \text{H})(4.21 \times 10^{-6}\ \text{F})},$$

$$T = 2.945 \times 10^{-3}\ \text{s}.$$

Hence $t = \tfrac{1}{4}(2.945 \times 10^{-3}\ \text{s}) = 7.36 \times 10^{-4}\ \text{s} = 736\ \mu\text{s}.$

## 38-57

(a) With $U_E = \frac{1}{2}U_B$, Eq. 37 yields $U = U_B + U_E = 2U_E + U_E = 3U_E$. By Eq. 34, in terms of charge this relation becomes

$$U_E = \frac{1}{3}U,$$

$$q^2/2C = \frac{1}{3}(q_m^2/2C),$$

$$q = q_m/\sqrt{3}.$$

(b) If we set $\phi = 0$ in Eq. 41, we have

$$q = q_m\cos\omega t,$$

which has $q = q_m$ at $t = 0$. Setting $q = q_m/\sqrt{3}$ gives

$$q_m/\sqrt{3} = q_m\cos\omega t,$$

$$\omega t = 54.74° = 0.304\pi \text{ rad},$$

$$(2\pi/T)t = 0.304\pi \rightarrow t = 0.152T.$$

## 38-60

(a) The situation described does not correspond to either the current or the charge being at a maximum. Therefore, we must find the energy by evaluating both terms in Eq. 37. With $L = 0.0248$ H, $C = 7.73 \times 10^{-6}$ F, $i = 0.00916$ A, and $q = 3.83 \times 10^{-6}$ C, we find that

$$U = \frac{1}{2}Li^2 + q^2/2C = 1.040 \text{ } \mu J + 0.9488 \text{ } \mu J = 1.99 \text{ } \mu J.$$

(b) By Eq. 34,

$$q_m = [2UC]^{1/2} = [2(1.99 \times 10^{-6} \text{ J})(7.73 \times 10^{-6} \text{ F})]^{1/2},$$

$$q_m = 5.55 \times 10^{-6} \text{ C} = 5.55 \text{ } \mu C.$$

(c) Use Eq. 31:

$$i_m = [2U/L]^{1/2} = [2(1.99 \times 10^{-6} \text{ J})/(0.0248 \text{ H})]^{1/2},$$

$$i_m = 0.0127 \text{ A} = 12.7 \text{ mA}.$$

(d) At $t = 0$, Eq. 41 reduces to $q = q_m\cos\phi$. Hence,

$$\phi = \cos^{-1}(q/q_m) = \cos^{-1}(3.83/5.55) = \pm46.4°.$$

Since the capacitor is charging at $t = 0$, we must have $di/dt > 0$ at $t = 0$. But, by Eq. 42, at this time

$$i = -\omega q_m\cos\phi.$$

Only $\phi = -46.4°$ gives $i > 0$ at $t = 0$.
(e) With the capacitor discharging at $t = 0$, we must have the phase angle $\phi = +46.4°$ so that $i < 0$ at this instant; see the equation for the current in (d) above.


## 38-65

The maximum energy present is determined by the amplitude of the charge oscillations; this amplitude is the non-oscillatory part of Eq. 50, and we can write it as

$$q = q_m e^{-Rt/2L}.$$

In this expression, $q$ is the amplitude and $q_m$ is the amplitude at $t = 0$; this notation is different from that in Eq. 41, for example. Similarly, $U = q^2/2C$ and $U_m = q_m^2/2C$; hence,

$$U = U_m e^{-Rt/L}.$$

Putting $U = \frac{1}{2}U_m$, canceling the common factor $U_m$, and taking natural logarithms of both sides of the resulting equation gives ultimately

$$t = (L/R)\ln 2.$$

## 38-67

If we use the approximation $\omega' = \omega$, the time required for the 50 oscillations is

$$t = 50T = 50(2\pi/\omega) = 100\pi\sqrt{LC} = 0.510 \text{ s.}$$

The decrease in the maximum charge on the capacitor is described by Eq. 50 without the oscillatory term; i.e., by

$$q = q_m e^{-Rt/2L}.$$

Taking natural logarithms of both sides gives

$$Rt/2L = \ln(q_m/q).$$

Now we have $t/2L = (0.510 \text{ s})/2(0.22 \text{ H}) = 1.16 \ \Omega^{-1}$. Also, we desire that $q = 0.99q_m$. Putting these quantities into the previous equation and solving for the resistance yields $R = 0.0087 \ \Omega$.

See Problem 65 above, where we show that

$$U = U_m e^{-Rt/L}.$$

The energy loss per cycle is

$$\Delta U \approx \left(\frac{dU}{dt}\right)(\Delta t).$$

Differentiating the expression for $U$ above, we find

$$\frac{dU}{dt} = -\left(\frac{R}{L}\right)(U_m e^{-Rt/L}) = -\left(\frac{R}{L}\right)U.$$

For one cycle $\Delta t = 2\pi/\omega' \approx 2\pi/\omega$. Therefore

$$\frac{\Delta U}{U} = -\left(\frac{R}{L}\right)\left(\frac{2\pi}{\omega}\right) = -\frac{2\pi R}{\omega L}.$$

Evidently, the minus sign is omitted in the result quoted by HRK. Their result concerns the absolute value of the energy loss per cycle.

CHAPTER 39

39-2

(a) Use Eq. 6:

$$X_L = \omega L = 2\pi\nu L,$$

$$1280 \ \Omega = 2\pi\nu(0.0452 \ H),$$

$$\nu = 4510 \ Hz = 4.51 \ kHz.$$

(b) Now turn to Eq. 11; put $\omega = 2\pi\nu$ and solve for the capacitance $C$; we find

$$C = [2\pi\nu X_c]^{-1},$$

$$C = [2\pi(4510 \ s^{-1})(1280 \ \Omega)]^{-1} = 2.76 \times 10^{-8} \ F = 27.6 \ nF.$$

(c) The inductive reactance is directly proportional to the frequency, so that if the frequency is doubled, then

$$X_{L,new} = 2X_L = 2.56 \ k\Omega.$$

On the other hand, the capacitive reactance is inversely proportional to the frequency, so that

$$X_{C,new} = \tfrac{1}{2}X_C = 640 \ \Omega.$$

39-5

(a) By Eqs. 13 and 11,

$$i_m = (V_c)_{max}/X_c = \omega C(V_c)_{max} = \omega C\mathscr{E}_m,$$

$$i_m = (377 \ s^{-1})(4.15 \times 10^{-6} \ F)(25 \ V) = 0.0391 \ A = 39.1 \ mA.$$

(b) See Table 1; the current $i$ and the potential across the capacitor are 90° out of phase. Therefore, when the current is a maximum, $V_c$ is zero. By the loop rule, this means that the generator emf is zero also, since the circuit contains only the generator and the capacitor.
(c) By Eq. 1,

$$\omega t = \sin^{-1}(\mathscr{E}/\mathscr{E}_m) = \sin^{-1}(-13.8/25) = -33.5°; \ 213.5°.$$

We must choose $\omega t = -33.5°$ in order to have, as specified,

$$d\mathscr{E}/dt = \omega\mathscr{E}_m\cos\omega t > 0.$$

Therefore, by Eq. 2,

$$i = i_m \sin(\omega t - \phi),$$

$$i = (39.1 \text{ mA})\sin[-33.5° - (-90°)] = 32.6 \text{ mA}.$$

(d) With $i > 0$ and $\mathscr{E} < 0$, it follows that $P_{gen} = i\mathscr{E} < 0$, so the generator is taking energy from the rest of the circuit (the current is flowing in the direction opposite to that which the generator, acting alone, would cause the current to flow).

## 39-13

By Eq. 5,

$$(V_L)_{max} = i_m \omega L.$$

At resonance $\omega = 1/\sqrt{(LC)}$, so that

$$(V_L)_m - i_m \left(\frac{1}{\sqrt{LC}}\right) L - i_m \sqrt{\frac{L}{C}} .$$

Also at resonance $X_L = X_C$ so that, by Eq. 20, $Z = R$. Therefore, under resonance conditions, Eq. 21 yields

$$i_m = \mathscr{E}_m/R = (10 \text{ V})/(9.6 \text{ } \Omega) = 1.04 \text{ A}.$$

Putting this with $L = 1.2$ H and $C = 1.3 \times 10^{-6}$ F into the voltage equation above yields $(V_L)_{max} = 1000$ V $> \mathscr{E}_m$, which answers the query in the affirmative.

## 39-14

The resistance can be found from Eq. 24, but first calculate the needed reactances:

$$X_L = 2\pi v L = 2\pi(941 \text{ Hz})(0.0883 \text{ H}) = 522.1 \text{ } \Omega,$$

$$X_C = [2\pi v C]^{-1} = [2\pi(941 \text{ Hz})(937 \times 10^{-9} \text{ F})]^{-1} = 180.5 \text{ } \Omega.$$

By Eq. 24,

$$R = (X_L - X_C)\cot\phi = (522.1 \text{ } \Omega - 180.5 \text{ } \Omega)\cot 75° = 91.5 \text{ } \Omega.$$

## 39-17

(a) Using the given data, we can write

$$\mathscr{E} = 125\sin\omega t,$$

$$i = 3.2\sin(\omega t + 56.3°),$$

i.e., $\phi = -56.3°$ (see Eq. 2). By Eq. 21,

$$Z = \mathcal{E}_m/i_m = (125 \text{ V})/(3.2 \text{ A}) = 39.1 \ \Omega.$$

(b) Combining Eqs. 20 and 24 and then employing a trig identity, we find that

$$Z = R\sqrt{1 + \tan^2\phi} = \frac{R}{\cos\phi}.$$

Hence,

$$R = Z\cos\phi = (39.1 \ \Omega)\cos(-56.3°) = 21.7 \ \Omega.$$

(c) With $\phi < 0$, we have a predominately capacitive circuit.

## 39-19

(a) For the inductor,

$$(V_L)_{max} = i_m X_L = 2(V_R)_{max} = 2i_m R,$$

$$X_L = 2R.$$

As for the capacitor,

$$(V_C)_{max} = i_m X_C = (V_R)_{max} = i_m R,$$

$$X_C = R.$$

Eq. 24 now yields

$$\tan\phi = (2R - R)/R = 1 \rightarrow \phi = 45°; \ -135°.$$

The circuit is predominately inductive ($X_L > X_C$) so that $\phi = 45°$ (i.e., $\phi > 0$).
(b) By Eq. 20, we find $Z = R\sqrt{2}$. Therefore, Eq. 21 gives

$$i_m = \mathcal{E}_m/Z = \mathcal{E}_m/[R\sqrt{2}],$$

$$R = \mathcal{E}_m/[i_m\sqrt{2}] = (34.4 \text{ V})/[(0.32 \text{ A})\sqrt{2}] = 76.0 \ \Omega.$$

## 39-20

(a) At resonance, by Eq. 22, $\omega = 1/\sqrt{(LC)} = 229$ Hz.
(b) Also at resonance $Z = R$ so that $i_m = \mathcal{E}_m/R = (31.3 \text{ V})/(5.12 \ \Omega) = 6.11$ A.

(c) To find $\omega_1$ and $\omega_2$, put $i = \frac{1}{2}i_m = \mathcal{E}_m/2R$ in Eq. 19. Then square both sides of the resulting equation and rearrange to obtain

$$(LC)^2\omega^4 - (3R^2C^2 + 2LC)\omega^2 + 1 = 0.$$

Substitute the values of $L$, $R$, $C$ to get

$$3.636 \times 10^{-10}\omega^4 - 3.8166 \times 10^{-5}\omega^2 + 1 = 0.$$

If we write $x = \omega^2$, then we have a quadratic equation which can be solved by the quadratic formula. Then take $\sqrt{x}$ to obtain $\omega$. One finds $\omega_1 = 234$ rad/s and $\omega_2 = 225$ rad/s.
(d) The fractional half-width is $(234 - 225)/229 = 0.039$, using the value of $\omega$ from (a) in the denominator above.


## 39-26

(a) Evidently $X_c = 0$; Eq. 20 yields

$$Z = \sqrt{12.2^2 + 2.3^2} = 12.4 \ \Omega.$$

(b) Using the value of the impedance from (a), the average power can be computed from

$$\overline{P} = i_{rms}^2 R = (\frac{\mathcal{E}_{rms}}{Z})^2 R = (\frac{120}{12.4})^2(12.2) = 1.14 \text{ kW}.$$

(c) By Eq. 32,

$$i_{rms} = \sqrt{\frac{\overline{P}}{R}} = \sqrt{\frac{1140 \text{ W}}{12.2 \ \Omega}} = 9.67 \text{ A}.$$


## 39-30

(a) The antenna is "in tune" with incoming waves with a frequency equal to the resonant frequency of the antenna circuit. Using Eq. 22 we quickly find that $\omega = 6.712 \times 10^8$ rad/s; the corresponding frequency is $\nu = \omega/2\pi = 107$ MHz.
(b) Since we are at resonance,

$$i_{rms} = \mathcal{E}_{rms}/R = (9.13 \ \mu V)/(74.7 \ \Omega) = 122 \text{ nA}.$$

This is the maximum value of the rms current.
(c) We will need the capacitive reactance:

$$X_c = [\omega C]^{-1} = [(6.712 \times 10^8 \text{ rad/s})(0.27 \times 10^{-12} \text{ F})]^{-1} = 5509 \text{ } \Omega.$$

The rms potential difference (maximum value) across the capacitor is now found to be

$$V_{c,rms} = i_{rms}X_c = (122 \times 10^{-9} \text{ A})(5509 \text{ } \Omega) = 672 \text{ } \mu V.$$

## 39-37

With the switch open, the circuit elements are $R$, $C$, $L$. With the current leading the generator emf, $\phi = -20°$ (see Eq. 2). Eq. 24 yields

$$\tan(-20°) = (X_L - X_c)/R,$$

$$\tan 20° = (X_c - X_L)/R.$$

Evidently $X_c > X_L$ (capacitive circuit, consistent with $\phi < 0$). With the switch in position 1, the circuit contains two equal capacitors in parallel; hence $C_{eq} = 2C$ and $X_{c,eq} = \frac{1}{2}X_c$. Eq. 24 gives

$$\tan 10° = (X_L - \tfrac{1}{2}X_c)/R.$$

Finally, with the switch in position 2 the resistor and one of the capacitors are effectively removed from the circuit, which becomes an $LC$ circuit. Therefore, from Eq. 20 with $R = 0$, the impedance becomes $Z = X_c - X_L$. Eq. 21 yields

$$Z = X_c - X_L = \mathscr{E}_m/i_m = (170 \text{ V})/(2.82 \text{ A}) = 60.28 \text{ } \Omega.$$

Using this result in the second equation gives

$$R = (X_c - X_L)/\tan 20° = (60.28 \text{ } \Omega)/\tan 20° = 165.6 \text{ } \Omega.$$

Putting this into the third equation results in

$$X_L - \tfrac{1}{2}X_c = R\tan 10° = (165.6 \text{ } \Omega)\tan 10° = 29.20 \text{ } \Omega.$$

We know have two equations in the two unknowns $X_c$ and $X_L$. Solving these we get $X_c = 179.0 \text{ } \Omega$ and $X_L = 118.7 \text{ } \Omega$. With a frequency of 60 Hz, $\omega = 2\pi\nu = 377$ rad/s. By Eqs. 6 and 11, one finds $C = 14.8 \text{ } \mu F$ and $L = 315$ mH.

## 39-39

(a) We have immediately by Eq. 42,

$$V_s = V_p(N_s/N_p) = (120 \text{ V})(10/500) = 2.4 \text{ V.}$$

(b) For the secondary windings,

$$i_s = V_s/R_s = (2.4 \text{ V})/(15 \text{ }\Omega) = 0.16 \text{ A}.$$

The current in the primary windings can be found from Eq. 44:

$$i_p = i_s(N_s/N_p) = (0.16 \text{ A})(10/500) = 0.0032 \text{ A} = 3.2 \text{ mA}.$$

## 40-1

We encounter $\epsilon_0$ for the first time in Eq. 3 of Chapter 27; its value is given there as

$$\epsilon_0 = 8.8542 \text{ X } 10^{-12} \text{ C}^2/\text{N} \bullet \text{m}^2.$$

As for $\mu_0$, it is found in Eq. 6 in Chapter 35 in the form

$$\mu_0 = 4\pi \text{ X } 10^{-7} \text{ T} \bullet \text{m/A}.$$

For the numerical value in Eq. 1, we find

$$\mu_0\epsilon_0 = 1.1127 \text{ X } 10^{-17},$$

$$\frac{1}{\sqrt{\mu_0\epsilon_0}} = 2.998 \text{ X } 10^8,$$

as asserted. For the units,

$$[\mu_0][\epsilon_0] = (\text{T} \bullet \text{m/A})(\text{C}^2/\text{N} \bullet \text{m}^2) = \text{C}^2 \bullet \text{T}/\text{N} \bullet \text{A} \bullet \text{m} = \text{C} \bullet \text{T} \bullet \text{s/N} \bullet \text{m},$$

since A = C/s. From the units of the equation $F_B = qvB$, we conclude that

$$\text{N} = \text{C(m/s)T} \quad \rightarrow \quad \text{C} \bullet \text{T} = \text{N} \bullet \text{s/m}.$$

Using this, we find

$$\text{C} \bullet \text{T} \bullet \text{s/N} \bullet \text{m} = \text{s}^2/\text{m}^2,$$

and therefore the units of the speed of light $c$ are predicted by Eq. 1 to be

$$(\text{C} \bullet \text{T} \bullet \text{s/N} \bullet \text{m})^{-1/2} = \text{m/s},$$

as they must be.

## 40-3

There are different expressions for the magnetic field $B$ at points with $r < R$ and points at $r > R$. The maximum value of the field is at $r = R$. This maximum, which can be found by setting $r = R$ in either of the two expressions referred to above is

$$B_{max} = \tfrac{1}{2}\mu_0\epsilon_0 R(dE/dt).$$

There will be a point with $r < R$ at which $B = \frac{1}{2}B_{max}$. To find it, set $B = \frac{1}{2}B_{max}$ using the appropriate expression for $B$ given in Sample Problem 1. We find

$$\tfrac{1}{2}\mu_0\epsilon_0 r(dE/dt) = \tfrac{1}{2}[\tfrac{1}{2}\mu_0\epsilon_0 R(dE/dt)],$$

$$r = \tfrac{1}{2}R = \tfrac{1}{2}(5 \text{ cm}) = 2.5 \text{ cm}.$$

Now do the analogous calculation using the expression for $B$ valid at points exterior to $R$; we obtain

$$\frac{1}{2r}(\mu_0\epsilon_0 R^2)(dE/dt) = \tfrac{1}{2}[\tfrac{1}{2}\mu_0\epsilon_0 R(dE/dt)],$$

$$r = 2R = 2(5 \text{ cm}) = 10 \text{ cm}.$$

## 40-4

The displacement current $i_d$ is, by Eq. 6,

$$i_d = \epsilon_0 d\Phi_E/dt = \epsilon_0 A(dE/dt).$$

Now let $x$ be the plate separation. Therefore $E = V/x$, with $V$ the potential difference across the plates. If the plates are fixed, so that $x$ does not change, then

$$i_d = \epsilon_0 A\frac{d(V/x)}{dt} = \frac{\epsilon_0 A}{x}\frac{dV}{dt}.$$

But for a parallel-plate capacitor with $\kappa_e = 1$, we have, by Eq. 7 of Chapter 31, $C = \epsilon_0 A/x$. Using this, the expression above becomes

$$i_d = C\frac{dV}{dt},$$

as we were to prove.

## 40-11

(a) We have $r < R$, so the magnetic field is given by

$$B = \mu_0 i_d r/2\pi R^2;$$

that is, by the result of Sample Problem 5, Chapter 35 with $i = i_d$ (this Sample Problem is appropriate since it concerns a uniform current passing through a circular area). In terms of the

287

displacement current density

$$j_d = i_d/\pi R^2,$$

this expression becomes

$$B = \tfrac{1}{2}\mu_0 j_d r.$$

Now, numerically,

$$j_d = 1.87 \text{ mA/cm}^2 = (1.87 \times 10^{-3} \text{ A})/(1 \times 10^{-4} \text{ m}^2) = 18.7 \text{ A/m}^2,$$

so we get for the magnetic field

$$B = \tfrac{1}{2}(4\pi \times 10^{-7} \text{ T} \bullet \text{m/A})(18.7 \text{ A/m}^2)(0.053 \text{ m}) = 623 \text{ nT}.$$

(b) Use the result of Problem 6(a):

$$j_d = \epsilon_0(dE/dt),$$

$$18.7 \text{ A/m}^2 = (8.85 \times 10^{-12} \text{ C}^2/\text{N} \bullet \text{m}^2)(dE/dt),$$

$$dE/dt = 2.11 \times 10^{12} \text{ V/m} \bullet \text{s}.$$

40-13

From Sample Problem 1, at $r = R$,

$$B = \tfrac{1}{2}\mu_0 \epsilon_0 R(dE/dt),$$

so that

$$B_{max} = \tfrac{1}{2}\mu_0 \epsilon_0 R(dE/dt)_{max}.$$

For sinusoidal oscillations in the electric field,

$$(dE/dt)_{max} = \omega E_m = 2\pi\nu E_m = 2\pi\nu[V_m/d],$$

$$(dE/dt)_{max} = 2\pi(60 \text{ Hz})[(162 \text{ V})/(0.0048 \text{ m})] = 1.272 \times 10^7 \text{ V/m} \bullet \text{s}.$$

Therefore, the maximum value of the magnetic field, at the location $r = R$, is

$$B_{max} = \tfrac{1}{2}(4\pi \times 10^{-7})(8.85 \times 10^{-12})(32.1 \times 10^{-3})(1.272 \times 10^7),$$

$$B_{max} = 2.27 \times 10^{-12} \text{ T} = 2.27 \text{ pT}.$$

You may wish to substitute the units in the equation for the magnetic field above (we omitted them for reasons of space), and verify that they reduce to tesla T.

40-19

With $\omega = 2\pi\nu$, we can use the given equation to find the cavity radius $a$:

$$a = 2.41c/\omega = 2.41c/2\pi\nu,$$

$$a = 2.41(3 \times 10^8 \text{ m/s})/2\pi(60 \text{ s}^{-1}) = 1.92 \times 10^6 \text{ m} = 1920 \text{ km}.$$

Evidently, this result is independent of the length of the cavity, for the length does not appear in the quoted formula for the frequency.

# CHAPTER 41

## 41-3

(*a*) Use the relation $\nu\lambda = c$, being careful with the units; we have

$$\nu = c/\lambda = (3 \times 10^8 \text{ m/s})/(0.067 \times 10^{-15} \text{ m}) = 4.5 \times 10^{24} \text{ Hz}.$$

(*b*) By the same equation, we have immediately

$$\lambda = c/\nu = (3 \times 10^8 \text{ m/s})/(30 \text{ s}^{-1}) = 1 \times 10^7 \text{ m}.$$

## 41-5

The angular frequency of the oscillator is given by Eq. 44 in Chapter 38. But $\omega = 2\pi\nu$ and, for light, $\nu = c/\lambda$. Combining these relations gives

$$\omega - \frac{1}{\sqrt{LC}} - 2\pi\nu - 2\pi\left(\frac{c}{\lambda}\right),$$

$$L - \left(\frac{\lambda}{2\pi c}\right)^2\left(\frac{1}{C}\right).$$

Numerically,

$$\lambda/2\pi c = (550 \times 10^{-9} \text{ m})/2\pi(3 \times 10^8 \text{ m/s}) = 2.92 \times 10^{-16} \text{ s},$$

so that

$$L = (2.92 \times 10^{-16} \text{ s})^2/(17 \times 10^{-12} \text{ F}) = 5.0 \times 10^{-21} \text{ H}.$$

## 41-11

The energy is given by (power)(time). Paying attention to the SI prefixes (see Table 2, Chapter 1), we find

$$E = Pt = (100 \times 10^{12} \text{ J/s})(1 \times 10^{-9} \text{ s}) = 100 \text{ kJ}.$$

In this problem, $E$ is energy (not electric field). Note that we do not need the wavelength for this calculation. One might wonder why it was given; in general HRK does not provide extraneous data in its problems, but there are some exceptions of which this problem is one.

(a) With the lamp radiating uniformly in all directions, we have at the two locations (see Sample Problem 1)

$$P = 4\pi r_1^2 I_1 = 4\pi r_2^2 I_2.$$

Let position 2 be closer to the lamp by the distance a. Also, set the intensity at this closer position equal to $nI_1$ ($n > 1$). We now have

$$r_1^2 I_1 = (r_1 - a)^2 (nI_1),$$

$$\pm r_1 = (r_1 - a)\sqrt{n},$$

$$r_1 = a \frac{\sqrt{n}}{\sqrt{n} \pm 1}.$$

If we pick the + sign in this last equation, we find $r_1 < a$, which is impossible for then position 2 would be "past" the lamp. Hence, we use the - sign to obtain

$$r_1 = a \frac{\sqrt{n}}{\sqrt{n} - 1} = (162 \text{ m}) \frac{\sqrt{1.5}}{\sqrt{1.5} - 1} = 883 \text{ m}.$$

(b) We are not actually told the intensity, in W/m², at either location, only the ratio for the two locations. As a result, the intensity $I_1$ drops out, and our result is correct regardless of the power output of the lamp, provided it radiates uniformly in all directions.

41-19

(a) By Eq. 6,

$$B_m = E_m/c = (1.96 \text{ V/m})/(3 \times 10^8 \text{ m/s}) = 6.53 \text{ nT}.$$

(b) By Eq. 18,

$$I = E_m^2/2\mu_0 c,$$

$$I = (1.96 \text{ V/m})^2/2(4\pi \times 10^{-7} \text{ T} \cdot \text{m/A})(3 \times 10^8 \text{ m/s}),$$

$$I = 5.10 \times 10^{-3} \text{ W/m}^2 = 5.10 \text{ mW/m}^2.$$

(*c*) See Sample Problem 1; we find that

$$P = I[4\pi r^2] = (0.0051 \text{ W/m}^2)[4\pi(11.2 \text{ m})^2] = 8.04 \text{ W}.$$

## 41-22

The diameter of the beam at the Moon's surface is

$$d = r\theta = (3.82 \times 10^8 \text{ m})(0.88 \times 10^{-6} \text{ rad}) = 336 \text{ m}.$$

(We found the distance to the Moon in Appendix C.) The area of the beam at the Moon's surface is

$$A = \tfrac{1}{4}\pi d^2 = \tfrac{1}{4}\pi(336 \text{ m})^2 = 8.87 \times 10^4 \text{ m}^2.$$

Hence, the intensity of the beam, by Eq. 29 in Chapter 19, is

$$I = P/A = (3850 \text{ W})/(8.87 \times 10^4 \text{ m}^2) = 43.4 \text{ mW/m}^2.$$

(We did not write $I = P/4\pi r^2$ because the laser does not radiate uniformly in all directions.)

## 41-25

(*a*) By Eq. 18,

$$E_m^2 = 2\mu_0 cI = 2(4\pi \times 10^{-7} \text{ T} \bullet \text{m/A})(3 \times 10^8 \text{ m/s})(7.83 \times 10^{-6} \text{ W/m}^2),$$

$$E_m = 7.684 \times 10^{-2} \text{ V/m}.$$

(*b*) Use Eq. 6:

$$B_m = E_m/c = (0.07684 \text{ V/m})/(3 \times 10^8 \text{ m/s}) = 2.56 \times 10^{-10} \text{ T}.$$

(*c*) By Sample Problem 1,

$$P = I[4\pi r^2],$$

$$P = (7.83 \times 10^{-6} \text{ W/m}^2)[4\pi(11.3 \times 10^3 \text{ m})^2] = 12.6 \text{ kW}.$$

## 41-30

(*a*) Sighting along the resistor in the direction of the current flow, **E** points directly away from the observer, while **B** is directed transverse and at a right angle to the resistor axis in the clockwise sense. By the right-hand rule **E** X **B** and therefore also **S** are directed radially in toward the axis.
(*b*) Since **E** and **B** are at 90° to each other,

$$S = EB/\mu_0,$$

by Eq. 13. But $E = V/l$, where $V$ is the potential drop along the distance $l$ (see Eq. 13 in Chapter 30). Also, by Ohm's law, $V = iR$. Putting these relations together, the equation above becomes

$$S = iRB/\mu_0 l.$$

In terms of the resistivity $\rho$, Eq. 13 in Chapter 32 tells us that $R = \rho l/\pi a^2$. Furthermore, the magnetic field at the surface of a current-carrying wire, by Sample Problem 5 in Chapter 35, can be written in our present notation as $B = \mu_0 i/2\pi a$. With these relations, our equation for the magnitude of the Poynting vector becomes

$$S = i^2\rho/2\pi^2 a^3.$$

The integral desired involves $\mathbf{S} \cdot d\mathbf{A}$ over the cylindrical surface of the wire, considering a piece of length $l$. The angle between $\mathbf{S}$ and the element of area $d\mathbf{A}$ is 180°, and the magnitude of $\mathbf{S}$ is the same at all points on the surface. Thus

$$\int \mathbf{S} \cdot d\mathbf{A} = -SA = -(i^2\rho/2\pi^2 a^3)(2\pi a l) = -i^2(\rho l/\pi a^2) = -i^2 R.$$

41-33

You absorb an amount of energy

$$U = Pt = IAt,$$

$$U = (1100 \text{ W/m}^2)(1.3 \text{ m}^2)[(2.5 \text{ h})(3600 \text{ s/h})] = 12.87 \text{ MJ}.$$

Hence, by Eq. 20, you receive

$$p = U/c = (12.87 \times 10^6 \text{ J})/(3 \times 10^8 \text{ m/s}) = 0.0429 \text{ kg} \cdot \text{m/s},$$

of linear momentum.

41-42

Let $f$ be the fraction of the incident beam that is reflected. The radiation pressure due to the part of the beam energy that is absorbed is, by Sample Problem 2,

$$p_a = (1 - f)S/c.$$

Here $S$ is the magnitude of the Poynting vector of the incident beam. The radiation pressure due to the reflected part of the incident beam is

$$p_r = 2(fS)/c.$$

The factor of 2 is present for the same reason as the factor of 2 is present in Eq. 21. The total radiation pressure is

$$p = p_a + p_r = (1 + f)S/c.$$

For a plane wave with energy flux $S$, an amount of energy $SAt$ crosses an area $A$ normal to the beam in time $t$. But in this same time the wave travels a distance $ct$. Hence, $SAt$ is the energy contained in a cylindrical volume of base area $A$ and length $ct$, so that the energy density $u$ in the wave is

$$u = SAt/Act = S/c.$$

In terms of the energy density $u$, then, the radiation pressure found above is

$$p = (1 + f)u = u + fu.$$

The first term on the right is the energy density of the incident beam and the second is the energy density of the reflected beam. Since energy is a scalar, the total radiation energy density just outside the surface is $u + fu$, rather than $u - fu$, even though the incident and reflected beams are moving in opposite directions. Thus, the assertion of the problem has been established.


## 41-44

The momentum carried off by the laser beam in time $t$ is, by Eq. 20, $p = U/c$, since the process is just the reverse of absorption. Of course, $U$ is just the energy carried by the beam emitted in the time $t$. If $P$ is the power of the laser beam, then $U = Pt$ and $p = Pt/c$. Hence the force exerted by the beam on the ship is

$$F = dp/dt = P/c = ma.$$

Hence, the speed $v$ reached in time $t$, assuming the spaceship started from rest, is $v = at$ so we have

$$v = Pt/mc,$$

$$v = (10,000 \text{ W})(86,400 \text{ s})/(1500 \text{ kg})(3 \text{ X } 10^8 \text{ m/s}),$$

$$v = 1.92 \text{ X } 10^{-3} \text{ m/s} = 1.92 \text{ mm/s}.$$


## 41-46

For a perfectly reflecting cylinder the radiation pressure, by Sample Problem 2, is

$$p = 2I/c.$$

But $p = F/A$, so the force on the cylinder is

$$F = 2IA/c.$$

The intensity of the laser beam is $I = P/A_{beam}$, where $P$ is the power output of the laser. Therefore, the upward force on the cylinder due to the beam is

$$F = 2PA/A_{beam}c.$$

The downward acting weight of the cylinder is

$$W = Mg = \rho Vg = \rho AHg.$$

For the cylinder to hover, these forces must be equal in magnitude; setting the expressions equal and solving for the desired height $H$ we get

$$H = 2P/A_{beam}\rho cg = 2P/\tfrac{1}{4}\pi d^2\rho cg,$$

where we introduced the beam diameter $d$. Upon substituting the data, in SI base units of course, we find $H = 490$ nm.


## 41-49

(a) Let $r$ = radius, $\rho$ = density of the particle at a distance $x$ from the Sun, which has a mass $M$ and power output $P$. The gravitational force on the particle due to the Sun is

$$F_{grav} = GMm/x^2 = GM(4\pi r^3\rho/3)/x^2.$$

From Sample Problem 2, the force due to radiation pressure is

$$F_{rad} = SA/c = (P/4\pi x^2)(\pi r^2)/c,$$

assuming that the Sun radiates uniformly in all directions. The critical particle radius $r = R_0$ occurs when these forces just balance. Setting $F_{grav} = F_{rad}$, solving for $r$ and then relabeling this radius $R_0$, we find

$$R_0 = 3P/16\pi\rho cGM.$$

(b) To evaluate this radius numerically, extract data on the Sun from Appendix C; we find $P = 3.9 \times 10^{26}$ W and $M = 1.99 \times 10^{30}$ kg. From Chapter 16, $G = 6.67 \times 10^{-11}$ N•m²/kg². We also have $\rho = 1000$ kg/m³ and $c = 3 \times 10^8$ m/s. With these we find $R_0 = 585$ nm.

CHAPTER 42

## 42-2

Each vibration occurs over a distance equal to the wavelength $\lambda$. Hence, the number # of vibrations taking place during a time $t$, during which the light wave travels a distance $ct$, is

$$\# = ct/\lambda,$$

$$\# = (3 \times 10^8 \text{ m/s})(430 \times 10^{-12} \text{ s})/(520 \times 10^{-9} \text{ m}) = 248{,}000.$$

## 42-8

Consider Eq. 5, specifically the middle version. Substituting $v = c/\lambda$, $v_0 = c/\lambda_0$, this equation can be written in the form

$$\frac{1}{\lambda} = \frac{1}{\lambda_0}(1 + u/c)^{\frac{1}{2}}(1 - u/c)^{-\frac{1}{2}};$$

we have canceled the common factor of $c$ and use the exponent $\frac{1}{2}$ in place of the square-root sign. For low speeds $u/c \ll 1$. Now use the BINOMIAL THEOREM (Appendix H) on each of the factors involving $u/c$, identifying the $x$ in Appendix H with $u/c$. Use the first version of the BINOMIAL THEOREM with the first factor and the second version with the second factor and keep only the first two terms of each. That is, write

$$(1 + u/c)^{\frac{1}{2}} = 1 + \tfrac{1}{2}u/c,$$

$$(1 - u/c)^{-\frac{1}{2}} = 1 + \tfrac{1}{2}u/c.$$

Making these substitutions in the first equation gives

$$\frac{1}{\lambda} = \frac{1}{\lambda_0}(1 + \tfrac{1}{2}u/c)^2 \approx \frac{1}{\lambda_0}(1 + u/c).$$

In the last step we dropped the term $\tfrac{1}{4}(u/c)^2$ as being very small compared with the other terms. Rewriting our last equation gives

$$\lambda_0 = \lambda(1 + u/c),$$

$$\Delta\lambda/\lambda = u/c,$$

where $\Delta\lambda = \lambda_0 - \lambda$.

## 42-11

From Fig. 1 we pick $\lambda_0 = 475$ nm for blue light. Now use Eq. 5 but put $u/c = -0.2$, since the rocketship is receding and Eq. 5 is written for source and observer approaching. We obtain

$$\frac{1}{\lambda} = \frac{1}{475 \text{ nm}} \sqrt{\frac{1 - 0.2}{1 + 0.2}},$$

$$\lambda = 582 \text{ nm}.$$

Turning again to Fig.1, we find that a wavelength of 582 nm corresponds to a yellow or perhaps yellow-orange color. (The color demarcations are not sharp, so there may be some uncertainty in the color assignments.)

## 42-14

Eq. 5 applies directly since you are approaching the light:

$$\frac{1}{\lambda} = \frac{1}{\lambda_0} \sqrt{\frac{1 + u/c}{1 - u/c}}.$$

Square both sides and solve for $u/c$ to find

$$u/c = \frac{x - 1}{x + 1}, \quad x = (\lambda_0/\lambda)^2.$$

In our present case, $x = (620/540)^2 = 1.318$, so we get $u/c = 0.137$, or $u = 4.11 \times 10^7$ m/s.

## 42-15

One edge of the Sun approaches at speed $u = R\omega$, and the other edge recedes at the same speed. Here $R$ is the radius of the Sun and $\omega = 2\pi/T$ is the angular frequency of rotation; see Eq. 12 in Chapter 11. As instructed, turn to Appendix C for needed data on the Sun; we find $R = 6.96 \times 10^8$ m and $T = 26$ d at the equator of the Sun (see the fine print). From these data, we calculate the linear speed of rotation:

$$u = R[2\pi/T] = (6.96 \times 10^8 \text{ m})[2\pi/(26 \text{ d})(86,800 \text{ s/d})],$$

giving $u = 1950$ m/s $< c$. Evidently, we can use the low speed version of the Doppler shift formula, as presented in Problem 8; that is

$$\Delta\lambda = \lambda(u/c),$$

$$\Delta\lambda = (553 \text{ nm})(1950/3 \text{ X } 10^8) = 0.0036 \text{ nm}.$$

Hence the Doppler shift is ±0.0036 nm, the + sign for light from the receding edge and the - sign for the approaching edge.

## 42-24

(a) We have $u/c = 0.05$. The aberration equation is given in Problem 22(a). For $\theta' = 0°$, this equation gives $\theta = 0°$ also.
(b) With $u/c = 0$ and $\theta' = 45°$, the aberration equation gives $\theta = 43°$.
(c) In this case $\theta' = 90°$; with $u/c = 0.05$ again, the aberration equation yields $\theta = 87°$.

## 42-27

For the first source the angle $\theta$ in Eq. 18 is 90°. We cannot ignore the Lorentz factor (as was done in Sample Problem 3) since $u/c = 0.717$ is not very small compared with unity (i.e., with one). Since $\cos 90° = 0$, we obtain

$$\nu = (188 \text{ MHz})\sqrt{1 - 0.717^2} = 131 \text{ MHz}.$$

For the second source we want to have $\nu = 131$ MHz also. The rest frequency for this source is $\nu_0 = 162$ MHz. Eq. 18 yields

$$131 \text{ MHz} = (162 \text{ MHz})\frac{\sqrt{1 - 0.717^2}}{1 - 0.717\cos\theta}.$$

Solving this equation we find $\theta = 78.9°$.

## 43-1

(a) The angle $\theta_1$ is the angle of reflection and this equals the angle of incidence (see Eq. 1); that is, $\theta_1 = 38°$.
(b) Apply Snell's law, Eq. 2, to find the angle of refraction $\theta_2$:

$$n_1\sin\theta_1 = n_2\sin\theta_2,$$

$$(1.58)\sin38° = (1.22)\sin\theta_2,$$

$$\theta_2 = 52.9°.$$

## 43-6

When the pipe is evacuated, the light beam travels the distance $L$ at the speed of light in a vacuum $c$, for a travel time of $L/c$. With air in the pipe the travel time is $L/v$, where the speed of light in air is $v = c/n$, by Eq. 3. Here $n$ is the index of refraction of air; its value is given in Table 1. Hence, the difference in travel times is

$$\Delta t = \frac{L}{v} - \frac{L}{c} = \frac{L}{c/n} - \frac{L}{c} = \frac{L}{c}(n - 1),$$

$$\Delta t = \frac{1610 \text{ m}}{3 \times 10^8 \text{ m/s}}(1.00029 - 1) = 1.56 \text{ ns}.$$

Since $v < c$, the travel time is greater when the pipe contains air.

## 43-12

Using Eq. 3 for the speed of light in the glass block, the desired travel time is

$$t = AB/v = n(AB)/c.$$

The distance $AB$ is given by

$$AB = L/\cos\phi,$$

where $\phi$ is the angle of refraction at $A$ (the angle between the dashed normal and the light beam in the glass block). Therefore, the travel time is

$$t = nL/c\cos\phi.$$

To find $\phi$, use the law of refraction at A; we find

$$n_{air}\sin\theta = n\sin\phi,$$

$$(1)\sin 24° = 1.63\sin\phi,$$

$$\phi = 14.45°.$$

We can now evaluate numerically the formula above for the travel time to get

$$t = (1.63)(0.547 \text{ m})/(3 \times 10^8 \text{ m/s})\cos 14.45° = 3.07 \text{ ns}.$$

<u>43-14</u>

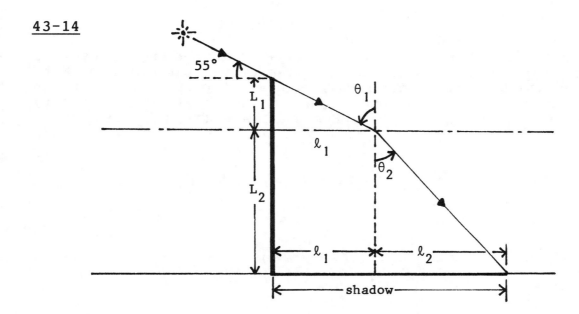

The length $s$ of the shadow is (see sketch above)

$$s = l_1 + l_2 = L_1\tan\theta_1 + L_2\tan\theta_2.$$

In this equation $L_1$ = 64 cm, $L_2$ = 200 cm - 64 cm = 136 cm, and $\theta_1$ = 90° - 55° = 35°. The angle $\theta_2$ must be found from the law of refraction; using $n = 1$ for air and $n = 1.33$ for water (Table 1), we find

$$(1)\sin 35° = (1.33)\sin\theta_2,$$

$$\theta_2 = 25.5°.$$

Therefore, the length of the shadow is

$$s = (64 \text{ cm})\tan 35° + (136 \text{ cm})\tan 25.5° = 110 \text{ cm}.$$

From the small right triangle at the top of the sketch, we have

$$\tan\theta_2 = x/d_{app}.$$

Since the light rays are close to the normal, the angle $\theta_2$ is very small. If we express the angle in radians, then

$$\tan\theta_2 \approx \theta_2,$$

by the last of the TRIGONOMETRIC EXPANSIONS found in Appendix H. In this case we have

$$d_{app} = x/\theta_2.$$

By the law of refraction,

$$n\sin\theta_1 = (1)\sin\theta_2.$$

Again, employ a small-angle approximation; this time use

$$\sin\theta \approx \theta,$$

applied to each of the angles in the equation above to get

$$n\theta_1 = \theta_2.$$

With this, the formula for the apparent depth becomes

$$d_{app} = x/n\theta_1.$$

Now consider the large right triangle in the sketch; we find that

$$\tan\theta_1 = x/d \approx \theta_1.$$

Solve this last expresssion for $x$ and substitute into the equation for the apparent depth to obtain finally

$$d_{app} = \theta_1 d/n\theta_1 = d/n.$$

43-23

Cerenkov radiation is emitted if the speed $v$ of the particle exceeds the speed $c/n$ of light in the material. We want

$$v_{muon} > c/n > v_{pion},$$

in order that the muons emit Cerenkov radiation and the pions do

301

not. The corresponding range of values of $n$ is

$$c/v_{muon} < n < c/v_{pion}.$$

Thus, we need to find $c/v$ for each particle. We are given the momentum $p$ and mass $m$ for each (in fact, the momenta are equal). The relativistic momentum is given in magnitude by (see Eq. 23 of Chapter 21)

$$p = \frac{mv}{\sqrt{1 - (v/c)^2}}.$$

This equation can be inverted to solve for $c/v$, obtaining

$$\frac{c}{v} = \frac{\sqrt{1 + (p/mc)^2}}{p/mc}.$$

For the muon, $mc = 106$ MeV/$c$ so that $p/mc = 145/106 = 1.368$. The equation above then gives $c/v_{muon} = 1.24$. For the pion, $mc = 135$ MeV/$c$, $p/mc = 145/135 = 1.074$ and $c/v_{pion} = 1.37$. Thus we must have $1.24 < n < 1.37$ for the muons, but not the pions, to emit Cerenkov radiation as they pass through this material.

43-26

The image is 10 cm behind the mirror, the same distance that the object is in front of the mirror. Since you are 30 cm in front of the mirror you are 40 cm from the image, so that 40 cm is the distance for which you must focus your eyes.

43-34

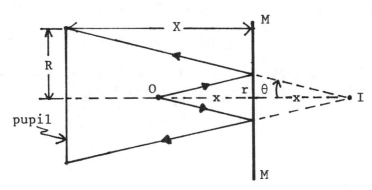

In the sketch the pupil is a distance $X$ from the mirror MM. The object O is a distance $x$ from the mirror, and therefore we know that the image I is a distance $x$ behind the mirror. Consider the rays shown, which leave the object at the greatest possible angle and still enter the pupil. They define the radius $r$ of the region of the mirror that is used in viewing the image. (The diagram can

302

be rotated about the line OI, the line $r$ sweeping out a circular region.) From the triangles in the sketch we have

$$\tan\theta = r/x = R/(X + x),$$

$$r = Rx/(X + x) = (0.25)(10)/(24 + 10) = 0.0735 \text{ cm}.$$

(All quantities are expressed in cm; clearly the diagram is not drawn to scale.) Finally, the desired area on the mirror is

$$A = \pi r^2 = \pi(0.0735 \text{ cm})^2 = 0.0170 \text{ cm}^2.$$

## 43-35

Let $P$ be the power output of the source. With only the screen present, the intensity at the center of the screen, call it $I_1$, is

$$I_1 = P/4\pi d^2,$$

assuming that the source radiates uniformly in all directions. With the mirror also present, we still receive radiation at the center of the screen of this intensity directly from the source. We also receive radiation from the image of the source formed by the mirror. This image is at a distance $d$ behind the mirror, so that it is $3d$ from the screen. Hence, the intensity at the center of the screen with the mirror present, call it $I_2$, is

$$I_2 = I_1 + P/4\pi(3d)^2 = I_1 + (1/9)I_1 = (10/9)I_1.$$

## 43-39

(a) By Eq. 13,

$$\lambda_n = \lambda/n = (612 \text{ nm})/(1.51) = 405 \text{ nm}.$$

(b) By Eq. 15, the optical path length is

$$L = nL_1 = (1.51)(1.57 \text{ μm}) = 2.37 \text{ μm}.$$

(c) The number of vibrations of the light beam traveling the distance in a vacuum is

$$\#_{vac} = L_1/\lambda = (1570 \text{ nm})/(612 \text{ nm}) = 2.565.$$

Similarly, the number of vibrations by the light beam traveling the same distance through the material medium is

$$\#_{med} = L_1/\lambda_n = (1570 \text{ nm})/(405 \text{ nm}) = 3.877.$$

The difference is $\Delta\# = 1.312$ vibrations, for a phase difference of 0.312 vibrations or $(0.312)(360°) = 112°$.

(*a*) Let $n_1$ be the index of
refraction of the medium in
which the prism is immersed.
The incident light strikes face
*ab* at an angle of incidence of 0°
so there is no refraction there.
The angle of incidence at the
hypotenuse of the prism is
90° - $\phi$. For this to be the
critical angle for total internal
reflection at *ac*, the angle of
refraction must be 90°. By Snell's law,

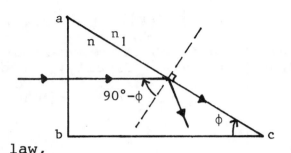

$$n\sin(90° - \phi) = n\cos\phi = n_1\sin90°,$$

$$\phi = \cos^{-1}(n_1/n).$$

For air $n_1$ = 1; with $n$ = 1.52, we get $\phi$ = 48.9°.
(*b*) For the prism immersed in water put $n_1$ = 1.33; we now get $\phi$ =
29.0°.

(*a*) Under the assumption of the problem, if $\theta$ is the critical angle
for total internal reflection, the rays emitted at angles with
respect to the vertical greater than $\theta$ are reflected back into the
water and do not escape. Only those rays emitted into directions
lying inside a cone of semiangle $\theta$ about the vertical will escape.
Imagine the source surrounded by a sphere of radius $r < h$. If $A$ is
the area intercepted by the cone on the sphere, then the desired
ratio is

$$f = A/4\pi r^2,$$

since the source is presumed, in the absence of contrary
information, to radiate uniformly in all directions. Now this area
is given by

$$A = 2\pi r^2(1 - \cos\theta),$$

a rather unfamiliar formula. By Eq. 16, the condition for the onset
of total internal reflection is

$$\sin\theta = 1/n,$$

using $n$ = 1 for air. We can solve for $\cos\theta$ from this using the
trigonometric identity

$$\sin^2\theta + \cos^2\theta = 1.$$

If we do this, put the resulting expression for $\cos\theta$ into the formula for A and then the expression for A into that for the fraction $f$, we find

$$f = \tfrac{1}{2}\left(1 - \sqrt{1 - 1/n^2}\right).$$

(b) The index of refraction of water is 1.33, so the ratio in this case is $f = 0.17$.

## 43-48

(a) As the light enters the fiber, it is refracted according to

$$(1)\quad \sin\theta = n_1\sin\phi,$$

using $n = 1$ as the index of refraction for air. The refracted ray strikes the cladding at an angle of incidence of $90° - \phi$. The critical condition for total internal reflection at the cladding, Eq. 16, becomes

$$n_1\cos\phi = n_2.$$

This can be rewritten in the form

$$\sin\phi = \sqrt{1 - \cos^2\phi} = \sqrt{1 - (n_2/n_1)^2}.$$

Hence, by the first equation,

$$\sin\theta = n_1\sqrt{1 - (n_2/n_1)^2} = \sqrt{n_1^2 - n_2^2}.$$

(b) Substituting $n_1 = 1.58$ and $n_2 = 1.53$ gives $\theta = 23.2°$.

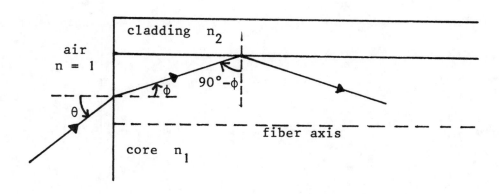

305

In Eq. 16 for the critical angle for total internal reflection, we identify $n_1$ with the quartz and $n_2$ with air and use $n_2 = 1$. From Fig. 4, for blue light (400 nm) $n_1 = 1.470$; for red light (700 nm) $n_1 = 1.455$. Eq. 16 now gives $\theta_{c,blue} = 42.9°$ and $\theta_{c,red} = 43.4°$. We get total internal reflection at angles of incidence $\theta > \theta_c$. Therefore, if $\theta > \theta_{c,red}$, both the red and blue light undergo total internal reflection and the beam appears white. If $\theta < \theta_{c,blue}$, neither beam undergoes total internal reflection; both are reflected, however, and the reflected beam will therefore appear white. However, if

$$\theta_{c,blue} < \theta < \theta_{c,red},$$

then the blue light undergoes total internal reflection but the red light does not; some of the red light is refracted out into the air. Hence the internally reflected beam is deficient in red light with no loss of blue light and appears bluish. We conclude that for $42.9° < \theta < 43.4°$, the internally reflected beam appears bluish. We cannot get a reddish beam at any angle of incidence.

44-1

By Eq. 5, $m = -i/o = 2.7$, so that

$$\frac{1}{o} = -\frac{2.7}{i}.$$

Substitute this into Eq. 1 to find

$$\frac{1}{i} + \frac{1}{o} = \frac{2}{r} = \frac{2}{35},$$

$$\frac{1}{i} - \frac{2.7}{i} = \frac{2}{35},$$

$$i = -29.8 \text{ cm.}$$

Hence, the object (face) distance from the mirror is

$$o = -i/2.7 = -(-29.8 \text{ cm})/(2.7) = 11.0 \text{ cm.}$$

44-3

(c) Since $f$ is positive, the mirror is concave. By Eq. 2, the radius of curvature is $r = 2f = 2(+20 \text{ cm}) = +40 \text{ cm}$. With $o = +30$ cm, we use Eq. 3 to find the image distance. Solving Eq. 3 for $i$ yields

$$i = fo/(o - f),$$

$$i = (20)(30)/(30 - 20) = +60 \text{ cm.}$$

Since $i$ is positive, we have a real image. By Eq. 5, we find the magnification to be

$$m = -i/o = -(+ 60 \text{ cm})/(+30 \text{ cm}) = -2.$$

With $m$ negative, the image is inverted so that: no, we do not have an upright image. Finally, note that numerically $r > o > f$, so that the ray diagram is close to that shown in Fig. 3a. In that figure, we see that the image is indeed inverted, about twice the size of the object, and properly located ($i > r$).

(a) Apply the law of refraction
at the point where the ray shown
on the sketch strikes the sphere
to find that

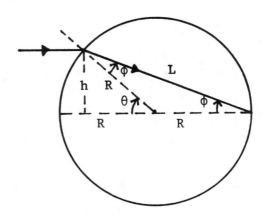

$$(1) \sin\theta = n\sin\phi.$$

By examining the various triangles
in the sketch, we see that

$$\sin\phi = h/L = h/(2R\cos\phi),$$

$$\sin\phi = \sin\theta/2\cos\phi.$$

Now substitute the result from the law of refraction to obtain

$$\sin\phi = (n\sin\phi)/2\cos\phi,$$

$$\cos\phi = \tfrac{1}{2}n.$$

Therefore,

$$\sin\phi = \sqrt{1 - \cos^2\phi} = \sqrt{1 - n^2/4},$$

and

$$\sin\theta = n\sin\phi = h/R = n\sqrt{1 - n^2/4}.$$

For any value of $n < 2$, there is only one value of $h$, the value
satisfying the equation above, such that the ray strikes the back
of the sphere. A parallel beam of light (containing rays over a
range of values of $h$) will not be brought to a focus at the back of
the sphere.
(b) If $n \approx 2$, then $h/R \approx 0$, indicating that a very narrow beam will
be brought to a focus at the back of the sphere.
(c) For the beam to be brought to a focus at the center of the
sphere, the refracted rays would have to pass along the normals.
This means an angle of refraction of 0°, which violates the law of
refraction for non-zero angles of incidence.

44-13

Solving Eq. 13 for the image distance gives

$$i = fo/(o - f).$$

Clearly, in looking at the Sun, we have $o \gg f$ so that this reduces to $i = f$ for all practical purposes. By Eq. 15, disregarding the minus sign,

$$m = i/o = f/o = d_i/D,$$

where $d_i$ is the diameter of the image and $D$ is the diameter of the Sun. From Appendix C we conclude that $D = 2(6.96 \times 10^8$ m$) = 1.392 \times 10^9$ m; also $o = 1.5 \times 10^{11}$ m (Earth-Sun distance). Therefore, the diameter of the image is

$$d_i = D(f/o),$$

$$d_i = (1.392 \times 10^9 \text{ m})(0.27 \text{ m}/1.5 \times 10^{11} \text{ m}) = 0.0025 \text{ m} = 2.5 \text{ mm}.$$

## 44-14

(a) Suppose we orient the lens as shown in the sketch and presume that the object is to the left of the lens. Then the region to the left of the lens is the V-side and the region to the right is the R-side. The radius of curvature of the flat side of the lens is $r_1 = \infty$. The center of curvature C of the convex side is in the V-region, so that $r_2 = -20$ cm. By Eq. 14,

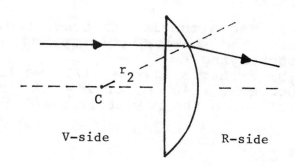

$$\frac{1}{f} = (1.5 - 1)\left[0 - \frac{1}{(-20 \text{ cm})}\right],$$

$$f = +40 \text{ cm}.$$

(b) With $o = 40$ cm, we have $o = f$. Eq. 13 now yields $i = \infty$; i.e., the light emerges from the lens parallel to the axis and is not brought to a focus.

## 44-19

The Newtonian form applies to converging lenses ($f > 0$). Begin with the Gaussian form, Eq. 13. In terms of $x$ and $x'$, the object and image distances are

$$o = f \pm x,$$

$$i = f \pm x'.$$

In these equations, the upper signs apply for a real image ($o > f$, $i > 0$), the lower signs for a virtual image ($o < f$, $i < 0$). Put these expressions into the Gaussian formula to obtain

$$\frac{1}{f \pm x} + \frac{1}{f \pm x'} = \frac{1}{f}.$$

Combine the two terms on the left into a single fraction and then cross-multiply with the right to obtain

$$2f^2 \pm xf \pm x'f = f^2 \pm xf \pm x'f + xx',$$

$$f^2 = xx',$$

in which $x$ and $x'$ both are positive.

## 44-22

(a) The situation as described, with the lens between the object and the image, corresponds to Fig. 14(a), so the lens is converging.
(b) With $m = -i/o = -1/2$, we have $o = 2i$. But $d = i + o = 40$ cm. Hence $3i = 40$ cm, or $i = 13.3$ cm. Therefore $o = 26.7$ cm.
(c) Using the result from (b) in Eq. 13, we find that $f = 8.89$ cm.

## 44-24

Solve Eq. 13 for the image distance $i$ to find that

$$i = fo/(o - f).$$

Let $x$ be the distance between object and image, so that

$$x = i + o.$$

(All quantities are positive for the optical arrangement under consideration.) Substituting the first equation for $i$ into the equation for $x$ gives

$$x = o^2/(o - f).$$

To find the minimum $x$, set $dx/do = 0$; one finds

$$dx/do = o(o - 2f)/(o - f)^2 = 0,$$

$$o = 2f.$$

Putting this result into the equation for $x$ gives for the minimum value of $x$

$$x = (2f)^2/(2f - f) = 4f.$$

This must be the minimum $x$, rather than the maximum, because we know that for $o \approx f$, the image distance $i \approx \infty$.

## 44-31

(a) We work in meters for all distances. Use Eq. 13 to find the distance $i_1$ of the image formed by the lens:

$$i_1 = o_1 f/(o_1 - f) = (1.12)(0.58)/(1.12 - 0.58) = 1.203 \text{ m}.$$

This distance is measured from the lens. Hence, this image is at a distance 1.97 m - 1.203 m = 0.767 m in front of the plane mirror. For the mirror, it is a real object because light diverges from it. The mirror form an image of this image at the same distance, 0.767 m, behind the mirror. This image is at a distance $o_2 = 1.203 + 2(0.767) = 2.737$ m from the converging lens. Hence, the final image is at a distance

$$i_2 = o_2 f/(o_2 - f) = (2.737)(0.58)/(2.737 - 0.58) = 0.736 \text{ m}$$

from the lens, on the side of the lens containing the original object. You might want to make a sketch of the situation.
(b) See Fig. 14(a) The image is real, since it is formed from a converging lens with $o_2 > f$.
(c) The first image formed by the lens is inverted. The mirror preserves this orientation in the image it forms. The final image is inverted with respect to the image formed by the mirror, and therefore upright compared with the original object.
(d) For the first image formed by the lens,

$$m_1 = -i_1/o_1 = -1.203/1.12 = -1.074.$$

The magnification of the plane mirror is +1. The magnification by the lens in forming the final image from the image in the mirror is

$$m_2 = -i_2/o_2 = -0.736/2.737 = -0.2689.$$

Hence the final magnification, relative to the original object, is

$$M = (-1.074)(+1)(-0.2689) = +0.289.$$

## 44-36

(a) Examining Fig. 25, we see that

$$s = 25 - 4.2 - 7.7 = 13 \text{ cm}.$$

(b) We want the image distance to be $i = s + f_{ob} = 13 + 4.2 = 17.2$ cm. With $f_{ob} = 4.2$ cm, Eq. 13 gives $o = 5.56$ cm. The distance from $F_1 = 5.56 - 4.2 = 1.4$ cm.
(c) By Eq. 15, $m = -i/o = -17.2/5.56 = -3.1$.
(d) From Eq. 29, $m_\theta = (25 \text{ cm})/f_{ey} = 25/7.7 = 3.2$.

(e) The final magnification is given by Eq. 31:

$$M = -(s/f_{ob})(25/f_{ey}) = -(13/4.2)(25/7.7) = -10.$$

## 44-42

(a) The mirror $M$ may be thought of as replaced by a lens of the same focal length. The mirror $M'$, being flat, does not affect the magnification, since for a flat mirror the magnification is +1. Hence

$$m_\theta = -f_{ob}/f_{ey},$$

as before, the lens also inverting the image.
(b) Considering the mirror to be spherical, Eq. 3 gives

$$i = f_{ob}o/(o - f_{ob}) = (16.8)(2000)/(2000 - 16.8) = 16.94 \text{ m.}$$

The magnification is $m = -i/o = -(16.94 \text{ m})/(2000 \text{ m}) = 0.00847$. Since the object's size is 1 m, the image size is 8.47 mm.
(c) The focal length is $f = r/2 = 5.0$ m. Thus, by the first equation, $m_\theta = 200 = (5 \text{ m})/f_{ey}$, so that $f_{ey} = 2.5$ cm.

## 45-5

For small angles, $\sin\theta \approx \theta$ (when $\theta$ is expressed in radians rather than degrees), so that Eq. 1,

$$d\sin\theta = m\lambda,$$

becomes

$$d\theta = m\lambda.$$

For changes in $\theta$ and consequent changes in $\lambda$, we have

$$d(\Delta\theta) = m(\Delta\lambda).$$

Combining this with the previous equation gives

$$\Delta\lambda = \lambda(\Delta\theta/\theta) = (589 \text{ nm})(0.10\theta/\theta) = 59 \text{ nm}.$$

Hence the desired wavelength is

$$\lambda_{new} = \lambda + \Delta\lambda = 589 + 59 = 648 \text{ nm}.$$

## 45-10

The geometric path lengths from the slits to the central part of the screen, with the mica in place, are equal. The phase difference arises because one beam traverses a distance $x$ in mica as the other is traversing the same distance in air. The phase difference so introduced must be $7(2\pi \text{ rad})$, since $2\pi$ rad corresponds to one wavelength. Hence,

$$\frac{2\pi x}{\lambda_n} - \frac{2\pi x}{\lambda_a} = 7(2\pi),$$

$$x(\frac{1}{\lambda/n} - \frac{1}{\lambda}) = 7,$$

$$x = 7\lambda/(n - 1) = 7(550 \text{ nm})/(1.58 - 1) = 6640 \text{ nm}.$$

## 45-17

Examining Fig. 4, we see that for the rays from the two slits, the slit-screen distances to the fringe in question are

$$r_1{}^2 = D^2 + (y - \tfrac{1}{2}d)^2,$$

$$r_2{}^2 = D^2 + (y + \tfrac{1}{2}d)^2.$$

Now the fractional error should be small, so for $y$, use the approximate formula (see Sample Problem 2):

$$y = \lambda mD/d = (589 \text{ nm})(10)(40 \text{ mm})/(2 \text{ mm}) = 0.1178 \text{ mm}.$$

Using this, we find for the slit-screen distances

$$r_1{}^2 = 40^2 + (0.1178 - 1)^2 \rightarrow r_1 = 40.0097273 \text{ mm},$$

$$r_2{}^2 = 40^2 + (0.1178 + 1)^2 \rightarrow r_2 = 40.0156154 \text{ mm}.$$

Hence the difference in these distances is $\Delta r = 5.8881 \times 10^{-3}$ mm. In Eq. 1, $d\sin\theta$ is the approximate value of $\Delta r$; we find that

$$d\sin\theta = m\lambda = (10)(589 \times 10^{-6} \text{ mm}) = 5.89 \times 10^{-3} \text{ mm}.$$

Therefore, the fractional error sought is

$$\text{error} = (5.89 - 5.8881)/(5.8881) = 3.2 \times 10^{-4}.$$

## 45-23

(a) The optical path difference $S_2D - S_1D$ equals the geometric path difference since the paths from both sources to the detector are in the same medium i.e., air. For maxima, this path difference must equal an integral number of wavelengths; by the Pythagorean theorem, this condition becomes

$$\sqrt{d^2 + x^2} - x = n\lambda,$$

$n = 1, 2, 3, \ldots$ . The value $n = 0$ is not possible since $d \neq 0$. Solving this equation for $x$ gives

$$x = (d^2 - n^2\lambda^2)/2n.$$

Putting $n = 3, 2, 1$, together with $d = 4.17$ m, $\lambda = 1.06$ m gives $x_3 = 1.21$ m, $x_2 = 3.22$ m, and $x_1 = 8.13$ m.
(b) For completely destructive interference the waves must arrive at the detector exactly 180° out of phase and also with equal amplitudes. The first requirement can be met but the second cannot since the waves, being spherical, experience a $1/d^2$ fall-off in intensity and therefore a $1/d$ fall-off in amplitude, and the distances $d$ are necessarily different for the two waves. (It is implied in the problem statement that the sources radiate waves of the same amplitude, since they are excited by the same oscillator.)

The intensity $I$ is given by Eqs. 11 and 12:

$$I = 4I_0\cos^2\tfrac{1}{2}\phi,$$

with $\phi = 2\pi d\sin\theta/\lambda \approx 2\pi\theta d/\lambda$ for small $\theta$. At a point where the intensity is one-half the maximum of $4I_0$, we have

$$2I_0 = 4I_0\cos^2\tfrac{1}{2}\phi,$$

$$\cos^2\tfrac{1}{2}\phi = 1/2.$$

The smallest positive $\phi$ satisfying this equation is $\phi = \pi/2$. Hence, the first half-intensity point occurs where $\theta$ has the value given from

$$\pi/2 = 2\pi\theta d/\lambda,$$

$$\theta = \lambda/4d.$$

A symmetrical half-intensity point falls at $\theta = -\lambda/4d$, with the $\theta = 0$ maximum between them. Thus the half-width of this central maximum is $\Delta\theta = 2(\lambda/4d) = \lambda/2d$.

The electric field components of the two waves are

$$E_1 = E_0\sin\omega t,$$

$$E_2 = 2E_0\sin(\omega t + \phi),$$

with $\phi = 2\pi d\sin\theta/\lambda$. The sum is to be expressed in the form

$$E = E_1 + E_2 = E_\theta\sin(\omega t + \beta).$$

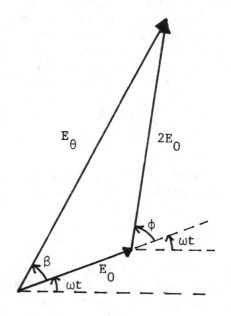

The intensity of this wave is proportional to $E_\theta^2$. Apply the law of cosines (see Appendix H under TRIANGLES, the last formula) to the triangle formed by $E_1$, $E_2$, $E_\theta$ in the phasor diagram to get

$$E_\theta^2 = E_0^2 + (2E_0)^2 - 2(E_0)(2E_0)\cos(\pi - \phi) = E_0^2[5 + 4\cos\phi].$$

Hence,

$$I = I_0[5 + 4\cos\phi] = I_0[1 + 8\cos^2(\tfrac{1}{2}\phi)],$$

using one of the TRIGONOMETRIC IDENTITIES in Appendix H in the last step. At the center of the slit ($\phi = 0$) we have $I = I_m = 9I_0$ so we have finally

$$I = (I_m/9)[1 + 8\cos^2(\pi d\sin\theta/\lambda)].$$

## 45-27

Both rays, one reflected from the top of the coating and the other from the coating-glass interface, experience a phase change of $\pi$ on reflection. Hence, the relative phase difference induced by the reflections is zero and only the difference in path lengths is effective. The conditions for maxima and minima given in Eqs. 16 and 17, which presume a relative phase change on relection of $\pi$, must be reversed. That is, for cancellation we must have

$$2dn = (m + \tfrac{1}{2})\lambda.$$

For the thinnest coating pick $m = 0$ to get

$$d = \lambda/4n = (620 \text{ nm})/4(1.25) = 124 \text{ nm}.$$

## 45-32

The ray striking the outer surface of the soap film from the air undergoes a phase change of $\pi$ on reflection. The ray that passes into the film and is reflected at the inner soap-air interface does not under a phase change on reflection. Hence there is a relative phase change of $\pi$ due to the reflections and the condition for a maximum is as given in Eq. 16. To see if there actually is a maximum (bright region), solve the equation for $m$; we find

$$2(1210 \text{ nm})(1.33) = (m + \tfrac{1}{2})(585 \text{ nm}),$$

$$m = 5.00.$$

Since we actually obtain an integer, the condition for a bright region is satisfied.

## 45-33

There is a phase change of $\pi$ on reflection from both the upper and lower surfaces of the oil. Thus, there is no relative phase change due to the reflections, so that Eqs. 16 and 17 for the maxima and minima are interchanged. For minima, then,

$$2dn = (m + \tfrac{1}{2})\lambda.$$

If we take $\lambda_1 = 485$ nm, $\lambda_2 = 679$ nm, then $m_1 = m_2 + 1$ (there being no minima between these). Since $2dn$ is the same for the two wavelengths, the equation above implies that

$$[(m_2 + 1) + \tfrac{1}{2}](485 \text{ nm}) = [m_2 + \tfrac{1}{2}](679 \text{ nm}),$$

$$m_2 = 2.$$

Now use the first equation applied to $\lambda_2 = 679$ nm:

$$2d(1.32) = (2 + \tfrac{1}{2})(679 \text{ nm}),$$

$$d = 643 \text{ nm}.$$

## 45-38

(a) There is a phase change of $\pi$ on reflection for both the ray reflected from the top of the oil drop and the ray reflected from the bottom of the oil drop. This means zero relative phase change on reflection, so we interchange Eqs. 16 and 17 for the conditions for maxima and minima. At the outer regions of the drop, the thickness $d \approx 0$ and since $m \geq 0$, we see that only the condition for a maximum, Eq. 17, can be satisfied, with $m = 0$. Hence the outer regions are bright.
(b) From Fig. 1 in Chapter 42, we pick $\lambda = 475$ nm for blue light. Eq. 17 yields for the thickness,

$$d = m\lambda/2n = (3)(475 \text{ nm})/2(1.2) = 594 \text{ nm}.$$

(c) As the thicker regions of the drop are examined, the fringes fall closer and closer together, since closely neighboring regions can differ in thickness by many wavelengths. Eventually, the finges cannot be distinguished by the unaided eye.

## 45-39

There is no phase change on reflection for the ray reflected from the bottom surface of the top plate, but there is a phase change of $\pi$ for the ray reflected from the top surface of the bottom plate. Hence the condition of maxima is that of Eq. 16, with $n = 1$ (air film); i.e.,

$$2d = (m + \tfrac{1}{2})\lambda.$$

If a bright fringe appears at the end, it would have an order number

$$m = 2d/\lambda - \tfrac{1}{2} = 2(48,000 \text{ nm})/(680 \text{ nm}) - \tfrac{1}{2} = 140.7.$$

This indicates that a bright fringe does not lie at the end ($m$ must be an integer; 140.7 is not close enough). The maximum nearest the end must be the $m = 140$ maximum. Since there is an $m = 0$ bright fringe near the other end of the plates (where the air wedge has a thickness of $\lambda/4$), there are $140 + 1 = 141$ bright fringes in all. Note that the 120 mm length of the plates does not enter directly; of course this dimension does affect the thickness of the fringes.

From Sample Problem 6 we have for the radii of the rings

$$r^2 = (m + \tfrac{1}{2})\lambda R/n,$$

where we have put $\lambda/n$ in place of $\lambda$ to account for the medium between the lens and the plate having an index of refraction $n \neq 1$. For air between lens and plate put $n = 1$ to find

$$(1.42/2)^2 = (10 + \tfrac{1}{2})\lambda R.$$

(We note that the radius is one-half the given diameter.) With the liquid having displaced the air, we have

$$(1.27/2)^2 = (10 + \tfrac{1}{2})\lambda R/n.$$

Solve for the quantity $\lambda R$ in the first equation and substitute the result into the second equation to find

$$n = (1.42/1.27)^2 = 1.25.$$

(a) The intensity is given by Eq. 13, which we can write as

$$I = I_m\cos^2\tfrac{1}{2}\phi.$$

We find the phase difference $\phi$ from the top equation, right column, on p. 953 of HRK; set the path difference $= 2d$, replace $\lambda$ by $\lambda/n$ to find that

$$\phi = 4\pi nd/\lambda = 4\pi(1.38)(100 \text{ nm})/\lambda = (1734 \text{ nm})/\lambda.$$

With $\lambda = 450$ nm, this gives $\phi = 3.853$ rad $= 221°$. The first equation now gives $I = 0.123I_m$, for an 88% reduction.
(b) With $\lambda = 650$ nm, we find $\phi = 2.668$ rad $= 153°$; in this case $I = 0.055I_m$, for a 95% reduction.

Let $L$ be the length of the chamber.The phase difference between the two rays arises from the passage of light in one arm through a length $2L$ in air of density less (once the pumping starts) than the air in the other arm. The density affects the index of refraction. If the lengths of the arms are not changed during the pumping process then the phase difference due to the differences in air density is given by

$$\phi = 2\pi[\text{optical path difference}]/\lambda,$$

where the optical paths are $n_1(2L)$ and $n_2(2L)$. If we assume perfect pumping, then put $n_2 = 1$ for the arm with the chamber, and $n_1 = n$ in the other arm, where $n$ is the density of air under normal atmospheric pressure. We now have

$$\phi = 4\pi L(n - 1)/\lambda.$$

Each fringe shift corresponds to a phase change of $2\pi$, so the number $N$ of fringes that pass across the field of view during removal of the air from the chamber is

$$N = \phi/2\pi = 2L(n - 1)/\lambda.$$

Substituting the data and solving for $n$ gives

$$60 = 2(5 \times 10^7 \text{ nm})(n - 1)/(500 \text{ nm}),$$

$$n = 1.0003.$$

## CHAPTER 46

### 46-3

(a) Let the slit-screen distance be $D$, and let $y$ be the distance from $P_0$ (which locates the central maximum) to any feature. Then, from Fig. 7,

$$\theta = \tan^{-1}(y/D) = \tan^{-1}(1.62 \text{ cm}/216 \text{ cm}) = 0.430°.$$

(b) By Eq. 3, using the angle found in (a), we have

$$a = m\lambda/\sin\theta,$$

$$a = (2)(441 \text{ nm})/(\sin 0.430°) = 1.18 \times 10^5 \text{ nm} = 118 \ \mu m.$$

### 46-7

The linear distance $y$ between the central maximum and either of the $m = 1$ minima is $\frac{1}{2}(5.20 \text{ mm}) = 2.6$ mm. With a slit-screen distance $D = 823$ mm, the diffraction angle $\theta$ of the first minmum is

$$\theta = \tan^{-1}(y/D) = \tan^{-1}(2.6/823) = 0.181°.$$

Hence, by Eq. 3 with $m = 1$,

$$a\sin 0.181° = (1)(546 \text{ nm}),$$

$$a = 1.73 \times 10^5 \text{ nm} = 173 \ \mu m.$$

### 46-10

Associate the wire diameter with the slit width $a$ in single-slit diffraction. By Eq. 3, the diffraction angle $\theta$ of either of the $m = 10$ minima is given by

$$a\sin\theta = m\lambda,$$

$$(1.37 \times 10^{-3} \text{ m})\sin\theta = (10)(632.8 \times 10^{-9} \text{ m}),$$

$$\theta = 0.2646°.$$

With the wire-screen distance $D = 2.65$ m, the linear distance $y$ from the central maximum to either of the minima in question is

$$y = D\tan\theta = (2650 \text{ mm})\tan 0.2646° = 12.2 \text{ mm}.$$

Hence, the distance between the two $m = 10$ minima, one on each side of the central maximum, is $2(12.2 \text{ mm}) = 24.4$ mm.

## 46-11

(*a*) The diffraction angle follows from

$$\theta = \tan^{-1}(y/D) = \tan^{-1}(0.0113 \text{ m}/3.48 \text{ m}) = 0.186°.$$

(*b*) Using Eq. 7 we find

$$\alpha = \pi a \sin\theta/\lambda,$$

$$\alpha = \pi(25.2 \times 10^3 \text{ nm})\sin 0.186°/(538 \text{ nm}) = 0.478 \text{ rad}.$$

(*c*) By Eq. 8,

$$I_\theta/I_m = [\sin^2(0.478 \text{ rad})]/(0.478)^2 = 0.926.$$

In evaluating the last expression on our calculator, we must either have the calculator ready to accept angles in radians, or first convert to degrees: $0.478 \text{ rad} = (0.478/\pi)(180°) = 27.4°$.

## 46-16

In the sketch shown, $A$ is the desired angle. Since the angles of a triangle total 180°, and the triangle is isosceles, we have

$$A = \tfrac{1}{2}(\pi - \phi) = \tfrac{1}{2}\pi - \alpha,$$

where

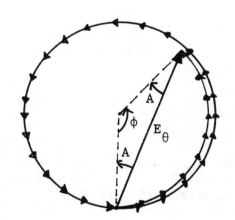

$$\alpha = \pi(a\sin\theta)/\lambda.$$

Now, at the first maximum beyond the central maximum,

$$a\sin\theta = (m + \tfrac{1}{2})\lambda,$$

where, in Problem 15, one finds that $m = 0.930$. Hence,

$$\alpha = \pi(m + \tfrac{1}{2}) = 1.43\pi \text{ rad} = 257.4°.$$

This means that $\phi = 2\alpha = 2(257.4°) = 514.8°$. We must subtract 360° for that part of the phasor chain that forms a closed circle, leaving $\phi = 514.8° - 360° = 154.8°$. Therefore, we actually have $\alpha = \tfrac{1}{2}\phi = \tfrac{1}{2}(154.8°) = 77.4°$, so that $A = 90° - 77.4° = 13°$.

## 46-17

(*a*) The angular separation inside the eye, which is filled with a fluid of index of refraction $n$, is, by Eq. 12,

$$\theta_R' = 1.22\lambda_n/d = 1.22(562 \text{ nm})/(5 \times 10^6 \text{ nm})n,$$

$$\theta_R' = (137 \text{ } \mu\text{rad})/n.$$

The angular separation outside the eye is $\theta_R$. But, by the law of refraction,

$$(1) \sin(\tfrac{1}{2}\theta_R) = n\sin(\tfrac{1}{2}\theta_R'),$$

which, for small angles, yields

$$\theta_R = n\theta_R' = 137 \text{ } \mu\text{rad}.$$

(b) The linear separation $\Delta x$ between the headlights and the distance $r$ to the car are related by

$$r = \Delta x/\theta_R = (1.42 \text{ m})/(137 \times 10^{-6}) = 10.4 \text{ km}.$$

## 46-19

The linear separation between the two points is found from

$$\Delta x = r\theta_R = r(1.22\lambda/d),$$

$$\Delta x = (3.82 \times 10^8 \text{ m})(1.22)(565 \times 10^{-9} \text{ m})/(5.08 \text{ m}),$$

$$\Delta x = 51.8 \text{ m}.$$

## 46-26

The angular diameter of the beam's central maximum is $\Delta\theta = 2\theta_R$. If the distance between the observatory and the shuttle is $r$ and the diameter of the central maximum is $D$, then

$$D = r(\Delta\theta) = 2r\theta_R = 2r(1.22\lambda/d),$$

and therefore,

$$d = 2.44r\lambda/D,$$

$$d = 2.44(354 \times 10^3 \text{ m})(500 \times 10^{-9} \text{ m})/(9.14 \text{ m}) = 4.73 \text{ cm}.$$

## 46-29

(a) The sketches on p. 323 illustrate the formation of a halo as the Moon shines through a cloud containing suspended water droplets. The ring will appear red if blue light is absent. Since the angle $\theta$ for the first minimum is given by the circular aperture diffraction formula $\sin\theta = 1.22\lambda/d$, blue light, as it has the smallest wavelength in the visible spectrum, will have its first minimum closer to the Moon (smallest $\theta$ gives the smallest $\phi$) than

any other color, giving the ring its reddish appearance.
(b) Since the rays MP and MO are virtually parallel, $\theta \approx \phi = 1.5(0.25°) = 0.375°$. Then, by (a),

$$d = 1.22\lambda_{blue}/\sin\theta,$$

$$d = 1.22(475 \text{ nm})/\sin 0.375° = 88.5 \ \mu m.$$

The wavelength of blue light was taken from Fig. 1 in Chapter 42.

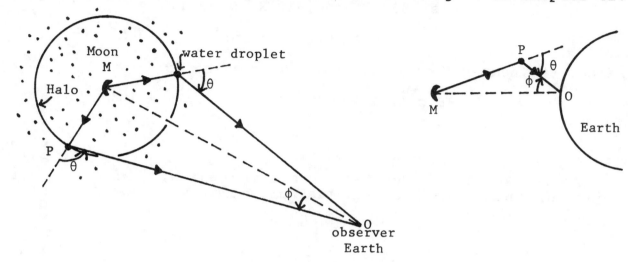

(c) A bluish ring will be seen at the scattering angle for the first minimum of red light. Since $\lambda_{red} = 700$ nm $\approx 1.5(\lambda_{blue})$, the radius of the blue ring will be 1.5 times that of the red ring, or about 2.3 times the apparent lunar radius. The intensity of the ring will be very low however. If water droplets of various sizes are present, variously-sized rings of these colors are present, giving the ring a whitish appearance due to the mixture of these colors.
(d) These halos are a diffraction effect, a rainbow being formed by refraction.

### 46-32

The location of the second minimum is given by $a\sin\theta = 2\lambda$; the interference fringes are located at angles given from $d\sin\theta' = m\lambda$. But $d = 11a/2$ (see Sample Problem 7) so that the interference fringes are given by $(11a/2)\sin\theta' = m\lambda$. Eliminating the slit width $a$ between the diffraction and interference equations gives $11\sin\theta' = m\sin\theta$. Hence, the $m = 11$ interference fringe falls at the same location ($\theta = \theta'$) as the second diffraction minimum and is not seen. There are 11 interference fringes inside the central maximum, so that the $m = 6$ fringe falls just outside the first diffraction minimum. Therefore, between the first and second diffraction minima will be found the $m = 6, 7, 8, 9, 10$ interference fringes, for five in all.

47-2

Find the ruling spacing from Eq. 1:

$$d\sin\theta = m\lambda,$$

$$d(\sin 33.2°) = 2(612 \text{ X } 10^{-7} \text{ cm}),$$

$$d = 2.235 \text{ X } 10^{-4} \text{ cm}.$$

Hence, the number $N$ of rulings on the grating is

$$N = (2.86 \text{ cm})/(2.235 \text{ X } 10^{-4} \text{ cm}) = 1.28 \text{ X } 10^{4}.$$

47-5

(a) Apply Eq. 1 to the two principal maxima; since they are adjacent, their order numbers must differ by 1, so we have

$$d\sin\theta_1 = m\lambda,$$

$$d\sin\theta_2 = (m + 1)\lambda.$$

Subtracting these equations gives

$$d(\sin\theta_2 - \sin\theta_1) = \lambda,$$

$$d(0.30 - 0.20) = 600 \text{ nm},$$

$$d = 6000 \text{ nm} = 6.0 \ \mu\text{m}.$$

(b) Since the fourth order is missing, the $m = 4$ maximum, angular position given by Eq. 1, must fall at the same position as the first diffraction minimum (given by Eq. 1 of Chapter 46). That is,

$$a\sin\theta = \lambda,$$

$$d\sin\theta = 4\lambda,$$

for the same value of $\theta$. Dividing the equations gives

$$a = d/4 = (6 \ \mu\text{m})/4 = 1.5 \ \mu\text{m}.$$

(c) The largest possible diffraction angle is 90°. From Eq. 1 we find the corresponding order number:

$$m = d\sin\theta/\lambda = (6000 \text{ nm})\sin 90°/(600 \text{ nm}) = 10.$$

Strictly speaking, the $m = 10$ maximum does not appear on the screen since its diffraction angle is 90° and the light travels parallel to the screen. Also, the $m = 8$ maximum falls at the same location as the second diffraction minimum, as can be verified from Eq. 1 and Eq. 3 of Chapter 46, as so does not appear. Therefore, the orders actually seen on the screen are those of $m = 0, 1, 2, 3, 5, 6, 7, 9$.

## 47-8

(a) The maxima and minima of $I$ vs. $\phi$ occur at those values of $\phi$ that satisfy $dI/d\phi = 0$; using the result from Problem 7, and taking the derivative we find

$$dI/d\phi = 0 = -(4/9)I_m\sin\phi(1 + 2\cos\phi).$$

Solutions are found for the following conditions:

$$\sin\phi = 0 \rightarrow \phi = \pm n\pi, \quad n = 0, 1, 2, \ldots;$$

$$\cos\phi = -\tfrac{1}{2} \rightarrow \phi = \pm 2\pi/3, \pm 4\pi/3, \pm 8\pi/3, \ldots .$$

For $\phi = 0, \pm 2\pi$, etc, we have $I = I_m$; these are the principal maxima. At $\phi = \pm\pi$, etc, we find that $I = I_m/9$ and these are the secondary maxima. All values of $\phi$ for which $\cos\phi = -\tfrac{1}{2}$ yield $I = 0$, the minima. To find the values of $\phi$ for which the intensity is one-half that of a principal maximum, set $I = \tfrac{1}{2}I_m$ in the formula displayed in Problem 7. One gets

$$\tfrac{1}{2}I_m = (1/9)I_m(1 + 4\cos\phi + 4\cos^2\phi),$$

$$\cos\phi = 0.5607 \rightarrow \phi = 0.9756 \text{ rad.}$$

But $\phi = 2\pi\sin\theta/\lambda$, so that if $\theta$ is small,

$$\theta = 0.1553\lambda/d.$$

The halfwidth is $\Delta\theta = 2\theta$ so we have

$$\Delta\theta = 0.3106\lambda/d = \lambda/3.22d.$$

(b) For the double-slit, the corresponding halfwidth is $\lambda/2d$. Evidently, the interference fringes narrow as the number of slits is increased, assuming the same slit separation.

## 47-11

(a) For $N \gg 1$, the phasors form the arc of circles. Let $E$ be the amplitude from any one slit; the total length of the phasors then is $NE$. At the $n$-th minimum the phasors form $n$ closed circles. At the $k$-th maximum they form $(k + \tfrac{1}{2})$ circles, each with circumference $\pi D_k$, where $D_k$ is the diameter of the circles. (We sketch the figure for the $k = 2$ maximum.) The total length of the phasors is still $NE$ so we have

$$(k + \tfrac{1}{2})\pi D_k = NE,$$

$$D_k/NE = 1/(k + \tfrac{1}{2})\pi.$$

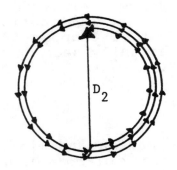

The quantity $D_k$ corresponds to $E_\theta$; recalling that intensity is proportional to the square of the amplitude, we have $I_k = AD_k^2$ and $I_{max} = A(NE)^2$, where $A$ is a constant. Squaring the equation above, then, yields

$$I_k/I_{max} = 1/(k + \tfrac{1}{2})^2\pi^2.$$

For the single slit,

$$I/I_{max} = \sin^2\alpha/\alpha^2.$$

At a maximum $\alpha \approx (k + \tfrac{1}{2})\pi$; using this approximation gives $\sin\alpha = \pm 1$ so the equation above becomes

$$I/I_{max} = 1/(k + \tfrac{1}{2})^2\pi^2.$$

This resembles the formula found first for the grating.
(b) For these secondary maxima we can put $k \approx \tfrac{1}{2}N$; also, for large $N$, we set $N + 1 = N$. Our formula for the intensity can now be written

$$I_k/I_{max} = 1/N^2(\tfrac{1}{2}\pi)^2 \approx 1/N^2.$$

(c) In the single slit formula, $I_{max}$ corresponds to the phasors from all the elements of the slit being lined up. In the grating, $I_{max}$ indicates that the phasors from all the slits (one phasor per slit since diffraction effects are ignored here) are lined up. From (a), the equations for the intensities are identical if the grating is considered one large slit of width $Nd$.


47-13

The separation between adjacent rulings is

$$d = (1 \text{ mm})/315 = 3.175 \times 10^{-3} \text{ mm} = 3175 \text{ nm}.$$

Set $\theta = 90°$ for the longest observable wavelength; with $m = 5$, Eq. 1 then yields

$$(3175 \text{ nm})\sin 90° = 5\lambda_{max},$$

$$\lambda_{max} = 635 \text{ nm}.$$

The shortest wavelength in the visible spectrum is 400 nm. This is observable in the fifth order since, by Eq. 1, it falls at a smaller angle than does a longer wavelength. Hence the observable visible wavelengths are those between 400 nm and 635 nm.

## 47-19

Eq. 1 tells us that, in any order, the longer wavelengths fall at the larger angles. Therefore, for possible overlap, examine the diffraction angles for 700 nm in the second order and 400 nm in the third order. For the second order spectrum,

$$d\sin\theta_{700} = 2(700 \text{ nm}) = 1400 \text{ nm}.$$

For the shorter wavelength in the third order,

$$d\sin\theta_{400} = 3(400 \text{ nm}) = 1200 \text{ nm}.$$

Comparing these two equations we see that, whatever the value of $d$,

$$\theta_{700} > \theta_{400},$$

and therefore the red region of the second order spectrum overlaps the blue region of the third order spectrum.

## 47-22

($a$) The number of rulings on the grating is

$$N = (620 \text{ mm}^{-1})(5.05 \text{ mm}) = 3131.$$

Combining Eqs. 10 and 11, we find

$$\lambda/\Delta\lambda = Nm,$$

$$\Delta\lambda = \lambda/Nm = (481 \text{ nm})/(3131)(3) = 0.0512 \text{ nm}.$$

($b$) The ruling separation is

$$d = (1 \text{ mm})/620 = 1.613 \times 10^{-3} \text{ mm} = 1613 \text{ nm}.$$

To find the highest observable order, use Eq. 1 with $\theta = 90°$; we find

$$m = (1613 \text{ nm})\sin 90°/(481 \text{ nm}) = 3.35.$$

Hence the $m = 3$ order itself is the highest observable order.

## 47-24

($a$) Combining Eqs. 10 and 11, one finds that

$$N = \lambda/m(\Delta\lambda).$$

Set $\Delta\lambda = 415.496$ nm $- 415.487$ nm $= 0.009$ nm. For $\lambda$ itself, use the average of the two wavelengths. Hence,

$$N = (415.491 \text{ nm})/(2)(0.009 \text{ nm}) = 23,100.$$

(b) The ruling spacing is

$$d = (4.15 \text{ cm})/(23,100) = 1.797 \text{ X } 10^{-4} \text{ cm.}$$

Again using the average of the wavelengths, we find the diffraction angle from Eq. 1:

$$(1797 \text{ nm})\sin\theta = (2)(415.491 \text{ nm}),$$

$$\theta = 27.5°.$$

## 47-27

(a) We do the calculations for the third order $m = 3$; for the other orders the calculations are similar. The ruling separation is

$$d = (76 \text{ X } 10^6 \text{ nm})/(40,000) = 1900 \text{ nm.}$$

By Eq. 1 we find

$$\theta = \sin^{-1}[m\lambda/d],$$

$$\theta = \sin^{-1}[(3)(589 \text{ nm})/(1900 \text{ nm})] = 68.43°.$$

The result of Problem 26 is now useful:

$$D = (\tan\theta)/\lambda = (\tan 68.43°)/(589 \text{ nm}),$$

$$D = 4.295 \text{ X } 10^{-3} \text{ rad/nm} = 0.246°/\text{nm},$$

using the conversion factor 1 rad $= (180°/\pi)$.
(b) By Eq. 11,

$$R = Nm = (40,000)(3) = 120,000.$$

## 47-35

The angle $\theta$ in Eq. 12 is measured from the crystal surface (see Fig. 18b) and therefore equals $90° - 51.3° = 38.7°$. For first order we have $m = 1$. Eq. 12 now yields

$$2d\sin\theta = m\lambda,$$

$$2(39.8 \text{ pm})\sin 38.7° = (1)\lambda,$$

$$\lambda = 49.8 \text{ pm.}$$

From Bragg's law, Eq. 12, we find the interplanar spacing $d$:

$$2d\sin\theta = m\lambda,$$

$$2d\sin63.8° = (1)(0.261 \text{ nm}),$$

$$d = 0.1454 \text{ nm}.$$

From Fig. 28 we see that the interplanar spacing $d$ (which is the perpendicular distance between the dashed lines) is one-half the diagonal of the square of edge length $a_0$; i.e.,

$$d = \tfrac{1}{2}(a_0\sqrt{2}).$$

Therefore,

$$a_0 = 2d/\sqrt{2} = 2(0.1454 \text{ nm})/\sqrt{2} = 0.206 \text{ nm}.$$

# CHAPTER 48

48-4

One polarizing sheet reduces the intensity of an incident unpolarized beam to one-half the incident intensity $I_0$; i.e., to $\frac{1}{2}I_0$; (see Fig. 4). The second sheet reduces the intensity of the now polarized beam transmitted by the first sheet to the extent given by Eq. 1. Therefore, the effect of the two sheets is to yield a transmitted intensity $I$ given by

$$I = (\tfrac{1}{2}I_0)\cos^2\theta.$$

For $I = \frac{1}{3}I_0$, we find that

$$\theta = \cos^{-1}(\sqrt{\tfrac{2}{3}}) = 35.3°.$$

48-7

The first sheet reduces the intensity to $\frac{1}{2}I_0$, where $I_0$ is the intensity of the incident unpolarized beam. Now use Eq. 1 three times in succession to account for transmission of the now linearly polarized beam through the three remaining sheets. The final transmitted intensity, therefore, is

$$I = (\tfrac{1}{2}I_0)(\cos^2 30°)(\cos^2 30°)(\cos^2 30°),$$

$$I = \tfrac{1}{2}I_0\cos^6 30° = \tfrac{1}{2}I_0(\tfrac{1}{2}\sqrt{3})^6,$$

$$I/I_0 = 27/128.$$

In the final step above we solved, as requested, for the fraction of the incident intensity that is transmitted.

48-10

Let $u$ be the intensity of the unpolarized component and $p$ the intensity of the polarized component of the incident beam. Regardless of the orientation of the Polaroid, the transmitted intensity of the unpolarized component is $\frac{1}{2}u$. When the characteristic polarizing direction of the Polaroid is in the plane of polarization of the linearly polarized component, the transmitted intensity of this component is $p$ (i.e., no reduction of intensity; $\theta = 0$ in Eq. 1). Hence the maximum possible intensity of the transmitted beam is

$$I_{max} = \tfrac{1}{2}u + p.$$

Rotating the Polaroid sheet by 90° reduces to zero the transmitted intensity of the polarized component ($\theta = 90°$ in Eq. 1). Thus, the minimum transmitted intensity is

$$I_{min} = \tfrac{1}{2}u.$$

We are told that $I_{max} = 5I_{min}$, so that

$$\tfrac{1}{2}u + p = 5(\tfrac{1}{2}u),$$

$$p = 2u.$$

The intensity of the original mixed beam is $p + u = 3u$. We see that this beam is 2/3 polarized and 1/3 unpolarized.

## 48-16

(a) In the absence of other information, assume that the incident beam is in air. At the polarizing angle the reflected and refracted rays are at 90° to each other. But the angle of reflection equals the angle of incidence $\theta_i$ and, since we are at the polarizing angle, we write $\theta_i = \theta_p$. Thus, with $\theta_r$ = angle of refraction, we have

$$\theta_p = 90° - \theta_r = 90° - 31.8° = 58.2°.$$

By Eq. 3,

$$n = \tan\theta_p = \tan 58.2° = 1.61.$$

(b) From (a), $\theta_p = 58.2°$.

## 48-17

For the o and e-rays to combine to form linearly polarized light, they must emerge from the crystal out of phase by either 0°, 180°, 540°, etc. A phase difference of 0° requires a plate of zero thickness. The thinnest plate, therefore, is that which induces a phase difference of 180° as the rays emerge from the plate. This corresponds to

$$N_e - N_o = m + \tfrac{1}{2}$$

in Sample Problem 3. Again, we take $m = 0$ for the thinnest plate. Thus, we have

$$\tfrac{1}{2} = x(\Delta n)/\lambda,$$

where $\Delta n = 0.022$, as found in Table 1. Solving for the plate thickness $x$ we get

$$x = \lambda/2(\Delta n),$$

$$x = (525 \text{ nm})/2(0.022) = 1.19 \times 10^4 \text{ nm} = 11.9 \ \mu\text{m},$$

certainly a thin plate.

(a) By Eq. 11, the rate of transfer of angular momentum is

$$dL/dt = (dU/dt)/\omega = P/\omega,$$

where $P$ is the power of the beam. Calculation of the angular frequency $\omega$ is straightforward:

$$\omega = 2\pi v = 2\pi c/\lambda,$$

$$\omega = 2\pi(3 \times 10^8 \text{ m/s})/(516 \times 10^{-9} \text{ m}) = 3.653 \times 10^{15} \text{ s}^{-1}.$$

Hence,

$$dL/dt = (106 \text{ W})/(3.653 \times 10^{15} \text{ s}^{-1}) = 2.90 \times 10^{-14} \text{ W} \cdot \text{s}.$$

(b) Now let $v$ and $\omega$ refer to the angular speed of the object (i.e., to the disk). Over a time $t$ the object absorbs an amount of angular momentum

$$L_{obj} = (dL/dt)t.$$

For $L_{obj}$ use Eq. 11 of Chapter 13, extracting the rotational inertia $I_{obj}$ from Fig. 9c in Chapter 12, so that

$$L_{obj} = I_{obj}\omega = (\tfrac{1}{2}MR^2)(2\pi v).$$

Substituting this into the first equation and solving for the desired time yields

$$t = \pi v MR^2/(dL/dt).$$

Numerically we have $M = 9.45 \times 10^{-6}$ kg, $R = 0.0026$ m, $v = 1.5 \text{ s}^{-1}$ and, from (a), $dL/dt = 2.90 \times 10^{-14}$ W$\cdot$s. Putting all this into the equation above gives $t = 10{,}380$ s $= 2$ h 53 min.

49-6

First convert the temperature to kelvins:

$$T = 500 + 273 = 773 \text{ K}.$$

Now Eq. 2 gives the total radiated power per unit area. Therefore, the power radiated by the fireplace is

$$P = IA = \epsilon A \sigma T^4,$$

$$P = (0.90)(0.5 \text{ m}^2)(5.67 \times 10^{-8} \text{ W/m}^2 \bullet \text{K}^4)(773 \text{ K})^4,$$

$$P = 9100 \text{ W} = 9.1 \text{ kW}.$$

49-9

(a) By Eq. 4, the initial temperature of the cavity radiator is

$$T_i = 2898/\lambda_{max,i} = (2898 \ \mu\text{m} \bullet \text{K})/(25 \ \mu\text{m}) = 115.9 \text{ K}.$$

The temperature of the radiator is now raised to $T_f$ such that

$$I(T_f) = 2I(T_i).$$

By Eq. 1, this can be written as

$$\sigma T_f^4 = 2\sigma T_i^4,$$

$$T_f = 2^{\frac{1}{4}} T_i = 2^{\frac{1}{4}}(115.9 \text{ K}) = 138 \text{ K}.$$

(b) For the final temperature, Eq. 4 can be solved for the wavelength at which the spectral radiancy has its maximum value:

$$\lambda_{max,f} = (2898 \ \mu\text{m} \bullet \text{K})/(138 \text{ K}) = 21.0 \ \mu\text{m}.$$

49-11

The oven emits energy at the rate $AI(T_o)$ into the room and also absorbs energy at the rate $AI(T_r)$ from the room. In these expressions, $T_o$ = oven temperature = $215 + 273 = 488$ K and $T_r$ = room temperature = $26.2 + 273 = 299.2$ K. This radiant energy is transferred through the area $A = 5.2 \text{ cm}^2 = 5.2 \times 10^{-4} \text{ m}^2$. The net power transferred from oven to room is

$$P_{net} = AI(T_o) - AI(T_r).$$

Invoke Eq. 2 and this becomes

$$P_{net} = \epsilon A \sigma (T_o{}^4 - T_r{}^4).$$

Substituting the data in SI base units yields

$$P_{net} = (1)(5.2 \times 10^{-4})(5.67 \times 10^{-8})(488^4 - 299.2^4) = 1.44 \text{ W}.$$

This seems small until we realize that an oven transfers most of its internal energy to the room as heat by conduction through the oven walls.

## 49-14

(a) The area of the emitting surface of the filament is

$$A = \pi D L = \pi (0.28 \times 10^{-3} \text{ m})(0.018 \text{ m}) = 1.583 \times 10^{-5} \text{ m}^2.$$

In this formula $D$ is the diameter and $L$ the length of the filament. If $T$ is the operating temperature, and assuming that the power rating of the bulb corresponds to the actual rate of emission of radiation, then

$$P = AI = \epsilon A \sigma T^4,$$

$$100 \text{ W} = (1)(1.583 \times 10^{-5} \text{ m}^2)(5.67 \times 10^{-8} \text{ W/m}^2 \bullet \text{K}^4) T^4,$$

$$T = 3250 \text{ K} = 2980°\text{C}.$$

(b) Assume that the specific heat of tungsten does not change with the temperature and is given by the value in Appendix D,

$$c = 134 \text{ J/kg} \bullet \text{K}.$$

Also, the mass of the filament is

$$m = \rho V = \rho (\tfrac{1}{4} \pi D^2 L).$$

In this expression $\rho$ is the density of tungsten; we shall assume that it does not vary with temperature (i.e., ignore thermal expansion) and its value is that found in Appendix D. Now, Eq. 4 of Chapter 25 implies that

$$dQ/dt = mc(dT/dt).$$

In this expression we shall identify $-dQ/dt$ with $P$, the rate of emission of radiant energy after the bulb is switched off. The filament then loses internal energy, thus the minus sign. Making this identification, and employing the expression for $P$ found in (a), we have

$$\epsilon \sigma A T^4 = -mc(dT/dt),$$

$$\epsilon \sigma (\pi D L) T^4 = -(\rho \tfrac{1}{4} \pi D^2 L) c(dT/dt),$$

$$dt = -a(dT/T^4),$$

where

$$a = \rho Dc/4\epsilon\sigma = 3.193 \times 10^9 \text{ K}^3 \cdot \text{s},$$

is a constant with the numerical value as above, which the reader can doubtless verify. The required cooling time is

$$t = -a\int_{T_i}^{T_f} dT/T^4 = \tfrac{1}{3}a\left[\frac{1}{T_f^3} - \frac{1}{T_i^3}\right].$$

Put $T_i = 3250$ K and $T_f = 3250 - 500 = 2750$ K to find $t = 20.2$ ms.

## 49-18

(a) We require that

$$R(400 \text{ nm}) = 3.5R(200 \text{ nm}).$$

Now turn to Eq. 6. The factor $2\pi c^2 h$ cancels out. Since

$$(400 \text{ nm})^5 = 2^5(200 \text{ nm})^5 = 32(200 \text{ nm})^5,$$

the factor $(200 \text{ nm})^5$ will also cancel. Let

$$x = hc/(400 \text{ nm})kT,$$

so that

$$hc/(200 \text{ nm})kT = 2x.$$

Taking cognizance of all these considerations, the requirement on the spectral radiancy expressed in the first equation reduces to

$$\frac{1}{32(e^x - 1)} = (3.5)\frac{1}{e^{2x} - 1}.$$

Now put

$$y = e^x; \quad e^{2x} = (e^x)^2 = y^2.$$

335

Since $(y^2 - 1) = (y + 1)(y - 1)$, the spectral radiancy equation, after a little algebra, yields

$$y = (32)(3.5) - 1 = 111.$$

Therefore,

$$x = \ln y = \ln(111) = hc/(400 \times 10^{-9} \text{ m})kT.$$

Put $h = 6.63 \times 10^{-34}$ J•s, $c = 3 \times 10^8$ m/s and $k = 1.38 \times 10^{-23}$ J/K to find that $T = 7650$ K.
(b) For the new condition it is necessary to write

$$3.5R(400 \text{ nm}) = R(200 \text{ nm}).$$

With the same definitions of $x$ and $y$ as in (a), we now get

$$y = (32/3.5) - 1 = 8.143,$$

$$x = \ln(8.143),$$

$$T = 17,200 \text{ K}.$$


## 49-21

(a) The Einstein temperature is defined by $T_E = h\nu/k$. Hence, in Eq. 14 for the molar heat capacity,

$$h\nu/kT = T_E/T.$$

If $T = T_E$, then $h\nu/kT = 1$. In this event Eq. 14 reduces to

$$C_V = \left[\frac{e}{(e - 1)^2}\right](3R) = 0.921(3R),$$

i.e., 92.1% of its classical value of 3R.
(b) In Eq. 13 for the molar internal energy, we can write

$$3N_A h\nu = 3N_A k T_E = 3RT_E.$$

Again, if $T = T_E$, then $h\nu/kT = 1$. Therefore, when this condition is satisfied,

$$E_{int} = 3RT/(e - 1) = 0.582(3RT),$$

or 58.2% of its classical value.

49-25

(a) From Appendix D, the molar mass of aluminum is $M = 27$ g/mol, so there are $n = m/M = 12/27 = 0.444$ mol of aluminum present. The classical value of the molar specific heat at constant volume is $C_v = 3R = 24.9$ J/mol·K. Under this assumption, the required heat amounts to

$$Q = n(\Delta E_{int}) = nC_v\Delta T,$$

$$Q = (0.444 \text{ mol})(24.9 \text{ J/mol·K})(100 \text{ K}) = 1110 \text{ J}.$$

(b) In Einstein's theory $Q = n(\Delta E_{int})$ still holds, but Eq. 13 must be used to evaluate the change in internal energy. As noted in Problem 21(b) above, $3N_Ah\nu = 3RT_E$ so that

$$3nN_Ah\nu = 3nRT_E = 3(0.444 \text{ mol})(8.31 \text{ J/mol·K})(290 \text{ K}) = 3210 \text{ J}.$$

Also $h\nu/kT = T_E/T$. With $T_i = 80$ K and $T_f = 180$ K, the required amount of heat predicted by Einstein's theory is

$$Q = (3210 \text{ J}) \left[ \frac{1}{e^{290/180} - 1} - \frac{1}{e^{290/80} - 1} \right] = 713 \text{ J},$$

certainly very different from the classical value.

49-29

Use the result of Problem 27(a) to calculate the energies of the two photons:

$$E_1 = (1240 \text{ eV·nm})/(375 \text{ nm}) = 3.307 \text{ eV},$$

$$E_2 = (1240 \text{ eV·nm})/(580 \text{ nm}) = 2.138 \text{ eV}.$$

The photon labeled 1 is absorbed, so the energy of the atom is increased by 3.307 eV. Photon 2 is emitted, so the atom loses 2.138 eV of energy. Hence, for the atom there is a net gain of 3.307 eV − 2.138 eV = 1.17 eV of energy.

49-36

Calculate the energy $h\nu$ of the photon in eV, as follows:

$$h\nu = (6.63 \times 10^{-34} \text{ J·s})(3.19 \times 10^{15} \text{ s}^{-1})/(1.6 \times 10^{-19} \text{ J/eV}),$$

$$h\nu = 13.22 \text{ eV}.$$

Eq. 18 can now be used to find the maximum kinetic energy of the emitted photoelectrons; we obtain

$$K_{max} = h\nu - \phi = 13.22 \text{ eV} - 2.33 \text{ eV} = 10.9 \text{ eV}.$$

## 49-39

(a) Calculate the energy of the photon from Problem 27(a), to obtain

$$E = h\nu = hc/\lambda = (1240 \text{ eV} \bullet \text{nm})/(410 \text{ nm}) = 3.02 \text{ eV}.$$

Now combine Eqs. 16 and 18 to find the stopping potential $V_0$:

$$h\nu = \phi + eV_0,$$

$$3.02 \text{ eV} = 1.85 \text{ eV} + eV_0,$$

$$V_0 = 1.17 \text{ V}.$$

(b) By Eq. 16,

$$K_{max} = eV_0 = e(1.17 \text{ V}) = 1.17 \text{ eV},$$

$$K_{max} = (1.17 \text{ eV})(1.6 \text{ X } 10^{-19} \text{ J/eV}) = 1.872 \text{ X } 10^{-19} \text{ J}.$$

Therefore, the speed $v$ of an electron with this amount of kinetic energy is found from

$$\tfrac{1}{2}mv^2 = K_{max},$$

$$\tfrac{1}{2}(9.11 \text{ X } 10^{-31} \text{ kg})v^2 = 1.872 \text{ X } 10^{-19} \text{ J},$$

$$v = 6.41 \text{ X } 10^5 \text{ m/s} = 641 \text{ km/s}.$$

## 49-43

Let $E$ be the energy of each photon. If $n$ of these photons arrive in time $t$ on an area $A$, then the rate at which they deliver energy is $P = nE/t$, and the associated intensity of the photon flux is

$$I = P/A = nE/tA.$$

The intensity is $I = 1380 \text{ W/m}^2$. Since this is in SI base units, the energy $E$ of each photon must be evaluated in joules; we find

$$E = hc/\lambda = (6.63 \text{ X } 10^{-34} \text{ J} \bullet \text{s})(3 \text{ X } 10^8 \text{ m/s})/(550 \text{ X } 10^{-9} \text{ m}),$$

$$E = 3.616 \text{ X } 10^{-19} \text{ J}.$$

Solve the first equation for the desired area $A$, to find

$$A = nE/tI.$$

Now put $n = N_A = 6.02 \text{ X } 10^{23}$, the Avogadro constant (also known as

Avogadro's number); we also have $t = 1$ min $= 60$ s, so that

$A = (6.02 \text{ X } 10^{23})(3.616 \text{ X } 10^{-19} \text{ J})/(60 \text{ s})(1380 \text{ W/m}^2) = 2.63 \text{ m}^2.$

### 49-49

(a) The energy $E$ of each photon, by Problem 27(a), is

$$E = (1240 \text{ eV} \bullet \text{nm}/589 \text{ nm})(1.6 \text{ X } 10^{-19} \text{ J/eV}),$$

$$E = 3.368 \text{ X } 10^{-19} \text{ J}.$$

The rate $R$ of emission of photons is therefore

$$R = P/E = (100 \text{ W})/(3.368 \text{ X } 10^{-19} \text{ J}) = 2.97 \text{ X } 10^{20} \text{ s}^{-1}.$$

(b) The photon flux $F$ at a distance $r$ from the lamp is given by

$$F = I/E,$$

where $I$ is the intensity (energy flux) at this same distance; dividing the energy flux by the energy per photon gives the photon flux. Now, for a source radiating uniformly in all directions,

$$I = P/A = P/4\pi r^2,$$

so that

$$F = P/4\pi r^2 E.$$

Put $F = 1/(\text{cm}^2 \bullet \text{s}) = 10,000/(\text{m}^2 \bullet \text{s})$; substitute also $P = 100$ W and the photon energy $E$ found in (a); solving for the required distance gives $r = 4.86 \text{ X } 10^7$ m $= 48,600$ km.
(c) In time $t$, the bulb emits $(P/E)t$ photons. At the end of this time interval, all of these photons are found within a sphere of radius $ct$ centered on the lamp. The photon number density is a function of distance $r$ from the lamp. Hence

$$(P/E)t - \int n dV - \int_0^{ct} n(4\pi r^2 dr).$$

You can verify that this integral equation is satisfied if

$$n = P/4\pi r^2 E c = F/c.$$

In the last step we used the relation for $F$ found in (b). After inserting the data in the appropriate SI base units, we find that the distance at which $n = 1 \text{ /cm}^3 = 1 \text{ X } 10^6 \text{ /m}^3$ is $r = 281$ m.
(d) With the equations given in (b) and (c), we find that at $r = 2$ m, the photon flux is $F = 5.91 \text{ X } 10^{18} \text{ /(m}^2 \bullet \text{s})$ and the photon number density is $n = 1.97 \text{ X } 10^{10} \text{ /m}^3$.

Ignore the violation of conservation of charge. Under the assumption that the photon gives all its energy to the initially stationary electron and so disappears, with the electron moving off with speed $v$, the conservation of momentum and of total energy applied to the photon-electron system, require that

$$E/c + 0 = 0 + \gamma mv,$$

$$E + mc^2 = 0 + \gamma mc^2.$$

Here $E$ is the energy of the incident photon and $\gamma$ is the Lorentz factor for the electron after the collision; see Eq. 15 in Chapter 21. By the vector nature of the law of conservation of momentum, the electron moves away along the same line on which the incident photon arrived. In the second equation above, transpose the rest energy of the electron to the other side and divide by $c$ to get

$$E/c = \gamma mc(1 - \gamma^{-1}).$$

This agrees with the first equation only if

$$v = c(1 - \gamma^{-1}),$$

$$v = c(1 - \sqrt{1 - v^2/c^2}).$$

This last relation requires that $v = 0$ or $v = c$. But $v = 0$ leads to $E = 0$ (by the first equation), so there is no photon to begin with. The second solution, $v = c$, is prohibited by the Special Theory (really Fact) of Relativity.

(a) The rest energy of an electron is 511 keV. Setting the photon energy $E = 511$ keV gives for the frequency

$$\nu = E/h = (511 \text{ keV})/(4.14 \times 10^{-18} \text{ keV·s}) = 1.23 \times 10^{20} \text{ s}^{-1}.$$

(b) The wavelength can be found from $\lambda = c/\nu$, or from the result in Problem 27(a); choosing the latter route gives

$$\lambda = (1240 \text{ eV·nm})/(511{,}000 \text{ eV}) = 0.00243 \text{ nm} = 2.43 \text{ pm}.$$

(c) The momentum of the photon is

$$p = E/c = 511 \text{ keV}/c.$$

In SI base units, this is

$$p = (511 \text{ keV})(1.6 \text{ X } 10^{-16} \text{ J/keV})/(3 \text{ X } 10^{8} \text{ m/s}),$$

$$p = 2.73 \text{ X } 10^{-22} \text{ kg} \cdot \text{m/s}.$$

## 49-53

The momentum of the photon is

$$p = h/\lambda = (6.63 \text{ X } 10^{-34} \text{ J} \cdot \text{s})/(589 \text{ X } 10^{-9} \text{ m}),$$

$$p = 1.126 \text{ X } 10^{-27} \text{ kg} \cdot \text{m/s}.$$

The photon has no rest mass. The mass of the atom is

$$m = M/N_A = (0.02299 \text{ kg/mol})/(6.02 \text{ X } 10^{23} \text{ mol}^{-1}),$$

$$m = 3.819 \text{ X } 10^{-26} \text{ kg}.$$

Presuming that the atom is moving slowly compared with the speed of light, momentum conservation gives for its reduction in speed upon absorbing the photon as

$$\Delta v = p/m = 2.95 \text{ cm/s},$$

upon substitution of the numerical values of $p$ and $m$ found above.

## 49-56

(a) Using the result of Problem 27(a),

$$E = (1240 \text{ eV} \cdot \text{nm})/(511,000 \text{ eV}) = 2.43 \text{ pm}.$$

(b) Find the wavelength of the scattered photon from Eq. 25. Now, in Sample Problem 8, we discover that

$$h/mc = 2.43 \text{ pm}.$$

Using this, Eq. 25 becomes

$$\lambda' = \lambda + (h/mc)(1 - \cos\phi),$$

$$\lambda' = 2.43 \text{ pm} + (2.43 \text{ pm})(1 - \cos 72°) = 4.11 \text{ pm}.$$

(c) Again use Problem 27(a), but this time to find the energy of the scattered photon:

$$E' = (1240 \text{ eV} \cdot \text{nm})/(4.11 \text{ X } 10^{-3} \text{ nm}) = 302 \text{ keV}.$$

Evidently the electron acquired 511 keV - 302 keV = 209 keV of kinetic energy as a result of the Compton collision.

The result of Problem 57 is useful here. In the equation cited there, $\Delta E = E - E'$, where $E$, $E'$ are the energies of the incident and of the scattered photon. We are told that $\Delta E/E = 0.10$. Also, $h\nu' = E'$, the energy of the scattered photon. But

$$E' = (1 - 0.10)E = (0.90)(215 \text{ keV}) = 193.5 \text{ keV}.$$

Furthermore, $mc^2 = 511$ keV. Taking all this into account, the equation given in Problem 57 reduces to

$$0.10 = (193.5/511)(1 - \cos\phi),$$

$$\phi = 42.6°.$$

50-5

(a) Assume that we are dealing with a nonrelativistic particle. Find the momentum of the proton from Eq. 1:

$$p = h/\lambda = (6.63 \times 10^{-34} \text{ J} \cdot \text{s})/(0.113 \times 10^{-12} \text{ m}),$$

$$p = 5.867 \times 10^{-21} \text{ kg} \cdot \text{m/s}.$$

The proton's speed follows from

$$v = p/m = (5.867 \times 10^{-21} \text{ kg} \cdot \text{m/s})/(1.67 \times 10^{-27} \text{ kg}),$$

$$v = 3.51 \times 10^{6} \text{ m/s} = 3510 \text{ km/s}.$$

(b) To find the accelerating potential, first calculate the kinetic energy of the proton. We obtain, working first in SI base units,

$$K = p^2/2m = (5.867 \times 10^{-21})^2/2(1.67 \times 10^{-27}) = 10.31 \text{ fJ},$$

$$K = (1.031 \times 10^{-14} \text{ J})/(1.6 \times 10^{-19} \text{ J/eV}) = 64.4 \text{ keV}.$$

If the proton was accelerated from rest, then the required accelerating potential $V$ can be found from

$$qV = K,$$

$$(e)V = 64.4 \text{ keV},$$

$$V = 64.4 \text{ kV}.$$

50-9

Using the extreme relativistic relation $p = E/c$, the de Broglie wavelength, assuming Eq. 1 is relativistically valid, is

$$\lambda = h/p = hc/E.$$

As pointed out in the problem statement, this is the same relation as for light (photons). This means that we can use the result of Problem 27(a) in Chapter 49 to evaluate the wavelength numerically:

$$\lambda = (1240 \text{ eV} \cdot \mu\text{m})/(32 \times 10^{9} \text{ eV}) = 3.9 \times 10^{-8} \text{ nm} = 0.039 \text{ fm}.$$

50-13

(a) The kinetic energy of the electrons after being accelerated (from rest) through a potential difference of 25 kV is $K = 25$ keV. The relation between kinetic energy $K$ and momentum $p$ is different in classical and relativistic physics. Classically $K = p^2/2m$, so that the momentum of the electrons is

$$(25 \text{ keV})(1.6 \text{ X } 10^{-16} \text{ J/keV}) = p^2/2(9.11 \text{ X } 10^{-31} \text{ kg}),$$

$$p = 8.537 \text{ X } 10^{-23} \text{ kg} \cdot \text{m/s}.$$

Therefore, the de Broglie wavelength of the electrons is

$$\lambda = h/p = (6.63 \text{ X } 10^{-34} \text{ J} \cdot \text{s})/(8.537 \text{ X } 10^{-23} \text{ kg} \cdot \text{m/s}) = 7.77 \text{ pm}.$$

(b) Relativistically, find the momentum from Eq. 32 in Chapter 21. For an electron $mc^2 = 511$ keV, so the total energy is $E = K + mc^2$ = 25 keV + 511 keV = 536 keV. We now have

$$E^2 = (pc)^2 + (mc^2)^2,$$

$$(536 \text{ keV})^2 = (pc)^2 + (511 \text{ keV})^2,$$

$$p = 161.8 \text{ keV}/c.$$

The de Broglie wavelength, if Eq. 1 holds relativistically, is

$$\lambda = h/p = (4.14 \text{ X } 10^{-18} \text{ keV} \cdot \text{s})/(161.8 \text{ keV}/c),$$

$$\lambda = (2.559 \text{ X } 10^{-20} \text{ s})(3 \text{ X } 10^8 \text{ m/s}) = 7.68 \text{ pm}.$$

Evidently, TV sets are marginally relativistic devices.


## 50-17

(a) The operating equation for the experiment is Eq. 2. From the text description, we see that in the original "setup", $\lambda = 165$ pm and $D = 215$ pm. Therefore, Eq. 2 predicts maxima at angles $\phi$ given from

$$m\lambda = D\sin\phi,$$

$$m(165 \text{ pm}) = (215 \text{ pm})\sin\phi,$$

$$\sin\phi = (0.767)m.$$

Clearly there is no solution except for $m = 1$, since higher values of $m$ would require $\sin\phi > 1$.
(b) We can use the formula quoted in Problem 2(b) to find the wavelength $\lambda = 0.158$ nm = 158 pm for $V = 50$ V. By Eq. 2,

$$(1)(158 \text{ pm}) = (215 \text{ pm})\sin\phi,$$

$$\phi = 47.3°.$$

We use the same value of $D$ since nothing was said about changing the diffracting crystal.

The variation in the angle of diffraction $\theta$ with the kinetic energy $K$ of the incident neutrons can be found by differentiating Bragg's law (see Sample Problem 4), $2d\sin\theta = m\lambda$ to get

$$2d\cos\theta\left(\frac{d\theta}{dK}\right) = m\left(\frac{d\lambda}{dK}\right).$$

For wavelength, use the de Broglie wavelength of the neutrons; we actually need $d\lambda/dK$, so differentiate

$$\lambda = \frac{h}{\sqrt{2mK}},$$

to get, after a little algebra (not too much),

$$\frac{d\lambda}{dK} = -\tfrac{1}{2}\,\frac{\lambda}{K}.$$

Putting this into the second equation gives

$$2d\cos\theta\left(\frac{d\theta}{dK}\right) = -\tfrac{1}{2}\,\frac{m\lambda}{K} = -\tfrac{1}{2}\,\frac{2d\sin\theta}{K},$$

$$\frac{d\theta}{dK} \approx \frac{\Delta\theta}{\Delta K} = -\tfrac{1}{2}\,\frac{\tan\theta}{K},$$

$$\Delta\theta = \tfrac{1}{2}(\tan\theta)\frac{\Delta K}{K},$$

ignoring the minus sign. Now this gives $\Delta\theta$ in radians. To convert to degrees, note that

$$1 \text{ rad} = (180/\pi)^\circ.$$

Using this conversion factor, the factor of $\frac{1}{2}$ in the last equation for $\Delta\theta$ is effectively changed to $90/\pi$, so that we obtain the result quoted in the problem.

50-21

Eq. 5 applies here, but introduce $\nu$ in place of $\omega$ so that we get the spread of frequencies in Hz; with this change, we have

$$(\Delta\omega)(\Delta t) \approx 1,$$

$$(2\pi\Delta\nu)(\Delta t) \approx 1,$$

$$(2\pi\Delta\nu)(0.23 \text{ s}) \approx 1,$$

$$\Delta\nu \approx 0.69 \text{ Hz}.$$

50-23

Use the energy-time uncertainty relationship, given by Eq. 7. To obtain the uncertainty in energy in eV, pick the value of $h$ as expressed in eV•s. Writing = rather than $\approx$, we find

$$(\Delta E)(\Delta t) = h/2\pi,$$

$$(\Delta E)(8.7 \times 10^{-12} \text{ s}) = (4.14 \times 10^{-15} \text{ eV•s})/2\pi,$$

$$\Delta E = 7.6 \times 10^{-5} \text{ eV} = 76 \ \mu\text{eV}.$$

50-31

The energy of the particle in any state is given by Eq. 16. In order to jump from the $n = 1$ state to the $n = 4$ state, the electron must absorb an amount of energy $\Delta E$ that added to $E_1$ yields $E_4$. That is,

$$\Delta E = E_4 - E_1 = (4^2 - 1^2)(h^2/8mL^2).$$

Substitute $h = 6.63 \times 10^{-34}$ J•s, $m = 9.11 \times 10^{-31}$ kg and the width $L = 253 \times 10^{-12}$ m; then convert the resulting energy in joules J to eV; we find $\Delta E = 88.3$ eV.

50-34

Numerically, we find that $h^2/8mL^2 = 6.031 \ \mu\text{eV}$. Now tabulate the five smallest distinct values of $\Sigma n^2$; we find

$$
\begin{aligned}
n_1{}^2 + n_2{}^2 + n_3{}^2 &= 1^2 + 1^2 + 1^2 = 3, \\
&= 1^2 + 2^2 + 1^2 = 6, \\
&= 1^2 + 2^2 + 2^2 = 9, \\
&= 1^2 + 1^2 + 3^2 = 11, \\
&= 2^2 + 2^2 + 2^2 = 12.
\end{aligned}
$$

Hence, the five lowest distinct energies are 18.1 $\mu$eV, 36.2 $\mu$eV, 54.3 $\mu$eV, 66.3 $\mu$eV, 72.4 $\mu$eV.

## 50-38

(a) Follow the method of Sample Problem 11, except that we are now interested in the region $0 \leq x \leq \frac{1}{3}L$. Also, we use the probability density of Eq. 14, suitably normalized with the constant $A$ as determined by Eq. 18 in Sample Problem 10. Taking account of all these machinations, we have

$$P_n(x) = (\frac{2}{L}) \sin^2 \frac{n\pi x}{L}.$$

Therefore, the desired probability is

$$P = \int P_n(x) \, dx = \frac{2}{L} \int_0^{\frac{1}{3}L} \sin^2 (\frac{n\pi x}{L}) \, dx.$$

In Sample Problem 10, we learn how to evaluate integrals of this kind; taking note of our different limits, we thereby obtain

$$P = \frac{1}{3} [1 - \frac{\sin(2\pi n/3)}{2\pi n/3}],$$

as asserted. Substituting values of the quantum number $n$, find that (b) for $n = 1$, $P = 0.196$; (c) $n = 2$ yields $P = 0.402$; (d) with $n = 3$ we get $P = \frac{1}{3}$. (e) The classical value is found in taking the limit $n \to \infty$. The factor $\sin(2\pi n/3) \leq 1$ always, but $1/2\pi n \to 0$ as $n \to \infty$. Hence, the classical value is $P = \frac{1}{3}$.

## 50-40

By Eq. 19, writing $=$ instead of $\approx$,

$$\ln T = -2kL.$$

We want the transmission probability to be $T = 1/100 = 0.01$. The value of $k$ found in Sample Problem 12 is $5.12 \times 10^9$ m$^{-1}$. Hence,

$$\ln(0.01) = -2(5.12 \times 10^9 \text{ m}^{-1})L,$$

$$L = 4.50 \times 10^{-10} \text{ m} = 0.45 \text{ nm}.$$

## 50-42

Take the natural logarithm of Eq. 19 (both sides, of course), and then square the result to remove the square root in $k$; we find

$$\ln T = -2kL,$$

$$(\ln T)^2 = 4k^2L^2,$$

$$(\ln T)^2 = 32\pi^2 m(U - E)L^2/h^2.$$

We want $T = 0.001$. We also have $m = 9.11 \times 10^{-31}$ kg, the mass of an electron, and $L = 700 \times 10^{-12}$ m. With $h = 6.63 \times 10^{-34}$ J•s, we find from the last equation that

$$U - E = 1.488 \times 10^{-19} \text{ J} = 0.930 \text{ eV},$$

$$E = 6 \text{ eV} - 0.93 \text{ eV} = 5.07 \text{ eV}.$$

51-4

(a) By Eq. 3,

$$\frac{1}{\lambda} = R(\frac{1}{m^2} - \frac{1}{n^2}) = (0.01097 \text{ nm}^{-1})(\frac{1}{1^2} - \frac{1}{3^2}),$$

$$\lambda = 102.6 \text{ nm}.$$

(b) The momentum of the photon is $p = E/c$. But $E = 1240/102.6 = 12.1$ eV. Hence $p = 12.1$ eV/$c$.

51-7

The first excited state corresponds to $m = 2$ in Eq. 3. Now see Sample Problem 1, from which we can conclude that, from the $m = 2$ state,

$$E_b = -E_2 = hcR/2^2.$$

But, Sample Problem 1 also shows us that

$$hcR = 13.6 \text{ eV}.$$

Hence, from the first excited state,

$$E_b = (13.6 \text{ eV})/2^2 = 3.40 \text{ eV}.$$

51-13

Consider a transition between the $n$ and $(n - p)$ levels, with $p$ an integer. The frequency of the photon involved is

$$\nu = \frac{1}{2}n^3\nu_0[\frac{1}{(n - p)^2} - \frac{1}{n^2}].$$

$$\nu = (\frac{1}{2}np\nu_0)\frac{2n - p}{(n - p)^2}.$$

For $n > p > 1$, $(2n - p)/(n - p)^2 \approx 2/n$, so that

$$v \approx (\tfrac{1}{2}npv_0)(2/n) = pv_0,$$

consistent with the correspondence principle.


## 51-17

The de Broglie wavelength (nonrelativistic case) is $\lambda = h/m_e v$. By Eq. 6,

$$v^2 = (Ze^2/4\pi\epsilon_0 m_e)r^{-1}.$$

Hence,

$$\lambda = 2\pi(a_0 r)^{\frac{1}{2}},$$

where $a_0$ is defined by Eq. 19. But Eq. 19 gives the orbital radii:

$$r = a_0 n^2.$$

Eliminating $r$ between the last two equations gives

$$\lambda = 2\pi a_0 n.$$

Finally, dividing the last two equations yields

$$\lambda/r = 2\pi/n.$$

From this last relation it is evident that $\lambda/r \to 0$ as $n \to \infty$.


## 51-20

In magnitude, the momentum $p_H$ of the recoiling hydrogen atom must equal the momentum $p$ of the emitted photon. The momentum of the photon is $p = E_{photon}/c$. Since the ground state has $n = 1$, Eq. 18 tells us that

$$E_{photon} = \Delta E = E_4 - E_1 = -13.6/4^2 - (-13.6/1^2) = 12.75 \text{ eV}.$$

Therefore,

$$p_H = p = 12.75 \text{ eV}/c,$$

$$p_H = (12.75 \text{ eV})(1.6 \times 10^{-19} \text{ J/eV})/(3 \times 10^8 \text{ m/s}),$$

$$p_H = 6.80 \times 10^{-27} \text{ kg} \cdot \text{m/s}.$$

The speed of the recoiling atom is

$$v = p_H/m = (6.80 \times 10^{-27} \text{ kg} \cdot \text{m/s})/(1.67 \times 10^{-27} \text{ kg}) = 4.07 \text{ m/s}.$$

(a) The photon energy is $E = (1240 \text{ eV} \cdot \text{nm})/(0.01 \text{ nm}) = 124 \text{ keV}$.
(b) We can use the result of Problem 65 in Chapter 49. The desired energy is $K_{max}$. For an electron, $mc^2 = 511 \text{ keV}$. We find that

$$K_{max} = (124)^2/[124 + \tfrac{1}{2}(511)] = 40.5 \text{ keV}.$$

(c) The kinetic energy of an electron moving at a speed $v = 0.1c$, calculated classically, is

$$K = \tfrac{1}{2}mv^2 = 4.1 \times 10^{-16} \text{ J} = 2.6 \text{ keV}.$$

This electron has much more kinetic energy than an electron in hydrogen, and still $K_{max} > K$. The photon-electron collision will probably result in the electron being ejected from the atom. There is no possibility of using such photon-electron interactions to see the electron making undisturbed orbital revolutions about the nucleus of the atom.

(a) In the Bohr theory the electron, carrying charge $-e$, revolves in a circular orbit of radius $r$. If $T$ is the period of its revolution, then the equivalent current $i$ due to its circular motion is $i = q/t = e/T$. The magnetic moment is, therefore,

$$\mu = iA = (e/T)(\pi r^2) = e\pi r^2/T.$$

This can be rewritten as

$$\mu = \tfrac{1}{2}e(2\pi r/T)r = \tfrac{1}{2}evr = \tfrac{1}{2}e(L/m),$$

where $L$ is the orbital angular momentum $L = mvr$. But $L$ is a quantized quantity; in Bohr's theory, $L = n\hbar$, so that

$$\mu = \tfrac{1}{2}e(n\hbar/m) = n(e\hbar/2m) = n\mu_B.$$

(b) The assignment by Bohr of angular momentum, embodied in Eq. 20, is not correct (i.e., disagrees with the result of quantum theory). Therefore, the result derived in (a) must be discarded. See the discussion on p. 1078 of HRK.

(a) In one orientation $\mu_z = +1\mu_B$, and in the other $\mu_z = -1\mu_B$. The energy difference between these orientations is, by Eq. 38 in Chapter 34,

$$\Delta U = \Delta(\mu \cdot \mathbf{B}) = (\Delta\mu_z)B = 2\mu_B B,$$

$$\Delta U = 2(57.88 \ \mu eV/T)(0.520 \ T) = 60.2 \ \mu eV.$$

(b) Set the photon energy $E$ equal to $\Delta U$; since $E = h\nu$, we find

$$\nu = E/h = (60.2 \ X \ 10^{-6} \ eV)/(4.14 \ X \ 10^{-15} \ eV \bullet s) = 14.5 \ GHz.$$

(c) The wavelength follows from

$$\lambda = hc/E = (1240 \ eV \bullet nm)/(60.2 \ X \ 10^{-6} \ eV) = 2.06 \ cm.$$

From Fig. 1 in Chapter 41, we see that this lies in the microwave region of the electromagnetic spectrum.


## 51-41

The maximum value of $m_1 = 1$, so we have $1 = 4$. For any value of $n$, the allowed values of $1$ extend from zero to $(n - 1)$ in integer steps. Therefore, all we can say about $n$ is that $n \geq 5$. Finally, we always have $m_s = \pm\frac{1}{2}$, since this is an intrinsic property of the electron.


## 51-48

First ignore spin. For each value of $n$ there are $n$ possible values of $1$ ranging from 0 to $(n - 1)$; for each $1$ there are $(21 + 1)$ possible values of $m_1$. Thus, the number $N$ of states at any $n$ is

$$N = \sum_0^{n-1} (21 + 1) = 2\sum_0^{n-1} 1 + n = 2(\sum_0^n 1 - n) + n = 2\sum_1^n 1 - n.$$

But,

$$\sum_1^n 1 = \frac{1}{2}n(n + 1),$$

since the average value of the terms in the sum is $\frac{1}{2}(n + 1)$ and there are $n$ terms. Hence,

$$N = 2[\tfrac{1}{2}n(n + 1)] - n = n^2.$$

With electron spin included, we have double the number of states (spin "up" and spin "down" for each of the states counted above), for a total of $2n^2$.

Since $\Delta r = 0.01a_0$ is much less than $1.00a_0$, we can consider the radial probability density $P_r(r)$ as being constant in the region between the spheres. In this event, the desired probability is

$$P = \int P_r(r)dr \approx [P_r(1.00a_0)](\Delta r).$$

By Eq. 41, $P_r(1.00a_0) = (4/a_0)e^{-2}$. Also, $\Delta r = 0.01a_0$. Thus,

$$P = (4/a_0)e^{-2}(0.01a_0) = 0.04e^{-2} = 5.4134 \times 10^{-3}.$$

Alternatively, we can do an exact calculation using the result of Problem 56(a). The probability that the electron lies inside a sphere of radius $1.01a_0$ is

$$P(1.01) = 1 - e^{-2.02}(5.0602),$$

obtained by putting $x = r/a_0 = 1.01$ in the formula quoted in the problem. Similarly, the probability of finding the electron inside a sphere of radius $a_0$ is

$$P(1.00) = 1 - e^{-2}(5).$$

The probability that the electron lies between the spheres is

$$\Delta P = 5e^{-2} - 5.0602e^{-2.02} = 5.4132 \times 10^{-3}.$$

We see that the approximation adopted in the first calculation is quite good.

## 51-54

Adopting the *Hint*, we write Eq. 40 as

$$\psi^2(r) = 1/\pi a_0^3,$$

and is justified since the radius $r_p$ of the proton is very much less than $a_0$, so that $2r/a_0 \ll 1$ in the exponent of $\psi^2$. But now $\psi^2$ is effectively a constant throughout the region occupied by the proton, so the probability $P$ of finding the electron inside this region is given to a close approximation by

$$P = \int \psi^2 dV = \psi^2(\Delta V) = (1/\pi a_0^3)(4\pi r_p^3/3) = (4/3)(r_p/a_0)^3.$$

With $r_p = 1.1 \times 10^{-15}$ m and $a_0 = 5.29 \times 10^{-11}$ m, this formula gives $P = 1.2 \times 10^{-14}$.

## 51-57

The formula quoted in Problem 56(a) gives the probability of finding the electron inside a sphere of radius $r$; this is a monotonically increasing function of $r$. Using the formula, evaluate the probability that the electron is inside a sphere of radius $r = 2a_0$ by setting $x = 2$:

$$P(2) = 1 - 13e^{-4}.$$

Similarly, the probability of finding the electron inside a sphere of radius $a_0$ is found by putting $x = 1$ to get

$$P(1) = 1 - 5e^{-2}.$$

The probability $\Delta P$ of finding the electron inside the larger sphere but outside the smaller plus the probability $P(1)$ of finding the electron inside the smaller sphere must equal the probability $P(2)$ of finding the electron inside the larger sphere. Therefore,

$$\Delta P = P(2) - P(1) = 5e^{-2} - 13e^{-4} = 0.439.$$

51-62

The radial probability density is given by Eq. 44. Note that $P_r(r) = 0$ at $r = 2a_0$; this fixes the upper limit on the integral for the probability $P$. Writing $x = r/a_0$, we find

$$P = \int_0^{2a_0} P_r(r)\, dr = \frac{1}{8} \int_0^2 x^2 (2 - x)^2 e^{-x} dx.$$

Since

$$\int x^2 (2 - x)^2 e^{-x} dx = -(x^4 + 4x^2 + 8x + 8) e^{-x},$$

as you should verify, we find, upon evaluating between the limits, that

$$P = 1 - 7e^{-2} = 0.0527.$$

51-64

(a) We conclude from Fig. 19 that the energy difference $\Delta E$ between the two upper levels must equal the difference in the energies $E_1$ and $E_2$ of the two photons emitted, since they both represent transitions from their respective upper level to the same lower level. But $E_1 = 1240/588.995 = 2.10528$ eV and $E_2 = 1240/589.592 = 2.10315$ eV, so that $\Delta E = 0.00213$ eV.
(b) Following the argument of Problem 51-38(a) above, we conclude that

$$\Delta E = 2\mu_B B,$$

$$0.00213 \text{ eV} = 2(5.79 \times 10^{-5} \text{ eV/T})B,$$

$$B = 18.4 \text{ T.}$$

## 51-65

(a) We must calculate the reduced mass $M$ of the muonic atom; this is given by

$$M = mm_p/[m + m_p].$$

In terms of the mass $m_e$ of the electron, we have $m = 207m_e$ and the proton mass $m_p = 1840m_e$. With these, we find $M = 186.1m_e$. From Eq. 19, we see that the Bohr radii are inversely proportional to $m_e$, and hence to $M$ when we substitute $M$ for $m_e$. Putting $n = 1$, we find for the first orbit in the muonic atom, $r_1 = (52.92 \text{ pm})/(186.1) = 0.284$ pm.

(b) The ionization energy is the negative of $E_1$, as given by Eq. 18. For hydrogen, this equals 13.6 eV. Examining Eq. 18, we see that $E_1$ is directly proportional to $m_e$. Hence, for the muonic atom, the ionization energy is $(186.1)(13.6 \text{ eV}) = 2.53$ keV.

(c) The most energetic photon corresponds to the Lyman series limit. In hydrogen, this wavelength is 91.2 nm (see Table 1). The photon wavelengths are inversely proportional to the Rydberg constant $R$. By Eq. 17, $R$ is directly proportional to $m_e$. Therefore, in the muonic atom, the desired wavelength is $(91.2 \text{ nm})/(186.1) = 0.490$ nm.

## CHAPTER 52

52-4

(a) The cutoff wavelength $\lambda_{min}$ is characteristic of the energy of the incident electrons, not of the properties of the target material upon which they fall. This cutoff wavelength is equal to the wavelength of a photon with an energy equal to the kinetic energy of the incident electrons, in this case 35 keV. Therefore, by Problem 1,

$$\lambda_{min} = 1240/35 = 35.4 \text{ pm}.$$

(b) The $K_\beta$ line results from an $M$ to $K$ energy level transition; see Fig. 3. The energy of the emitted photon is the difference in the energies of the atom with the affected electron in the $K$ and $M$ levels (or hole in the $M$ and $K$ levels). Thus,

$$hc/\lambda_{K\beta} = E_K - E_M = 25.51 \text{ keV} - 0.53 \text{ keV} = 24.98 \text{ keV},$$

$$\lambda_{K\beta} = 1240/24.98 = 49.6 \text{ pm}.$$

(c) The $K_\alpha$ line corresponds to an $L$ to $K$ electron transition (or a $K$ to $L$ hole transition). Hence,

$$hc/\lambda_{K\alpha} = E_K - E_L = 25.51 \text{ keV} - 3.56 \text{ keV} = 21.95 \text{ keV},$$

$$\lambda_{K\alpha} = 1240/21.95 = 56.5 \text{ pm}.$$

52-9

The initial kinetic energy of the electron is 50 keV, and its energy after each of the collisions is 25 keV, 12.5 keV, zero. Hence, the energies of the three photons are 50 - 25 = 25 keV, 25 - 12.5 = 12.5 keV, 12.5 - 0 = 12.5 keV ($h\nu = \Delta K$; see Fig. 2). These photons have wavelengths $\lambda = 1240/25,000 = 0.0496$ nm = 49.6 pm and $1240/12,500 = 0.0992$ nm = 99.2 pm. There are two of the longer wavelength photons.

52-10

(a) For these lines to be produced, an electron must be removed from the $K$ shell, creating a hole there. This requires 69.5 keV of energy, or an accelerating potential for electrons of (69.5 keV)/$e$ = 69.5 kV.
(b) By Problem 1,

$$\lambda_{min} = 1240/V = 1240/69.5 = 17.8 \text{ pm}.$$

(b) For the $K_\alpha$ line, $E = 69.5 - 11.3 = 58.2$ keV, and this is the energy of the photon. Therefore

356

$$\lambda_{K\alpha} = (1240 \text{ eV} \cdot \text{nm})/(58,200 \text{ eV}) = 0.0213 \text{ nm} = 21.3 \text{ pm}.$$

For $K_\beta$, $E = 69.5 - 2.3 = 67.2$ keV, so that

$$\lambda_{K\beta} = (1240 \text{ eV} \cdot \text{nm})/(67,200 \text{ eV}) = 0.0185 \text{ nm} = 18.5 \text{ pm}.$$

## 52-14

The energy $E$ of the photon is equal to the change in the kinetic energy $K$ of the electron. From Eqs. 27 and 15 in Chapter 21, we see that

$$E = K_i - K_f = (\gamma_i - \gamma_f)mc^2.$$

Setting $v_i/c = 2.73/3 = 0.91$, we find from Eq. 15 in Chapter 21 that $\gamma_i = 2.412$. Therefore the equation above becomes, upon substitution of this and the other data,

$$43.8 \text{ keV} = (2.412 - \gamma_f)(511 \text{ keV}),$$

$$\gamma_f = 2.326.$$

Inverting Eq. 15 of Chapter 21 to solve for the final speed $v_f$ of the electron yields

$$v_f = [1 - (1/\gamma_f)^2]^{\frac{1}{2}}c,$$

$$v_f = [1 - (1/2.326)^2]^{\frac{1}{2}}(3 \times 10^8 \text{ m/s}) = 2.71 \times 10^8 \text{ m/s}.$$

## 52-16

From Sample Problem 3, we see that

$$\lambda_{Nb}/\lambda_{Ga} = [(Z_{Ga} - 1)/(Z_{Nb} - 1)]^2.$$

In either Appendix D or Appendix E we find that $Z_{Ga} = 31$ and $Z_{Nb} = 41$. Thus,

$$\lambda_{Nb}/\lambda_{Ga} = [30/40]^2 = 9/16.$$

## 52-18

See Eq. 19 in Chapter 51, but note that the numerical value of $a_0$ given there of $a_0 = 52.92$ pm applies only if $Z = 1$. Bearing this in mind, we can write for the orbit radii in the Bohr model of "hydrogen-like" atoms,

$$r_n = (a_0/Z)n^2.$$

The ground state orbit has $n = 1$; therefore, the radius of this orbit is

$$r_1 = (52.92 \text{ pm}/92)(1)^2 = 0.575 \text{ pm},$$

since, for uranium (U), $Z = 92$.

## 52-23

(a) We do the calculations for sodium (Na). On the Bohr model the ionization energy is the negative of the energy $E$ given in Eq. 5; that is,

$$E_{ion} = (13.6 \text{ eV}/n^2)Z_{eff}^2,$$

where $Z_{eff}$ is the charge that must be assigned to the nucleus-closed electron shell to force the measured ionization energy to agree with the Bohr theory. From Fig. 5, we see that the valence electron is in a $3s$ state; that is, $n = 3$. Hence,

$$5.14 \text{ eV} = (13.6 \text{ eV}/3^2)Z_{eff}^2,$$

$$Z_{eff} = 1.84.$$

(b) For sodium (Na), $Z = 11$. The fraction sought, therefore, is given by $1.84/11 = 0.167$.
Now you repeat the problem for potassium.

## 52-29

Solve Eq. 8 for the temperature $T$ by taking natural logarithms of both sides, to find

$$\ln[n(E_2)/n(E_1)] = -(E_2 - E_1)/kT,$$

$$T = (E_1 - E_2)/k\ln[n(E_1)/n(E_2)].$$

In our case $E_2 - E_1 = 3.2$ eV; also $n(E_1)/n(E_2) = 41$, as you can verify. Therefore,

$$T = (3.2 \text{ eV})/(8.62 \times 10^{-5} \text{ eV/K})(\ln41) = 10,000 \text{ K}.$$

## 52-35

(a) We use not only the equation for the radius of the central disk given in Problem 34, but also the revelation contained there that this central disk contains only 84% (not 100%) of the incident power. Intensity is power per unit area and therefore the intensity $I$ of the beam at the target is

$$I = \frac{P}{A} = \frac{0.84 P_{laser}}{\pi R^2} = \frac{0.84 P_{laser}}{\pi (1.22 f\lambda/d)^2}.$$

We are instructed to set $f = r = 3000$ km $= 3 \times 10^6$ m, the distance to the target. Also $\lambda = 2.95$ $\mu$m $= 2.95 \times 10^{-6}$ m, $d = 3.72$ m and $P_{laser}$ = 5.3 MW $= 5.3 \times 10^6$ W. With these data, the equation above gives for the intensity at the target $I = 1.68 \times 10^5$ W/m$^2$ = 0.168 MW/m$^2$. This falls far short of the 120 MW/m$^2$ needed to destroy the missile.

(b) To find the wavelength to which the laser should be tuned (if possible), use the same equation above but set $I = 120 \times 10^6$ W/m$^2$. Inserting all the other data except wavelength, and solving for the wavelength, gives $\lambda = 1.10 \times 10^{-7}$ m $= 0.11$ $\mu$m.

(c) In this situation in which the laser wavelength remains at 2.95 $\mu$m, again set $I = 120 \times 10^6$ W/m$^2$, but leave the distance $r$ to the target as the unknown. The other quantities are unchanged. Solving for the distance gives $r = 1.12 \times 10^5$ m $= 112$ km for the range of this device.

# CHAPTER 53

## 53-3

The number density of gold atoms is given by

$$n_{Au} = \rho N_A / M,$$

$$n_{Au} = (19.3 \text{ g/cm}^3)(6.02 \times 10^{23} \text{ mol}^{-1})/(197 \text{ g/mol}),$$

$$n_{Au} = 5.90 \times 10^{22} \text{ cm}^{-3}.$$

With each atom contributing just one charge carrier, the number density of the charge carriers is

$$n = (1 \text{ carrier/atom})(5.90 \times 10^{22} \text{ atoms/cm}^3),$$

$$n = 5.90 \times 10^{28} \text{ m}^{-3}.$$

## 53-9

By Problem 3 (see above), the number density of charge carriers is $n = 5.90 \times 10^{28}$ m$^{-3}$. We could now use Eq. 5 to calculate the Fermi energy. But, if we peek ahead in the problems to Problem 13, we see that Eq. 5 has been put into a form convenient for calculation; that is, the constants in Eq. 5 have been evaluated once and for all. Using this result from Problem 13, then, we have

$$E_F = An^{\frac{2}{3}},$$

$$E_F = (3.65 \times 10^{-19} \text{ m}^2 \bullet \text{eV})(5.90 \times 10^{28} \text{ m}^{-3})^{\frac{2}{3}} = 5.53 \text{ eV}.$$

## 53-14

(a) Inverting Eq. 6 to solve for the exponent term gives

$$(E - E_F)/kT = \ln[1/p(E) - 1].$$

In our present case $p(E) = 0.910$; we find that $\ln[1/0.910 - 1] = -2.314$. Now evaluate $kT$; we obtain

$$kT = (8.62 \times 10^{-5} \text{ eV/K})(1050 \text{ K}) = 0.09051 \text{ eV}.$$

Therefore we have

$$(E - 7.06 \text{ eV})/(0.09051 \text{ eV}) = -2.314,$$

$$E = 6.85 \text{ eV}.$$

(b) Using the result of Problem 1, the density of states at this energy is found to be

$$n(E) = CE^{\frac{1}{2}},$$

$$n(6.85 \text{ eV}) = (6.81 \text{ X } 10^{27} \text{ m}^{-3} \bullet \text{eV}^{-3/2})(6.85 \text{ eV})^{\frac{1}{2}},$$

$$n(6.85 \text{ eV}) = 1.78 \text{ X } 10^{28} \text{ m}^{-3} \bullet \text{eV}^{-1}.$$

(c) Combining the results of (a) and (b) via Eq. 3 yields the density of occupied states at this energy; we have

$$n_o(E) = p(E)n(E),$$

$$n_o(6.85 \text{ eV}) = p(6.85 \text{ eV})n(6.85 \text{ eV}),$$

$$n_o(6.85 \text{ eV}) = (0.910)(1.78 \text{ X } 10^{28} \text{ m}^{-3} \bullet \text{eV}^{-1}),$$

$$n_o(6.85 \text{ eV}) = 1.62 \text{ X } 10^{28} \text{ m}^{-3} \bullet \text{eV}^{-1}.$$

<u>53-18</u>

Use the result of Problem 13 to find the number density of charge carriers (free electrons); we find

$$n = (E_F/A)^{3/2},$$

$$n = (11.66 \text{ eV}/3.65 \text{ X } 10^{-19} \text{ m}^2 \bullet \text{eV})^{3/2},$$

$$n = 1.806 \text{ X } 10^{29} \text{ electrons/m}^3.$$

The number density of aluminum atoms is

$$n_{Al} = \rho/m_{Al} = \rho N_A/M,$$

$$n_{Al} = (2700 \text{ kg/m}^3)(6.02 \text{ X } 10^{23} \text{ mol}^{-1})/(0.027 \text{ kg/mol}),$$

$$n_{Al} = 6.02 \text{ X } 10^{28} \text{ atoms/m}^3.$$

Evidently, for aluminum

$$n/n_{Al} = (1.806 \text{ X } 10^{29} \text{ electrons/m}^3)/(6.02 \text{ X } 10^{28} \text{ atoms/m}^3),$$

$$n/n_{Al} = 3 \text{ electrons/atom};$$

that is, each atom contributes 3 of its electrons to the pool of charge carriers.

<u>53-19</u>

Calculate the Fermi energy from the result of Problem 13. The number density of free electrons is given by

$$n = ZN/V,$$

where $Z = 26$ for iron, $N$ = the number of (completely ionized) iron atoms in the star and $V$ = the volume of the star. For the volume we have $V = 4\pi R^3/3$. The mass $m$ of an iron atom can be found from $m = M/N_A$. We use $M = 56$ g/mol for iron (the bulk value of $M = 55.847$ g/mol given in Appendix D is not appropriate for stars, although there is little numerical difference). If $M*$ is the mass of the star, then

$$N = M*N_A/M.$$

Putting all this together we find

$$n = 3ZM*N_A/4\pi R^3 M = 5.137 \times 10^{35} \text{ m}^{-3}.$$

(You should verify the numerical value; be sure to put $M = 0.056$ kg/mol.) Using this result in the formula given in Problem 13, we find $E_F = 234$ keV.


53-23

Rearranging the final equation in Problem 21 for the temperature, we get

$$T = 2E_F f/3k,$$

$$T = 2(4.71 \text{ eV})(0.013)/3(8.62 \times 10^{-5} \text{ eV/K}),$$

$$T = 474 \text{ K} = 201°C.$$


53-26

(a) The number $N$ of conduction electrons in the penny is equal to the number of copper atoms present, since each copper atom contributes one conduction electron. From Sample Problem 2, we learn that in copper $n = 8.49 \times 10^{28}$ m$^{-3}$ (this is the number density of conduction electrons). The volume of the penny is given by

$$V = m/\rho = (0.0031 \text{ kg})/(8960 \text{ kg/m}^3) = 3.46 \times 10^{-7} \text{ m}^3.$$

Therefore,

$$N = nV = (8.49 \times 10^{28} \text{ m}^{-3})(3.46 \times 10^{-7} \text{ m}^3) = 2.94 \times 10^{22}.$$

The average energy of these electrons, by Problem 25, is $(3/5)E_F$. Therefore the total energy potentially available is

$$E_{total} = N(3E_F/5).$$

Inserting the value of $N$ and of $E_F = (7.06 \text{ eV})(1.6 \times 10^{-19} \text{ J/eV})$ gives $E_{total} = 20$ kJ.
(b) The time is $t = E_{total}/P = (20,000 \text{ J})/(100 \text{ J/s}) = 200$ s.

(a) With 2 electrons contributed by each atom, the number density of free electrons is

$$n = 2n_{Zn} = 2(\rho N_A/M),$$

$$n = 2(7133 \text{ kg/m}^3)(6.02 \text{ X } 10^{23} \text{ mol}^{-1})/(0.06537 \text{ kg/mol}),$$

$$n = 1.31 \text{ X } 10^{29} \text{ m}^{-3}.$$

(b) By the result of Problem 13,

$$E_F = (3.65 \text{ X } 10^{-19} \text{ m}^2 \bullet eV)(1.31 \text{ X } 10^{29} \text{ m}^{-3})^{\frac{2}{3}} = 9.41 \text{ eV}.$$

(c) The Fermi speed $v_F$ is defined from $\frac{1}{2}mv_F^2 = E_F$. The mass of the electron is $m = 9.11 \text{ X } 10^{-31}$ kg. The Fermi energy was found in (b) above; it needs to be converted to joules. With this done, the defining formula gives $v_F = 1820$ km/s.

(d) At the Fermi speed, the de Broglie wavelength is $\lambda = h/mv_F = 400$ pm, the numerical value found using $h = 6.63 \text{ X } 10^{-34}$ J$\bullet$s, and $m$ and $v_F$ given above.

(a) First evaluate $kT$; we find

$$kT = (8.62 \text{ X } 10^{-5} \text{ eV/K})(289 \text{ K}) = 0.0249 \text{ eV}.$$

With the Fermi level at the middle of the gap, then at the bottom of the conduction band

$$E - E_F = \frac{1}{2}E_g = \frac{1}{2}(0.67 \text{ eV}) = 0.335 \text{ eV},$$

so that

$$(E - E_F)/kT = (0.335 \text{ eV})/(0.0249 \text{ eV}) = 13.5.$$

By Eq. 6, the occupancy probability is

$$p(E) = (e^{13.5} + 1)^{-1} = 1.4 \text{ X } 10^{-6}.$$

(b) Now set $p(E) = 3(1.4 \text{ X } 10^{-6}) = 4.2 \text{ X } 10^{-6}$. Turning to Eq. 6, we find that

$$(E - E_F)/kT = \ln[1/(4.2 \text{ X } 10^{-6}) - 1] = 12.4.$$

Therefore,

$$12.4 = (0.335 \text{ eV})/(8.62 \text{ X } 10^{-5} \text{ eV/K})T,$$

$$T = 313 \text{ K} = 40°C.$$

53-36

The number density of silicon atoms is

$$n_{Si} = \rho N_A/M = 4.99 \times 10^{22} \text{ cm}^{-3},$$

using $\rho = 2.33$ g/cm$^3$ and $M = 28.086$ g/mol; we found these values in Appendix D. From the result of Sample Problem 7, we can now find the number density of phosphorus atoms:

$$n_P = n_{Si}/(5 \times 10^6) = 9.98 \times 10^{15} \text{ cm}^{-3}.$$

The volume of the 1 g silicon sample is

$$V = m/\rho = (1 \text{ g})/(2.33 \text{ g/cm}^3) = 0.429 \text{ cm}^3.$$

Thus we have $(9.98 \times 10^{15} \text{ cm}^{-3})(0.429 \text{ cm}^3) = 4.28 \times 10^{15}$ atoms of phosphorus, for a mass of phosphorus of

$$m = (N/N_A)M,$$

$$m = (4.28 \times 10^{15}/6.02 \times 10^{23} \text{ mol}^{-1})(30.97 \text{ g/mol}) = 0.22 \text{ } \mu g.$$

Appendix D provided the molar mass of phosphorus.


53-41

(a) We have $kT = (8.62 \times 10^{-5} \text{ eV/K})(290 \text{ K}) = 0.0250$ eV. From Eq. 6,

$$\Delta E/kT = \ln[1/p(E) - 1],$$

where $\Delta E = E - E_F$. Putting $p(E) = 4.8 \times 10^{-5}$ and $kT = 0.025$ eV gives $\Delta E = 0.25$ eV. We are told to put $E = 0$ at the top of the valence band. Therefore, the bottom of the conduction band is at $E = E_g = 1.1$ eV. Hence, the donor level is at $E = E_d = 1.1$ eV $- 0.11$ eV $= 0.99$ eV. It follows that

$$\Delta E = 0.25 \text{ eV} = E_d - E_F = 0.99 \text{ eV} - E_F,$$

$$E_F = 0.74 \text{ eV}.$$

(b) At the bottom of the conduction band

$$\Delta E = E - E_F = 1.1 \text{ eV} - 0.74 \text{ eV} = 0.36 \text{ eV}.$$

Now use this and the value of $kT$ found in (a) in Eq. 6 to find the probability of occupancy as

$$p(E) = (e^{0.36/0.025} + 1)^{-1} = 5.6 \times 10^{-7}.$$

## 53-48

Evidently, photons with wavelengths shorter than 295 nm have energies sufficient to excite some electrons into jumping the gap. We conclude that a photon of wavelength 295 nm itself, if absorbed by an electron sitting at the top of the occupied band, will cause that electron to jump to the bottom of the next higher unoccupied band. (The photon disappears, the crystal appearing opaque at this and shorter wavelengths.) Thus, the energy of the photon must just equal the energy gap in the crystal. Hence

$$E_g = (1240 \text{ eV} \bullet \text{nm})/(295 \text{ nm}) = 4.20 \text{ eV}.$$

## 53-49

Calculate the energy of a photon of wavelength 140 nm; we find

$$E = (1240 \text{ eV} \bullet \text{nm})/(140 \text{ nm}) = 8.86 \text{ eV}.$$

This energy is greater than the band gap energy. Thus, electrons in the occupied band can absorb this photon and make the transition to the empty band above (in energy). The photon disappears, does not emerge from the other side of the crystal and therefore the crystal appears opaque at this wavelength.

# CHAPTER 54

## 54-3

The distance of closest approach (center to center) is

$$d = 6.98 \text{ fm} + 1.8 \text{ fm} = 8.78 \text{ fm} = 8.78 \times 10^{-15} \text{ m}.$$

Now see Sample Problem 1. With $q = 2e$, $Q = 79e$, we find, using SI base units, that

$$K_\alpha = \frac{1}{4\pi\epsilon_0}\frac{qQ}{d} = (8.99 \times 10^9)\frac{(2)(79)(1.6 \times 10^{-19})^2}{8.78 \times 10^{-15}},$$

$$K_\alpha = (4.14 \times 10^{-12} \text{ J})/(1.6 \times 10^{-13} \text{ J/MeV}) = 26 \text{ MeV}.$$

## 54-6

Using Eq. 1, we have immediately

$$A = (R/R_0)^3 = (3.6 \text{ fm}/1.2 \text{ fm})^3 = 27.$$

## 54-10

The number of nucleons in the nucleus equals the mass number $A$; therefore, the quantity sought is $E_B/A$. But

$$E_B = [Zm_H + Nm_n - m]c^2,$$

where $m_H$ is the atomic mass of hydrogen $^1$H and $m$ is the atomic mass of the nucleus, in this case $^{62}$Ni. We find the values of $m_H$ and $m_n$ below Eq. 4. For our nucleus, $Z = 28$ (see Appendix D) and $N = A - Z = 62 - 28 = 34$. Hence,

$$E_B = [28(1.007825 \text{ u}) + 34(1.008665 \text{ u}) - 61.928346 \text{ u}]c^2,$$

$$E_B = (0.585364)(uc^2) = (0.585364)(931.5 \text{ MeV}) = 545.267 \text{ MeV}.$$

Therefore, the average binding energy per nucleon is

$$E_B/A = (545.267 \text{ MeV})/62 = 8.79 \text{ MeV}.$$

## 54-11

(a) Call the new unit u'. By its definition,

$$1 \text{ u'} = 1.007825 \text{ u},$$

so that

$$1 \ u = (1/1.007825) \ u'.$$

Hence the mass $m$ of an atom of $^{12}C$ in the new unit is

$$m = 12 \ u = 12(1/1.007825) \ u' = 11.906830 \ u'.$$

Similarly, for $^{238}U$ we find

$$m = 238.050785 \ u = 238.050785(1/1.007825) \ u' = 236.202500 \ u'.$$

(b) We see that for the heavier atom, the approximate relation $m \approx A \ u$ fails to hold in the new mass unit. We found $m \approx 236 \ u'$ whereas $m \approx 238 \ u$. The use of $u'$ is awkward in at least this sense. Basically, the $^{1}H$ nucleus, with only one nucleon, is not "typical" enough of most nuclei to use as a basis for an atomic mass unit.

## 54-13

The abundances of the three isotopes must add up to 100%. Thus, if $f$ is the abundance of $^{25}Mg$, then the abundance of $^{26}Mg$ must be just $1 - 0.7899 - f = 0.2101 - f$. The bulk atomic mass, 24.305, is the average of the three isotope atomic masses weighted according to their abundances. That is,

$$m_{bulk} = \frac{f_{24}m_{24} + f_{25}m_{25} + f_{26}m_{26}}{f_{24} + f_{25} + f_{26}}.$$

The denominator, of course, equals 1. Substituting $m_{bulk} = 24.305 \ u$, $m_{24} = 23.985042 \ u$, $m_{25} = 24.985837 \ u$, $m_{26} = 25.982594 \ u$; also $f_{24} = 0.7899$, $f_{25} = f_{25}$, and $f_{26} = 0.2101 - f_{25}$, and then solving the equation for $f_{25}$ gives $f_{25} = 0.10005$. Hence $f_{26} = 0.110048$. We conclude that the abundances of the isotopes are 10.01% of $^{25}Mg$ and 11.00% of $^{26}Mg$.

## 54-18

The binding energy of a copper nucleus is

$$E_B = [Zm_H + Nm_n - m]c^2,$$

$$E_B = [29(1.007825) + 34(1.008665) - 62.929599](931.5 \ MeV),$$

$$E_B = 551.4 \ MeV.$$

To accuracy sufficient for calculating the number of atoms, the mass of a copper atom is

$$m = (62.93 \text{ u})(1.6605 \text{ X } 10^{-27} \text{ kg/u}) = 1.045 \text{ X } 10^{-25} \text{ kg}.$$

Therefore there must be

$$N = (0.003 \text{ kg})/(1.045 \text{ X } 10^{-25} \text{ kg}) = 2.871 \text{ X } 10^{22}$$

atoms of copper in the coin. The required energy is

$$E = NE_B = (2.871 \text{ X } 10^{22})(551.4 \text{ MeV})(1.6 \text{ X } 10^{-13} \text{ J/MeV}),$$

$$E = 2.53 \text{ X } 10^{12} \text{ J} = 2.53 \text{ TJ}.$$

### 54-20

By the conservation of total relativistic energy

$$m_n c^2 + m_H c^2 = m_d c^2 + K_d + E_\gamma,$$

where $E_\gamma$ is the energy of the gamma-ray photon (which, of course, has zero mass). By conservation of momentum, in magnitude

$$p_d = p_\gamma = E_\gamma/c.$$

Therefore

$$K_d = p_d^2/2m_d = E_\gamma^2/2m_d c^2,$$

$$K_d = (2.2233 \text{ MeV})^2/2(2.014102 \text{ u})(931.5 \text{ MeV/u}) = 0.001317 \text{ MeV}.$$

Therefore, the energy conservation equation yields

$$m_n = m_d - m_H + (K_d + E_\gamma)/c^2,$$

$$m_n = 2.014102 - 1.007825 + (2.2233 + 0.001317)/(931.5),$$

$$m_n = 1.0087 \text{ u}.$$

### 54-25

(a) By Eq. 8,

$$t_{\frac{1}{2}} = \ln2/\lambda = \ln2/(0.0108 \text{ h}^{-1}) = 64.2 \text{ h}.$$

(b) For $t = 3t_{\frac{1}{2}}$, $\lambda t = 3\ln2$, so that the desired fraction remaining is, by Eq. 6,

$$N/N_0 = e^{-3\ln2} = 0.125 = 1/8.$$

(c) In hours the elapsed time is $t = 240$ h. Now that $t$ is in the same time units that appear in the disintegration constant $\lambda$, we have $\lambda t = (0.0108 \text{ h}^{-1})(240 \text{ h}) = 2.592$. By Eq. 6,

$$N/N_0 = e^{-2.592} = 0.0749.$$

## 54-27

(a) The initial activity is

$$R_0 = \lambda N_0.$$

Calculate the disintegration constant from the half-life:

$$\lambda = \ln 2/t_{\frac{1}{2}} = \ln 2/(78.25 \text{ h})(3600 \text{ s/h}) = 2.4606 \times 10^{-6} \text{ s}^{-1}.$$

Since the mass of an atom of gallium is $A$ u, the number $N_0$ of atoms of gallium in the initially pure sample is

$$N_0 = (3.42 \text{ g})/(66.93 \text{ u})(1.66 \times 10^{-24} \text{ g/u}) = 3.078 \times 10^{22}.$$

Therefore, by the first equation,

$$R_0 = (2.4606 \times 10^{-6} \text{ s}^{-1})(3.078 \times 10^{22}) = 7.57 \times 10^{16} \text{ s}^{-1}.$$

(b) We now have

$$\lambda t = (2.4606 \times 10^{-6} \text{ s}^{-1})[(48 \text{ h})(3600 \text{ s/h})] = 0.4252.$$

The activity at this later time follows from Eq. 7 as

$$R = (7.57 \times 10^{16} \text{ s}^{-1})e^{-0.4252} = 4.95 \times 10^{16} \text{ s}^{-1}.$$

## 54-28

The exponent in Eq. 6 is $\lambda t = (\ln 2/t_{\frac{1}{2}})t = (\ln 2)t/t_{\frac{1}{2}}$. Therefore

$$e^{-\lambda t} = (e^{-\ln 2})^{t/t_{1/2}} = (2^{-1})^{t/t_{1/2}}.$$

Since $2^{-1} = 1/2$, putting this into the first form of the decay law yields

$$N = N_0 \left(\frac{1}{2}\right)^{t/t_{1/2}}.$$

This form of the decay law is useful when the half-life is given, for there is no need to compute the distintegration constant $\lambda$ in order to find the number of radioactive atoms that remain after a specified time $t$.

## 54-30

The disintegration constant is

$$\lambda = \ln 2/t_{\frac{1}{2}} = \ln 2/(12.7\ \text{h}) = 0.05458\ \text{h}^{-1}.$$

By Eq. 6, the number of $^{64}$Cu atoms remaining after 14 h is

$$N_{14} = N_0 e^{-0.05458(14)} = 0.46574 N_0.$$

Similarly, the number remaining after 16 h is

$$N_{16} = N_0 e^{-0.05458(16)} = 0.41758 N_0.$$

Therefore, the number that decayed during the two hour period must be given by

$$\Delta N = 0.04816 N_0.$$

Now, in terms of mass, $\Delta M = (m_{\text{Cu}})\Delta N$ and $M_0 = (m_{\text{Cu}})N_0$, so that

$$\Delta M = (0.04816)M_0 = (0.04816)(5.50\ \text{g}) = 0.265\ \text{g}.$$

## 54-33

The number of plutonium atoms present initially is

$$N_0 = (12\ \text{g})(6.02 \times 10^{23}\ \text{mol}^{-1})/(239\ \text{g/mol}) = 3.023 \times 10^{22}.$$

Also,

$$\lambda t = (\ln 2/24{,}100\ \text{y})(20{,}000\ \text{y}) = 0.5752.$$

Therefore, by Eq. 6, the number of plutonium atoms present after 20,000 y is

$$N = (3.023 \times 10^{22})e^{-0.5752} = 1.701 \times 10^{22}.$$

For each plutonium atom that decays, one $\alpha$ particle is produced, so the number of $\alpha$ particles created during the 20,000 y is

$$\Delta N = (3.023 - 1.701) \times 10^{22} = 1.322 \times 10^{22}.$$

This represent a mass $m$ of helium (recall that an $\alpha$ particle is a helium nucleus) of

$$m = (1.322 \times 10^{22}/6.02 \times 10^{23})(4\ \text{g}) = 0.0878\ \text{g}$$

As usual, Appendix D was our source for the value of the molar mass, this time of helium.

54-42

By conservation of momentum we must have, in magnitude, $p_\alpha = p_{Th}$. The definition of $Q$ is given in Eq. 42 of Chapter 10. Wuth $K_U = 0$, this becomes

$$Q = K_{Th} + K_\alpha.$$

Treat each particle classically, so that for each $K = p^2/2m$. Taking note of the momentum equality mentioned above, we find

$$Q = p_{Th}^2/2m_{Th} + K_\alpha = p_\alpha^2/2m_{Th} + K_\alpha = 2m_\alpha K_\alpha/2m_{Th} + K_\alpha,$$

$$Q = K_\alpha(1 + m_\alpha/m_{Th}),$$

$$Q = (4.196 \text{ MeV})(1 + 4.0026/234.04) = 4.268 \text{ MeV}.$$

54-46

If we call the anonymous nuclide X, then the overall reaction is

$$^A X_Z + {}^1 n_0 \rightarrow {}^0 e_{-1} + 2({}^4 He_2).$$

By the conservation of nucleon number

$$A + 1 = 0 + 2(4),$$

$$A = 7.$$

Conservation of charge yields

$$Z + 0 = -1 + 2(2),$$

$$Z = 3.$$

From the periodic table we learn that the element with $Z = 3$ is lithium Li. Therefore, the unknown nuclide is $^7$Li.

54-51

(a) As suggested, let $m'$ = mass of the nucleus of an atom. If we ignore the binding energy of the electrons, then the mass of an atom of atomic number $Z$ is

$$m = m' + Zm_e.$$

Therefore, if no neutrino is emitted, the $Q$ value of the reaction, in terms of atomic masses, follows from

$$Q = [m_C' - m_B' - m_e]c^2,$$

$$Q = [(m_C - 6m_e) - (m_B - 5m_e) - m_e]c^2,$$

$$Q = [m_C - m_B - 2m_e]c^2.$$

(b) By straightforward calculation using the last formula above,

$$Q = [11.011433 - 11.009305 - 2(0.0005486)](931.5 \text{ MeV}),$$

$$Q = 0.9602 \text{ MeV} = 960.2 \text{ keV}.$$

## 54-55

The relation between dose and dose equivalent is

$$\text{dose equivalent} = (\text{dose})(QF).$$

Therefore, in our case,

$$\text{dose} = (25 \text{ mrem})/(0.85 \text{ rem/rad}) = 29.4 \text{ mrad}.$$

By definition, 1 rad corresponds to an absorbed energy of $10^{-5}$ J/g, so that for any dose, the absorbed energy $E$ in joules is given by

$$E = (\text{dose in rad})(0.01 \text{ J/kg} \cdot \text{rad})(\text{mass in kg}).$$

The mass involved is the mass exposed to radiation, or 44 kg in this situation. Hence,

$$E = (0.0294 \text{ rad})(0.01 \text{ J/kg} \cdot \text{rad})(44 \text{ kg}) = 0.013 \text{ J} = 13 \text{ mJ}.$$

## 54-60

From the unit relations given in the problem, we have

$$1 \text{ } \mu\text{Sv} = 1 \times 10^{-6} \text{ Sv} = (1 \times 10^{-6})(100 \text{ rem}) = 1 \times 10^{-4} \text{ rem},$$

$$1 \text{ } \mu\text{Sv} = 0.1 \text{ mrem}.$$

Therefore, 12 $\mu$Sv/h = 1.2 mrem/h. The pilot is exposed for 20 h per week, or for $(20)(52) = 1040$ h per year (assuming no vacations). Hence, the annual equivalent dose is

$$\text{equivalent dose} = (1.2 \text{ mrem/h})(1040 \text{ h}) = 1250 \text{ mrem}.$$

## 54-62

The activity is

$$R = (4.6 \times 10^{-6} \text{ Ci})(3.7 \times 10^{10} \text{ s}^{-1}/\text{Ci}) = 1.702 \times 10^5 \text{ s}^{-1}.$$

We need the half-life in the same time unit (i.e., seconds):

$$t_{\frac{1}{2}} = (1.28 \times 10^9 \text{ y})(3.16 \times 10^7 \text{ s/y}) = 4.045 \times 10^{16} \text{ s}.$$

The corresponding disintegration constant is

$$\lambda = \ln2/t_{\frac{1}{2}} = \ln2/(4.045 \times 10^{16} \text{ s}) = 1.714 \times 10^{-17} \text{ s}^{-1}.$$

Thus, the number of atoms in the sample is

$$N = R/\lambda = (1.702 \times 10^5 \text{ s}^{-1})/(1.714 \times 10^{-17} \text{ s}^{-1}) = 9.93 \times 10^{21}.$$

The mass of the sample must be

$$m = Nm_K = (9.93 \times 10^{21})[(40 \text{ u})(1.66 \times 10^{-27} \text{ kg/u})],$$

$$m = 6.59 \times 10^{-4} \text{ kg} = 659 \text{ mg}.$$

## 54-68

By Eq. 15 we have

$$t(\ln2) = t_{\frac{1}{2}}\ln(1 + N_F/N_I),$$

$$(260 \times 10^6 \text{ y})(\ln2) = (4.47 \times 10^9 \text{ y})\ln(1 + N_F/N_I),$$

$$N_F/N_I = 0.0411.$$

But the mass of an atom can be expressed, to sufficent precision here, as $A$ u, where u is the atomic mass unit. Therefore

$$N_F = m_F/(A_F \text{ u}), \quad N_I = m_I/(A_I \text{ u}),$$

where $m_F$ and $m_I$ are the corresponding masses. It follows that

$$N_F/N_I = (m_F/m_I)(A_I/A_F),$$

$$0.0411 = (m_F/3.71 \text{ mg})(238/206),$$

$$m_F = 0.132 \text{ mg} = 132 \text{ } \mu\text{g}.$$

## 54-71

Compute $Q$ immediately from Eq. 18, using the general reaction of Eq. 17 as a guide to the notation, to get

$$Q = [58.933198 + 1.007825 - 58.934349 - 1.008665](931.5 \text{ MeV}),$$

$$Q = -1.855 \text{ MeV}.$$

Evidently this is an endothermic reaction: the proton must bring in a threshold kinetic energy to make the reaction go.

In the center of mass system, by definition,

$$Q = (K_b' + K_Y') - (K_a' + K_X'),$$

$K'$ = center of mass kinetic energy for each particle. For the threshold energy set $K_b' = K_Y' = 0$, so that

$$-Q = K_a' + K_X' = K_{cm} = K_{lab}\left(\frac{m_X}{m_X + m_a}\right),$$

by Problem 77. Hence, the threshold kinetic energy is

$$K_{lab} = K_{th} = (-Q)\left(\frac{m_X + m_a}{m_X}\right) = |Q|\left(\frac{m_X + m_a}{m_X}\right).$$

It is to be expected that $K_{th} > -Q$ because the mass deficiency must be supplied by the incoming particle's kinetic energy before the reaction can proceed at all; indeed, more than this is required since kinetic energy must be given to the products of the reaction, as seen in the lab, to satisfy conservation of momentum.

(a) The binding energy $E$ of the last neutron can be found by examining the removal reaction

$$^{91}\text{Zr} + E \rightarrow {}^{90}\text{Zr} + n.$$

Solving for $E$ and substituting energy equivalents gives

$E = (89.904703 + 1.008665 - 90.905644)(931.5 \text{ MeV}) = 7.19 \text{ MeV}.$

(b) Similarly, for the next neutron,

$$^{90}\text{Zr} + E \rightarrow {}^{89}\text{Zr} + n,$$

$E = (88.908890 + 1.008665 - 89.904703)(931.5 \text{ MeV}) = 12.0 \text{ MeV}.$

(c) The binding energy $E_B$ of $^{91}\text{Zr}$ is

$E_B = [40(1.007825) + 51(1.008665) - 90.905644](931.5 \text{ MeV}),$

which gives $E_B = 791.1$ MeV, or $791.1/91 = 8.69$ MeV/nucleon on the average. As expected, (a) is less and (b) is greater than this average binding energy per nucleon.

CHAPTER 55

<u>55-4</u>

Since an amount of energy $Q$ = 200 MeV is liberated, on the average, for each nucleus that undergoes fission, the energy $E$ released when $n$ nuclei undergo fission is

$$E = nQ.$$

If fissions occur at the rate $dn/dt$, then

$$dE/dt = P = (dn/dt)Q.$$

We must be careful with the energy units. We have $P$ = 2 W = 2 J/s, so express $Q$ in joules:

$$Q = (200 \text{ MeV})(1.6 \times 10^{-13} \text{ J/MeV}) = 3.2 \times 10^{-11} \text{ J}.$$

Therefore, the fission rate must be

$$dn/dt = P/Q = (2 \text{ J/s})/(3.2 \times 10^{-11} \text{ J}) = 6.25 \times 10^{10} \text{ s}^{-1}.$$

<u>55-6</u>

It is sufficiently accurate to use for the molar mass of plutonium $M$ = 239 g/mol in calculating the number $N$ of atoms of plutonium in 1 kg; therefore,

$$N = (1/0.239)(6.02 \times 10^{23}) = 2.519 \times 10^{24}.$$

Hence the energy released if all these nuclei undergo fission is

$$E = NQ = (2.519 \times 10^{24})(180 \text{ MeV}) = 4.534 \times 10^{26} \text{ MeV},$$

or $7.27 \times 10^{13}$ J if we multiply by $1.6 \times 10^{-13}$ J/MeV.

<u>55-10</u>

We will work with the third reaction as an example. Find the atomic number of Zr in the periodic table (we should, by now, know the $Z$ number of uranium U). Then write the reaction as

$$^{235}U_{92} + {}^{1}n_0 \rightarrow {}^{A}X_z + {}^{100}Zr_{40} + 2({}^{1}n_0).$$

Apply conservation of nucleon number to find

$$235 + 1 = A + 100 + 2(1),$$

$$A = 134.$$

Conservation of charge gives

$$92 + 0 = Z + 40 + 2(0),$$

$$Z = 52.$$

The element with $Z = 52$ is Tellurium (Te). Therefore $X = {}^{134}\text{Te}$.

## 55-11

Since $2(26) = 52$, no neutrons are emitted, for nucleon number must be conserved. It follows that

$$Q = (\Delta m)c^2,$$

$$Q = [51.940509 - 2(25.982593)](931.5 \text{ MeV}),$$

$$Q = -23.0 \text{ MeV}.$$

Since this is negative, ${}^{52}\text{Cr}$ will not undergo spontaneous fission.

## 55-15

(a) The overall reaction chain can be summarized as

$$^{238}\text{U}_{92} + {}^1\text{n}_0 \rightarrow {}^{140}\text{Ce}_{58} + {}^{99}\text{Ru}_{44} + n({}^0\text{e}_{-1}),$$

where, as usual, we found the $Z$ values in the periodic table. The conservation of charge yields

$$92 + 0 = 58 + 44 + n(-1),$$

$$n = 10.$$

(b) In ß⁻ decay, the electrons are accounted for when we use atomic masses. Thus

$$Q = (238.050784 + 1.008665 - 139.905433 - 98.905939)(931.5),$$

$$Q = 231 \text{ MeV}.$$

## 55-20

Assume that if the reactor consumes half its fuel in 3 y, then the lifetime on one fuel loading is 6 y. In this 6 y period, the reactor generates

$$E = Pt = (190 \times 10^6 \text{ J/s})(6 \text{ y})(3.16 \times 10^7 \text{ s/y}),$$

$$E = 3.602 \times 10^{16} \text{ J}.$$

Each fission, on the average, generates energy $Q = 200 \text{ MeV} = 3.2 \times 10^{-11}$ J. Therefore, the number $N$ of fissions occuring in the reactor over the 6 y is

$$N = E/Q = (3.602 \text{ X } 10^{16} \text{ J})/(3.2 \text{ X } 10^{-11} \text{ J}) = 1.123 \text{ X } 10^{27}.$$

The total mass of this number of $^{238}$U atoms is

$$m = (N/N_A)M,$$

$$m = (1.123 \text{ X } 10^{27}/6.02 \text{ X } 10^{23})(0.235 \text{ kg}) = 438 \text{ kg}.$$

## 55-24

We will use the relation given in Problem 23. We have $P(t) = 350$ MW and $P_0 = 1200$ MW. The neutron generation time is $t_{gen} = 1.3$ ms and $t = 2.6$ s $= 2600$ ms ($t$ and $t_{gen}$ must be in the same time units). Therefore, by Problem 23,

$$350 = (1200)k^{2600/1.3},$$

$$350 = (1200)k^{2000}.$$

Now take logarithms of both sides (either log or ln). We use log to get

$$\log(350) = \log(1200) + (2000)\log k,$$

$$k = 0.99938.$$

## 55-30

(a) From Problem 29, we conclude that the energy released in the bomb is

$$E = (0.066 \text{ Megaton})(2.6 \text{ X } 10^{28} \text{ MeV/Megaton}) = 1.716 \text{ X } 10^{27} \text{ MeV}.$$

Since 200 MeV is produced in each fission, the number $N_f$ of fissions that take place is

$$N_f = (1.716 \text{ X } 10^{27} \text{ MeV})/(200 \text{ MeV}) = 8.58 \text{ X } 10^{24}.$$

This is 4% of the number $N$ of $^{235}$U atoms in the bomb, so that

$$N = N_f/0.04 = 2.15 \text{ X } 10^{26}.$$

Therefore, the mass of uranium in the bomb is

$$m = (2.15 \text{ X } 10^{26}/6.02 \text{ X } 10^{23})(0.235 \text{ kg}) = 83.9 \text{ kg}.$$

(b) There are two fission fragments per fission, leading to 1.72 X $10^{25}$ fragments in all.
(c) Although 2.47 neutrons are produced, on average, in each fission, one of these neutrons must induce another fission to

377

keep the chain reaction going. Therefore, 1.47 are released to the environment per fission, or $(1.47)(8.58 \times 10^{24}) = 1.26 \times 10^{25}$ for the whole bomb. We have assumed that the bomb does not contain a source of neutrons other than those generated in the fission of the uranium.

## 55-34

Let $t$ stand for the present, and $N_5$, $N_8$ the present abundances of the isotopes. The abundances at the time sought ($t = 0$) are $N_{50}$, $N_{80}$. By the law of radioactive decay (see Problem 28 in Chapter 54),

$$N_5 = N_{50} \left(\frac{1}{2}\right)^{t/0.704},$$

and

$$N_8 = N_{80} \left(\frac{1}{2}\right)^{t/4.47},$$

where $t$ is in Gy. Now, we want $t$ such that $N_{50}/N_{80} = 0.03$; at present $N_5/N_8 = 0.0072$. Hence, dividing the equations gives

$$\frac{N_5}{N_8} = \frac{N_{50}}{N_{80}} \left(\frac{1}{2}\right)^{t\left[\frac{1}{0.704} - \frac{1}{4.47}\right]},$$

$$0.0072 = 0.03 \left(\frac{1}{2}\right)^{1.197t},$$

$$t = 1.72 \text{ Gy}.$$

## 55-35

In Sample Problem 5, the barrier height is taken to be $K$, where

$$K = \frac{1}{4\pi\epsilon_0} \frac{e^2}{4R},$$

where $R$ is the radius (in our case) of a proton. Put $1/4\pi\epsilon_0 = 8.99$ X $10^9$ N•m²/C², $e = 1.6$ X $10^{-19}$ C and $R = 0.8$ X $10^{-15}$ m to get $K = 0.45$ MeV (after converting from J to MeV) for the barrier height.

## 55-39

The number $N_d$ of deuterium atoms in 1 kg of deuterium is

$$N_d = (1 \text{ kg})/[(2.014)(1.66 \text{ X } 10^{-27} \text{ kg})] = 2.991 \text{ X } 10^{26}.$$

From the reaction equation, we see that two deuterium atoms are consumed in each fusion. Therefore, the number $N_f$ of fusions that take place is

$$N_f = \tfrac{1}{2}N_d = 1.496 \text{ X } 10^{26}.$$

The total energy released by all these fusions is

$$E = N_f Q = (1.496 \text{ X } 10^{26})[(3.27 \text{ MeV})(1.6 \text{ X } 10^{-13} \text{ J/MeV})],$$

$$E = 7.827 \text{ X } 10^{13} \text{ J}.$$

Therefore, the time $t$ sought is

$$t = E/P = (7.827 \text{ X } 10^{13} \text{ J})/(100 \text{ J/s})(3.16 \text{ X } 10^7 \text{ s/y}),$$

$$t = 24,800 \text{ y}.$$

## 55-44

(a) Since $E = mc^2$, we have

$$P = dE/dt = (dm/dt)c^2,$$

$$dm/dt = P/c^2 = (3.9 \text{ X } 10^{26} \text{ J/s})/(3 \text{ X } 10^8 \text{ m/s})^2,$$

$$dm/dt = 4.3 \text{ X } 10^9 \text{ kg/s}.$$

(b) The mass lost during the time $t = 4.5$ Gy is

$$\Delta M = (dm/dt)t,$$

$$\Delta M = (4.3 \text{ X } 10^9 \text{ kg/s})(4.5 \text{ Gy})(3.16 \text{ X } 10^{16} \text{ s/Gy}),$$

$$\Delta M = 6.11 \text{ X } 10^{26} \text{ kg}.$$

Hence, the fraction of mass lost is

$$\Delta M/M = (6.11 \text{ X } 10^{26} \text{ kg})/(2 \text{ X } 10^{30} \text{ kg}) = 3.1 \text{ X } 10^{-4}.$$

If we use $A = 4$ for helium, then the number $N_{He}$ of helium atoms in the star is

$$N_{He} = M_{star}/m_{He},$$

$$N_{He} = (4.6 \text{ X } 10^{32} \text{ kg})/[4(1.66 \text{ X } 10^{-27} \text{ kg})] = 6.93 \text{ X } 10^{58}.$$

Since 3 helium nuclei are consumed in each fusion, the number $N_f$ of fusions over the lifetime of the star in this phase is

$$N_f = \tfrac{1}{3}N_{He} = 2.31 \text{ X } 10^{58}.$$

The total energy $E$ released in all these fusions is

$$E = N_f Q,$$

$$E = (2.31 \text{ X } 10^{58})[(7.27 \text{ MeV})(1.6 \text{ X } 10^{-13} \text{ J/MeV})],$$

$$E = 2.69 \text{ X } 10^{46} \text{ J}.$$

Hence, the lifetime $t$ of the star during the helium burning phase is given by

$$t = E/P = (2.69 \text{ X } 10^{46} \text{ J})/[(5.3 \text{ X } 10^{30} \text{ J/s})(3.16 \text{ X } 10^7 \text{ s/y})],$$

$$t = 1.61 \text{ X } 10^8 \text{ y} = 0.161 \text{ Gy}.$$

(a) The rate $R_f$ at which proton-proton fusion cycles occur is given from

$$R_f = (dE/dt)/Q.$$

But $dE/dt = 3.9 \text{ X } 10^{26} \text{ J/s}$ and $Q = 26.7 \text{ MeV} = 4.272 \text{ X } 10^{-12} \text{ J}$. Substituting these numbers we get $R_f = 9.129 \text{ X } 10^{37} \text{ s}^{-1}$. Since 2 neutrinos are produced in each complete fusion cycle, the Sun emits neutrinos at the rate $R_n = 2R_f = 1.83 \text{ X } 10^{38} \text{ s}^{-1}$.
(b) The neutrinos are emitted isotropically by the Sun. The Earth presents an area $\pi R^2$ ($R$ = radius of the Earth) to the neutrinos, all of which must pass through a sphere of area $4\pi r^2$ ($r$ = the Earth-Sun distance). Therefore, the fraction $f$ of the solar neutrinos that strike the Earth is

$$f = \pi R^2/4\pi r^2 = 4.509 \text{ X } 10^{-10},$$

using numerical data from Appendix C. The rate $R_E$ at which solar neutrinos strike the Earth is

$$R_E = fR_n = (4.509 \text{ X } 10^{-10})(1.83 \text{ X } 10^{38} \text{ s}^{-1}) = 8.25 \text{ X } 10^{28} \text{ s}^{-1}.$$

One liter of water has a mass of $(1000 \text{ kg/m}^3)(0.001 \text{ m}^3) = 1$ kg. Of this, we are told that $1.5 \times 10^{-4}$ kg is HDO, or "heavy water." The molar mass of the HDO molecule is

$$M = 1.01 + 2.01 + 16 = 19.02 \text{ g/mol.}$$

Therefore, there are

$$N = [(1.5 \times 10^{-4} \text{ kg})/(0.01902 \text{ kg/mol})](6.02 \times 10^{23} \text{ mol}^{-1}),$$

$$N = 4.75 \times 10^{21},$$

molecules of HDO in the water. Since 2 molecules must participate in each fusion, the number $N_f$ of fusions is $N_f = \frac{1}{2}N = 2.375 \times 10^{21}$. The total amount of energy so available from the water is

$$E = N_f Q = (2.375 \times 10^{21})[(3.27 \text{ MeV})(1.6 \times 10^{-13} \text{ J/MeV})],$$

$$E = 1.24 \times 10^9 \text{ J.}$$

Therefore, if all the fusions take place over 1 day, the power generated is

$$P = E/t = (1.24 \times 10^9)/(86,400 \text{ s}) = 14.4 \text{ kW.}$$

# CHAPTER 56

## 56-1

(a) The gravitational force is given, in magnitude, by Eq. 1 in Chapter 16, and the electrostatic force, again in magnitude, by Eq. 4 in Chapter 27. In forming the ratio of these forces the terms $r^2$ will cancel. Note also that the mass of a positron equals that of the electron and their charges are both equal to $e$ in magnitude. Therefore, we have

$$F_{grav}/F_{elect} = 4\pi\epsilon_0 G(m/e)^2.$$

Numerically $4\pi\epsilon_0 = (8.99 \times 10^9 \text{ N} \cdot \text{m}^2/\text{C}^2)^{-1}$, $G = 6.67 \times 10^{-11} \text{ N} \cdot \text{m}^2/\text{kg}^2$, $m = 9.11 \times 10^{-31}$ kg and $e = 1.6 \times 10^{-19}$ C. With these numbers, we find $F_{grav}/F_{elect} = 2.4 \times 10^{-43}$.

(b) For the proton-antiproton pair, only the mass $m$ is different from the electron-positron pair. Since $m_p = 1840 m_e$, we now have

$$F_{grav}/F_{elect} = (1840)^2(2.4 \times 10^{-43}) = 8.1 \times 10^{-37}.$$

## 56-5

Use Eq. 32 in Chapter 21 to calculate the total energy $E_\pi$ of each created pion. From Table 4 we see that the rest energy of each pion is $mc^2 = 140$ MeV. Therefore,

$$E_\pi^2 = (pc)^2 + (mc^2)^2 = (358.3 \text{ MeV})^2 + (140 \text{ MeV})^2,$$

$$E_\pi = 384.7 \text{ MeV}.$$

Now apply conservation of total relativistic energy to the decay. The $\rho^0$ meson is at rest, so its kinetic energy is zero. Therefore,

$$E_{0\rho} = 2E_\pi = 2(384.7 \text{ MeV}) = 769 \text{ MeV}.$$

## 56-8

(a) The total energy of the tau is $E = mc^2 + K = 1784$ MeV + 2200 MeV = 3984 MeV. (We found the rest energy of the tau in Table 3, where the charge of the tau is also given as $+1e$.) We can now calculate the momentum of the tau from Eq. 32 of Chapter 21:

$$E^2 = (pc)^2 + (mc^2)^2,$$

$$(3984 \text{ MeV})^2 = (pc)^2 + (1784 \text{ MeV})^2,$$

$$p = 3562 \text{ MeV}/c,$$

$$p = (3562)(1.6 \times 10^{-13} \text{ J})/(3 \times 10^8 \text{ m/s}) = 1.90 \times 10^{-18} \text{ kg} \cdot \text{m/s}.$$

(b) The radius of the circular path is (see Section 34-3)

$$r = p/qB,$$

$$r = (1.90 \times 10^{-18} \text{ kg} \bullet \text{m/s})/(1.6 \times 10^{-19} \text{ C})(1.2 \text{ T}) = 9.9 \text{ m}.$$

## 56-11

We will examine decay (a). To begin, check relativistic energy. If the $\mu^-$ is at rest, its energy is rest energy which Table 3 gives as 105.7 MeV. The rest energy of the positron is 0.511 MeV and of the neutrinos is zero. Therefore, there is energy available to supply the necessary kinetic energy of the electron and energies of the neutrinos required to satisfy conservation of momentum. We have a problem, however, with charge conservation. The muon has negative charge, the positron (as its name implies) positive charge. The neutrinos have zero charge. Therefore, charge conservation is violated. Having found one violation, we must not assume that there are no others. The muon is a lepton, so we must examine conservation of $L_e$ and of $L_\mu$. Recall that the positron is an antielectron. In terms of $L_e$, the reaction reads

$$0 \rightarrow -1 + 0 + (-1),$$

so it is evident that conservation of electron lepton number is not obeyed. In terms of $L_\mu$ the reaction reads

$$+1 \rightarrow 0 + 1 + 0,$$

so that conservation of muon lepton number is satisfied. So we find violations of charge and of electron lepton number.
Now you examine decay (b).

## 56-15

We will look at reaction (a). In applying the conservation laws of charge, baryon number and strangeness, we can "cancel" one of the protons on each side. Having done this, then in terms of charge $Q$ we have

$$+1 = 0 + Q_x \rightarrow Q_x = +1.$$

For baryon number,

$$+1 = +1 + B_x \rightarrow B_x = 0.$$

And for strangeness,

$$0 = -1 + S_x \rightarrow S_x = +1.$$

Clearly the particle is not a baryon, so look at Table 4 (mesons). We see that only the $K^+$ has $Q = +1$ and $S = +1$. Now examine total energy, for which purpose we cannot cancel the two protons. Although the initial rest energy is less than the final rest energy, there are two particles initially, so the deficit can be supplied from the kinetic energies of the initial particles. Now its your turn to examine the other reactions.


## 56-22

We look at (a). The quark content of various particles are given in Fig. 4, but we need the $Q$ and $S$ numbers of the particles to locate them on the figure. In Fig. 3(b) we see that a $\Sigma^-$ has $Q = -1$, $S = -1$. The $\Sigma^-$ is a spin $\frac{1}{2}$ baryon; Fig. 4(b) gives the quark content as dds. In Table 9 we see that the neutron ($Q = 0$, spin $\frac{1}{2}$, $B = +1$, $S = 0$) has quark content udd. The pion must consist of a quark-antiquark combination. Fig. 3(a) gives $Q = -1$, $S = 0$ for the $\pi^-$. Turning to Table 8, we find the quark content to be $d\bar{u}$. Therefore, in terms of quarks the reaction is

$$dds \rightarrow udd + d\bar{u},$$

that is

$$d + d + s \rightarrow (u + d + d) + (d + \bar{u}),$$

which reduces to

$$s \rightarrow d + u + \bar{u}.$$

The other reactions await your attention.


## 56-25

By Hubble's law, Eq. 2,

$$v/c = Hd/c,$$

$$v/c = [67 \text{ km}/(s \bullet Mpc)](240 \text{ Mpc})/(3 \times 10^5 \text{ km/s}) = 0.0536.$$

This is small enough so that we can use the classical Doppler shift equation

$$\Delta\lambda/\lambda_0 = v/c,$$

$$\Delta\lambda/(656.3 \text{ nm}) = 0.0536,$$

$$\Delta\lambda = 35.2 \text{ nm}.$$

The galaxy is, presumably, receding so that the observed wavelength is $\lambda = \lambda_0 + \Delta\lambda = 690$ nm.

(a) First evaluate $kT = (8.62 \times 10^{-5}$ eV/K$)(2.7$ K$) = 233$ $\mu$eV. If there are 23% of the molecules in the excited state, and this is the only excited state, then 77% of the molecules must be in the ground state. Hence, by Eq. 8 in Chapter 52,

$$\frac{0.23}{1 - 0.23} = e^{-\Delta E/233},$$

$$\Delta E = 280 \text{ } \mu eV.$$

(We solved the equation by taking ln of both sides.)
(b) The photon wavelength is given from

$$\lambda = (1240 \text{ eV} \bullet \text{nm})/(280 \times 10^{-6} \text{ eV}) = 4.4 \times 10^6 \text{ nm} = 4.4 \text{ mm}.$$

56-31

(a) From Table 4 we extract the rest energy of the pion as 140 MeV. By Sample Problem 6, we see that the minimum temperature is

$$T = E/k = (140 \text{ MeV})/(8.62 \times 10^{-11} \text{ MeV/K}) = 1.6 \times 10^{12} \text{ K}.$$

(b) Use Eq. 3 to find the required age of the universe:

$$t = (1.5 \times 10^{10}/T)^2 = (1.5 \times 10^{10} \text{ s}^{\frac{1}{2}} \bullet \text{K}/1.6 \times 10^{12} \text{ K})^2,$$

$$t = 8.8 \times 10^{-5} \text{ s} = 88 \text{ } \mu s.$$